Glencoe Science
Features and Benefits

		Pages
Dynamic Instructional Strategies	**. . . present clear and comprehensive coverage of middle school science.**	
	■ Each engaging chapter opener includes a *Launch Lab* and *Foldables*™.	7, 35
	■ *National Geographic Visualizing* features illustrate important concepts in middle school science.	18, 48
A Strong Reading Strand	**. . . encourages active reading and learning for students of all reading levels. In the *Student Edition:***	
	■ as you read gives students a preview of learning objectives and vocabulary for each section;	8, 36
	■ *Reading Checks* help students check their reading comprehension; and	13, 39
	■ *Caption questions* ask students to interpret visuals.	13, 38
	■ *Reading Essentials, An Interactive Student Textbook* is designed to help struggling readers comprehend core content. It is written at a reading level of about two to three grades below the *Student Edition.*	
Meeting the Needs of All Students	**. . . facilitates understanding of science concepts for students of all learning levels. In the *Teacher Wraparound Edition:***	
	■ *Differentiated Instruction* strategies help meet the needs of students with learning disabilities and physical challenges, or create opportunities to enrich and extend students' learning;	10, 41
	■ *Daily Intervention* provides intervention strategies for struggling students; and	11, 42
	■ *Identifying Misconceptions* helps uncover and address common science misconceptions.	6F
	■ The *English/Spanish Glossary*, also in the *Student Edition,* helps English-language learners comprehend science terms.	195–201
Extensive Standardized Test Practice	**. . . gives students the opportunity to practice for state and national exams.**	
	■ Each chapter ends with a variety of standardized test practice questions, including *Multiple Choice, Short Response/Grid In*, and *Open Ended.*	150–151
A Variety of Labs	**. . . gets students excited about being involved in science. The *Student Edition* provides:**	
	■ *MiniLABs,* traditional labs, and *Design Your Own, Model and Invent,* and *Use the Internet* labs; and	13, 17, 54
	■ *Extra Try at Home Labs* provide opportunities for students to practice their science skills at home with adult supervision using materials from the kitchen, junk drawer, or backyard.	166–168
	■ *Virtual Labs* CD-ROM contains an interactive virtual lab for each chapter.	107
	■ *Video Labs* (VHS) reinforce lab techniques and safety skills, offer troubleshooting tips, and give expected outcomes.	111
	■ The *Science Lab Manual, Probeware Lab Manual,* and *Science Inquiry Lab Manual* provide additional opportunities to practice laboratory techniques.	
Multi-Level Review	**. . . presents multiple opportunities for all students to review and master content.**	
	■ Each section ends with a review that contains a *Summary* of the section's major concepts and a *Self Check* that has questions to assess student learning and practice math or science skills.	11, 42
	■ The *Study Guide* at the end of each chapter can preview, review, summarize, and visualize the chapter's main ideas.	29, 57
	■ *Study Guide* and *Reinforcement* help students grasp core content.	
Teacher Resources	**. . . provide innovative strategies to help new and experienced teachers.**	
	■ *Chapter Resources Fast File* ™ contains important reproducible masters.	34B
	■ Section Focus, Assessment, and Teaching transparencies accompany each chapter.	62C
	■ *Performance Assessment in the Science Classroom* has assessment guidelines, strategies, sample rubrics, and more.	
Online Resources	**. . . enrich the learning experience with the click of a mouse.**	
	■ For prescreened Web links, standardized test practice, self-check quizzes, chapter tests, *Vocabulary PuzzleMaker,* extra math practice, science career information, current science news, and *WebQuest* interactive projects, visit **booke.msscience.com**.	
	■ The complete interactive *Student Edition* is available at *The McGraw-Hill Learning Network* Web site, **mhln.com**.	
Technology	**. . . provides timesaving products to help teachers creatively engage their students.**	
	■ *MindJogger Videoquizzes* (VHS & DVD) provide a game-show style interactive quiz for each chapter.	
	■ Easy to edit *Interactive Chalkboard* Microsoft® PowerPoint® presentations include step-by-step lessons, an image bank, chapter and section review questions, standardized test practice, and transparencies.	
	■ *ExamView® Pro Testmaker* CD-ROM in English or Spanish allows you to customize assessments.	
	■ *TeacherWorks* CD-ROM is your all-in-one resource center that helps you plan and organize lessons.	
	■ *StudentWorks* CD-ROM solves the heavy backpack problem.	

SAFETY SYMBOLS

SAFETY SYMBOLS	HAZARD	EXAMPLES	PRECAUTION	REMEDY
DISPOSAL	Special disposal procedures need to be followed.	certain chemicals, living organisms	Do not dispose of these materials in the sink or trash can.	Dispose of wastes as directed by your teacher.
BIOLOGICAL	Organisms or other biological materials that might be harmful to humans	bacteria, fungi, blood, unpreserved tissues, plant materials	Avoid skin contact with these materials. Wear mask or gloves.	Notify your teacher if you suspect contact with material. Wash hands thoroughly.
EXTREME TEMPERATURE	Objects that can burn skin by being too cold or too hot	boiling liquids, hot plates, dry ice, liquid nitrogen	Use proper protection when handling.	Go to your teacher for first aid.
SHARP OBJECT	Use of tools or glassware that can easily puncture or slice skin	razor blades, pins, scalpels, pointed tools, dissecting probes, broken glass	Practice common-sense behavior and follow guidelines for use of the tool.	Go to your teacher for first aid.
FUME	Possible danger to respiratory tract from fumes	ammonia, acetone, nail polish remover, heated sulfur, moth balls	Make sure there is good ventilation. Never smell fumes directly. Wear a mask.	Leave foul area and notify your teacher immediately.
ELECTRICAL	Possible danger from electrical shock or burn	improper grounding, liquid spills, short circuits, exposed wires	Double-check setup with teacher. Check condition of wires and apparatus.	Do not attempt to fix electrical problems. Notify your teacher immediately.
IRRITANT	Substances that can irritate the skin or mucous membranes of the respiratory tract	pollen, moth balls, steel wool, fiberglass, potassium permanganate	Wear dust mask and gloves. Practice extra care when handling these materials.	Go to your teacher for first aid.
CHEMICAL	Chemicals can react with and destroy tissue and other materials	bleaches such as hydrogen peroxide; acids such as sulfuric acid, hydrochloric acid; bases such as ammonia, sodium hydroxide	Wear goggles, gloves, and an apron.	Immediately flush the affected area with water and notify your teacher.
TOXIC	Substance may be poisonous if touched, inhaled, or swallowed.	mercury, many metal compounds, iodine, poinsettia plant parts	Follow your teacher's instructions.	Always wash hands thoroughly after use. Go to your teacher for first aid.
FLAMMABLE	Flammable chemicals may be ignited by open flame, spark, or exposed heat.	alcohol, kerosene, potassium permanganate	Avoid open flames and heat when using flammable chemicals.	Notify your teacher immediately. Use fire safety equipment if applicable.
OPEN FLAME	Open flame in use, may cause fire.	hair, clothing, paper, synthetic materials	Tie back hair and loose clothing. Follow teacher's instruction on lighting and extinguishing flames.	Notify your teacher immediately. Use fire safety equipment if applicable.

Eye Safety
Proper eye protection should be worn at all times by anyone performing or observing science activities.

Clothing Protection
This symbol appears when substances could stain or burn clothing.

Animal Safety
This symbol appears when safety of animals and students must be ensured.

Handwashing
After the lab, wash hands with soap and water before removing goggles.

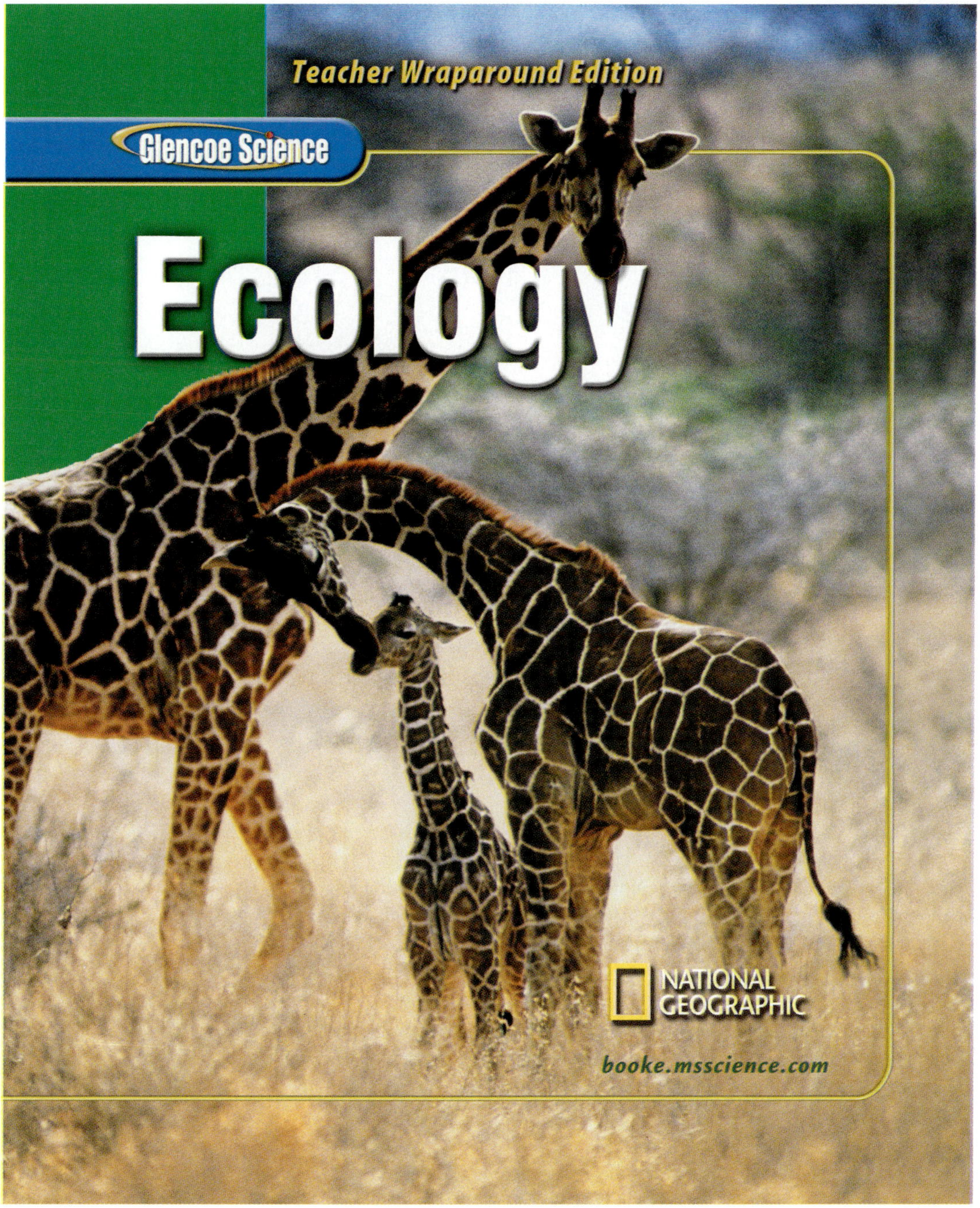

McGraw Hill Glencoe

New York, New York Columbus, Ohio Chicago, Illinois Peoria, Illinois Woodland Hills, California

Glencoe Science

Ecology

Three generations of reticulated giraffes are interacting on an African savanna. Females start breeding at around five years old, and gestation is about 15 months. Calves are born from a standing female, dropping six feet to the ground! They weigh 100 to 150 pounds, and are six feet tall at birth.

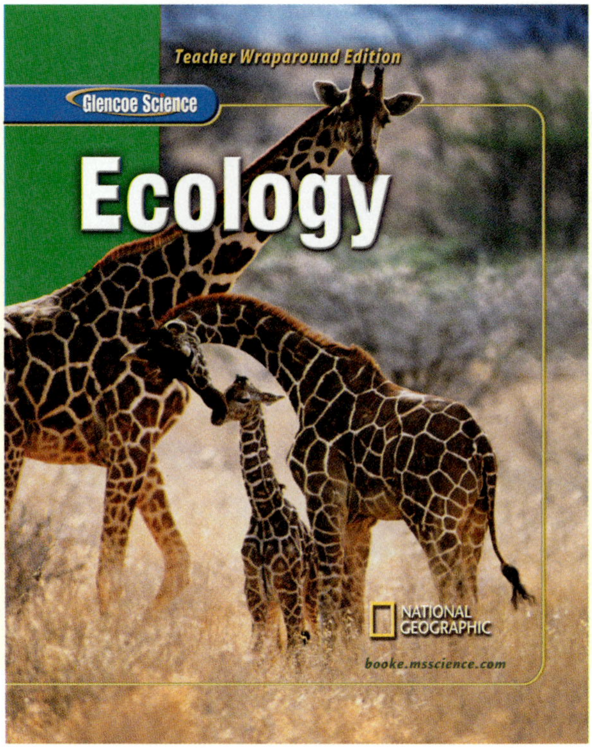

McGraw Hill Glencoe

The McGraw-Hill Companies

Send all inquiries to:
Glencoe/McGraw-Hill
8787 Orion Place
Columbus, OH 43240-4027

ISBN 0-07-861746-4 (Student Edition)
ISBN 0-07-861747-2 (Teacher Wraparound Edition)

Printed in the United States of America.

2 3 4 5 6 7 8 9 10 027/055 09 08 07 06 05

Authors

NATIONAL GEOGRAPHIC
Education Division
Washington, D.C.

Peter Rillero, PhD
Associate Professor of Science Education
Arizona State University West
Phoenix, AZ

Dinah Zike
Educational Consultant
Dinah-Might Activities, Inc.
San Antonio, TX

Series Consultants

CONTENT

Michael A. Hoggarth, PhD
Department of Life and Earth Sciences
Otterbein College
Westerville, OH

Dominic Salinas, PhD
Middle School Science Supervisor
Caddo Parish Schools
Shreveport, LA

MATH

Teri Willard, EdD
Mathematics Curriculum Writer
Belgrade, MT

READING

Elizabeth Babich
Special Education Teacher
Mashpee Public Schools
Mashpee, MA

SAFETY

Sandra West, PhD
Department of Biology
Texas State University-San Marcos
San Marcos, TX

ACTIVITY TESTERS

Nerma Coats Henderson
Pickerington Lakeview Jr. High School
Pickerington, OH

Mary Helen Mariscal-Cholka
William D. Slider Middle School
El Paso, TX

Science Kit and Boreal Laboratories
Tonawanda, NY

Series Reviewers

Maureen Barrett
Thomas E. Harrington Middle School
Mt. Laurel, NJ

Desiree Bishop
Environmental Studies Center
Mobile County Public Schools
Mobile, AL

Linda V. Forsyth
Retired Teacher
Merrill Middle School
Denver, CO

Amy Morgan
Berry Middle School
Hoover, AL

Darcy Vetro-Ravndal
Hillsborough High School
Tampa, FL

Glencoe Science 15-Book Series

Teach science your way!

With the 15 Life, Earth, and Physical Science titles in our modular series, you can select the science topics you want to cover and customize your science curriculum in any way you choose.

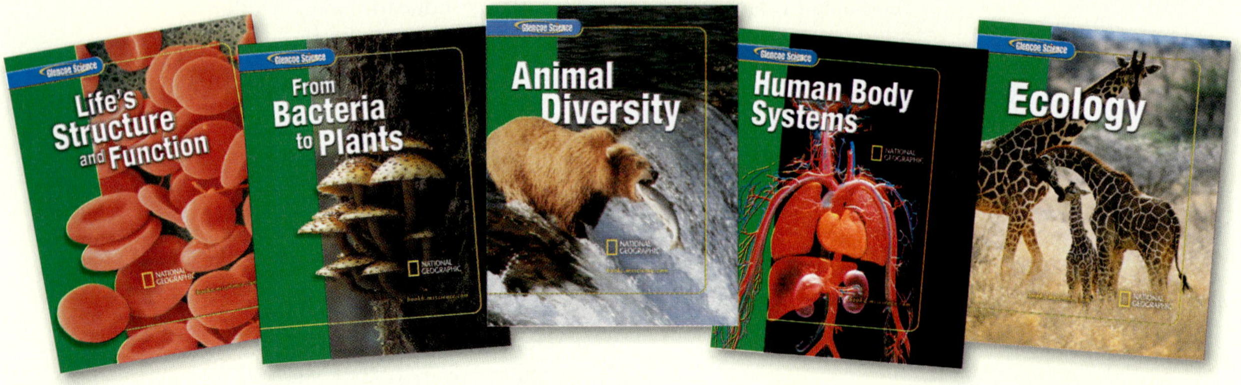

A Life's Structure and Function

1. Exploring and Classifying Life
2. Cells
3. Cell Processes
4. Cell Reproduction
5. Heredity
6. Adaptations over Time

B From Bacteria to Plants

1. Bacteria
2. Protists and Fungi
3. Plants
4. Plant Reproduction
5. Plant Processes

C Animal Diversity

1. Introduction to Animals
2. Mollusks, Worms, Arthropods, Echinoderms
3. Fish, Amphibians, and Reptiles
4. Birds and Mammals
5. Animal Behavior

D Human Body Systems

1. Structure and Movement
2. Nutrients and Digestion
3. Circulation
4. Respiration and Excretion
5. Control and Coordination
6. Regulation and Reproduction
7. Immunity and Disease

E Ecology

1. Interactions of Life
2. The Nonliving Environment
3. Ecosystems
4. Conserving Resources
5. Conserving Life

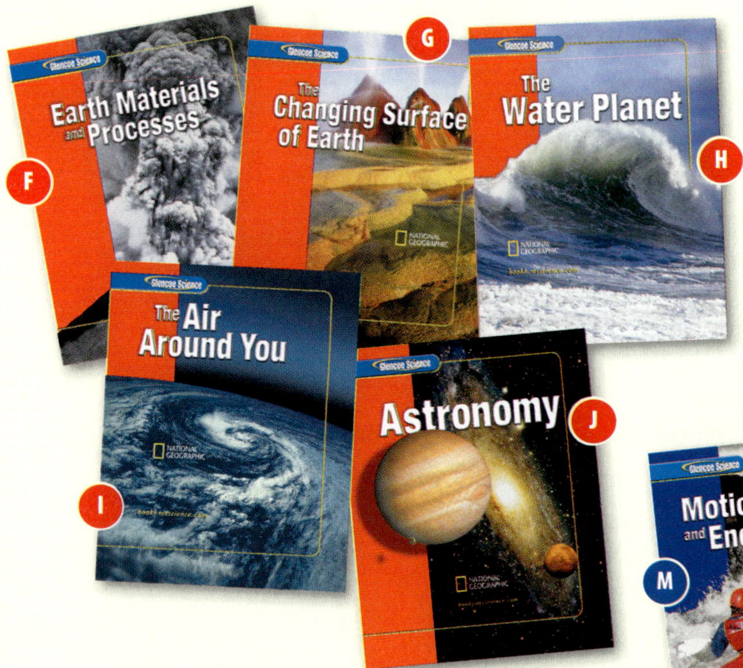

Multi-Level Review and Assessment

Each chapter provides five pages of review and testing to help you evaluate students' knowledge and ability to apply science concepts.

Section Review

- Summary pinpoints important concepts in the section.
- Skill-based questions promote critical thinking skills.

Study Guide

- Main idea summary of each section
- Concept mapping activity to help students visualize the main ideas

Chapter Review

- Using Vocabulary
- Checking Concepts
- Thinking Critically
- Performance Activities
- Applying Math

Standardized Test Practice

- Multiple Choice
- Short Response/ Grid In
- Open-Ended Questions

Dynamic Instruction

The consistent instructional strategies in each chapter strengthen students' learning—from the beginning of each chapter where students see "Chapter Preview," to the end where they have a chance to test the knowledge they have acquired and prepare for the next lesson.

Chapter Opener

- *Chapter Preview* introduces the main concepts.
- *Science Journal* promotes critical-thinking skills
- *Dinah Zike's Foldables*™ let students create interactive study guides.
- *Launch Labs* give students an opportunity to explore new ideas.

Section Opener

- *What You'll Learn* introduces main concepts.
- *Why It's Important* provides an answer to "Why do we have to learn this?"
- *Review Vocabulary* reviews a term that helps students better understand section content.
- *New Vocabulary* highlights new terms students will learn in the section.

Labs

- *Labs* allow students to design their own experiments or follow well-tested procedures, enabling them to learn and practice science processes.
- *MiniLABS* offer students quick and easy-to-do ways to clarify concepts and reinforce skills.

Assessment

Glencoe Science offers the Glencoe Assessment Advantage, a system designed to give you all the tools you need to prepare your students for success in any testing situation.

In the *Student Edition*

Section Review and **Applying Math** questions appear in every chapter.

Chapter Review questions help you evaluate students' knowledge and ability to apply science concepts.

Standardized Test Practice questions at the end of each chapter provide students with additional opportunities to practice their test-taking skills.

In the *Teacher Wraparound Edition*

Assessments located throughout the *Teacher Wraparound Edition* provide methods for assessing students' comprehension with Performance, Process, and Content exercises.

Teacher Classroom Resources

Performance Assessment in the Science Classroom

- Guidelines for assessing the performance of a task
- Reproducible activities for evaluating students
- Sample rubrics and checklists

***Fast File* Chapter Resources** provides six pages of assessement for every chapter including *Testing Concepts, Applying Concepts,* and *Writing Skills.*

Technology Support

MindJogger Videoquizzes are interactive video quizzes set in game show format. Each is designed for the full range of student learning styles.

Exam*View*® Pro Testmaker CD-ROM for Windows® and Macintosh® provides an easy way to create, edit, and customize your tests. Select your own test items by objective from two different levels of difficulty, or write and edit your own. Translate tests from English to Spanish and vice versa.

Rubrics

The following rubrics are sample scoring devices for short response and open-ended questions.

Short Response

Points	Description
2	The student demonstrates a thorough understanding of the science of the task. The response may contain minor flaws that do not detract from the demonstration of a thorough understanding.
1	The student has provided a response that is only partially correct.
0	The student has provided a completely incorrect solution or no response at all.

Open Ended

Points	Description
4	The student demonstrates a thorough understanding of the science of the task. The response may contain minor flaws that do not detract from the demonstration of a thorough understanding.
3	The student demonstrates an understanding of the science of the task. The response is essentially correct and demonstrates an essential but less than thorough understanding of the science.
2	The student demonstrates only a partial understanding of the science of the task. Although the student may have used the correct approach to a solution or may have provided a correct solution, the work lacks an essential understanding of the underlying science concepts.
1	The student demonstrates a very limited understanding of the science of the task. The response is incomplete and exhibits many flaws.
0	The student provides a completely incorrect solution or no response at all.

Time-Saving Teacher Resources

Glencoe Science provides an extensive array of support materials and resources designed to help you create and customize your science course quickly and easily.

FAST FILE Chapter Resources

For each chapter, Chapter Resources contain key reproducible masters along with additional teaching strategies, teacher support, and answer keys.

Teacher Wraparound Edition

The *Teacher Wraparound Edition* is your key to the teaching resources available. In addition to teaching strategies and suggestions, the *Teacher Wraparound Edition* provides a guide for all print and software materials available for each lesson.

Transparencies

Color Transparencies provides three types of transparencies for use while teaching each chapter. The *Section Focus Transparencies* are designed to generate interest and focus students' attention on the topic being presented in the section. The *Teaching Transparency* for each chapter addresses a major concept that will benefit from an extra visual learning aid. The *Assessment Transparency* for each chapter is set up to resemble standardized tests.

Exam*View*® Pro Testmaker CD-ROM

This CD-ROM will help you create, edit, and customize tests. In addition, it will help you create multiple versions of tests, translate tests from English to Spanish and vice versa, and build tests aligned with state standards.

Video Labs

These VHS cassettes contain step-by-step procedures for selected *Student Edition* labs. They also contain lab safety skills, teacher support, and troubleshooting advice.

TeacherWorks™

This CD-ROM is your all-in-one teacher resource center. Personalize a lesson plan, access resources from the *Teacher Wraparound Edition*, connect to the Internet, or make a to-do list. These are only a few of the many features that can assist you in the planning and organizing of your lessons.

StudentWorks™

This CD-ROM is a valuable resource for students to access content online and use online resources to continue learning chapter concepts.

Guided Reading Audio Program has been upgraded to StudentWorks Plus™

Interactive Chalkboard with Image Bank — PowerPoint® Presentations

This CD-ROM brings Microsoft® PowerPoint® presentations right to your door. With the large number of graphics provided, students can use a visual approach to learning chapter content.

Virtual Labs CD-ROM Program

The Virtual Labs CD-ROM contains a collection of labs that allow students to complete labs that are too expensive, take too long to complete, or might be too dangerous in a classroom laboratory.

Science Online

This website is a portal to hundreds of prescreened Internet sites (Web links) that correlate to content in the text. Visit this link to find interactive activities that review chapter concepts and to access the *Student Edition* online.

McGraw Hill Learning Network mhln.com

mhln.com is an online teaching and learning space for teachers, students, and parents.

Differentiated Instruction

Teaching Strategies

Following each suggested assessment and activity, ability levels are supplied to accommodate all students. For a key to the Teaching Strategies designations, see the C page before each chapter.

Identifying Misconceptions

These short, diagnostic, and perscriptive lessons target common science misconceptions.

Multiple Learning Styles

Look for these italicized designations under various activities to help you target your lessons to each student's preferred learning style.

- ***Kinesthetic*** learners learn through touch, movement, and manipulating objects.
- ***Visual-Spatial*** learners think in terms of images, illustrations, and models.
- ***Interpersonal*** learners understand and work well with other people.
- ***Intrapersonal*** learners can analyze their own strengths and weaknesses and may prefer to work on their own.
- ***Linguistic*** learners write clearly and easily understand the written word.
- ***Logical-Mathematical*** learners understand numbers easily and have highly-developed reasoning skills.

Daily Interventions

Found at the end of each chapter section, this feature is designed to intercept students who are struggling and prescribe a system to help them get back on track. *Reteach* provides reinforcement of the section's concepts through visual activities.

Differentiated Instruction

These activities present various teaching strategies designed to help you meet the special needs of students with learning disabilities, physical challenges, visual impairment, and hearing impairment. *Challenge* activities provide opportunities for students who excel to engage in activities and research projects that extend the chapter's concepts. English-language learners in the classroom will also find exercises that bridge the gap between language barriers and the chapter content.

Cultural Diversity

These readings provide insights into the unique ways in which people of different ethnicities and cultural heritage have approached science. The intent of these features is to build awareness and appreciation for the global community in which we live.

Inquiry-Based Science

The call for more inquiry-based science by the *National Science Education Standards* has been met by Glencoe Science.

Glencoe Science recognizes the importance of conducting inquiry-based science activities in the classroom. The process of doing inquiry models actual science practice, encouraging problem-solving strategies and developing critical thinking skills. Inquiry gets students actively involved in the learning process by allowing them to determine materials, procedures, or the topics and questions they want to investigate.

Inquiry can range from a very structured activity for those students who need more guidance to a more open-ended approach where students lead the investigations. Glencoe Science recognizes that the inquiry activities suggested will not look the same in every classroom. We encourage teachers to modify the suggested activities in a manner that best supports your students.

Glencoe also provides teachers with *Alternative Inquiry Labs,* teaching strategies or suggestions for making existing labs more inquiry-based.

Research-Based Learning Strategies

Glencoe Science incorporates the most current and applicable educational research on science learning and follows recommendations from the American Association for the Advancement of Science and the National Science Teachers Association. The following research-based strategies can be found throughout the text.

Learning Strategies

The following research-based strategies can be found throughout the text:

- **Using Prior Knowledge** Glencoe Science encourages students to use their prior knowledge to learn information because this adds relevance to the material. Students are referred back to other parts of the text or to their own real-life experiences.
- **Practicing Important Tasks** By offering students an opportunity to practice important tasks using a variety of labs and activities in the *Student Edition, Teacher Wraparound Edition,* ancillaries and technology, Glencoe Science makes learning fun and relevant for students.
- **Using Visuals to Communicate, Organize, and Reinforce Learning** High-quality art and photos throughout the text communicate concepts more efficiently and reinforce learning, while allowing students to organize information.
- **Motivating Students to Achieve** Active strategies and real-world experiences motivate students to achieve. Throughout Glencoe's programs, students are encouraged to apply their knowledge in ways that will motivate them to learn.
- **Developing Decoding and Reading Comprehension Strategies** Throughout the text, students are supplied with caption questions, reading checks, and other strategies to aid in comprehension.
- **Using Study Strategies** Through the use of highlighting, outlining, note-taking, summarizing, and other such strategies, students can monitor their own progress and organize information more effectively, thereby increasing their scientific literacy. These strategies are found throughout the text and ancillaries.

The use of these strategies within Glencoe Science will help teachers to achieve the goals set forth by the *National Science Education Standards.*

White Paper

The Glencoe Science White Paper outlines the educational strategies on which this program was based. This document provides specific examples from the *Student Edition, Teacher Wraparound Edition,* ancillary program, and technology resources, highlighting extensive use of educationally sound strategies that help students learn science.

Field Research and Testing

Feedback from students, teachers, curriculum supervisors, department chairpersons, parents, learning specialists, and science content experts was invaluable in the development of this program. The following pre-publication and post-publication research was conducted.

Prior to Publication

- Detailed classroom teacher and curriculum supervisor **surveys** were conducted by independently contracted researchers.
- A **nationwide panel** of science teachers, curriculum supervisors, and department chairpersons provided countless hours of feedback and assistance throughout program development.
- A wide range of **educator and content reviewers** provided in-depth reviews of and suggestions for manuscripts and pre-publication versions of the program.
- **Face-to-face interviews** with science teachers provided insight into teachers' day-to-day challenges.

After Publication

- Field tests were conducted in which students and teachers used a pre-publication manuscript in the classroom.
- Follow-up interviews, observations, and surveys of Glencoe Science users provide ongoing opportunities for program development and verification of program success.

Field-Test Results

- Field-test research indicates that test scores increased among students using Glencoe Science programs.
- Nine out of ten students earned higher scores after using Glencoe programs.
- Scores improved among both male and female students.
- Scores improved among both minority and non-minority students.
- Overall, the gap between the average pre-test score and a perfect score closed by 33 percent. Stated differently, on average, **scores increased 77 percent after students used the Glencoe program.**

KS Studios

National Education Standards

Correlation of *Glencoe Science* to the National Science Education Standards.

Content Standard	Chapter and Section
(UCP) Unifying Concepts and Processes	
1. Systems, order, and organization	A1-1, A1-2, A1-3, A1-4, A2-1, A2-2, A2-3, A3-1, A3-2, A3-3, A4-1, A4-2, A4-3, A5-1, A5-2, A5-3, A6-1, A6-2, A6-3, B1-1, B2-1, B2-2, B4-1, B4-2, B4-3, B5-1, B5-2, C1-1, C2-1, C2-2, C2-3, C2-4, C3-1, C3-2, C3-3, C3-4, C4-1, C4-2, C5-1, C5-2, D1-1, D1-2, D1-3, D2-1, D2-2, D3-1, D3-2, D3-3, D4-1, D4-2, D5-1, D5-2, D6-1, D6-2, D6-3, E1-1, E1-2, E1-3, E2-1, E2-2, E3-1, E3-2, E3-3
2. Evidence, models, and explanation	A1-1, A1-2, A1-3, A1-4, A2-1, A2-2, A2-3, A3-1, A3-2, A3-3, A4-1, A4-2, A4-3, A5-1, A5-2, A5-3, A6-1, A6-2, A6-3, B1-1, B2-1, B2-2, B3-2, B4-1, B4-2, B4-3, B5-1, B5-2, C1-1, C1-2, C1-3, C2-1, C2-2, C2-4, C3-1, C3-2, C3-3, C3-4, C4-1, C4-2, C5-1, C5-2, D1-1, D1-2, D1-3, D2-1, D3-1, D3-2, D3-3, D4-1, D4-2, D5-1, D5-2, D6-1, D6-2, D6-3, D7-1, D7-2, D7-3, E1-1, E1-2, E1-3, E2-1, E2-2, E3-1, E3-2, E3-3, E4-1, E4-2, E4-3, E5-1, E5-2
3. Change, constancy, and measurement	A1-1, A1-2, A1-3, A1-4, A2-1, A2-2, A3-1, A3-2, A3-3, A4-1, A4-3, A5-1, A5-2, A5-3, A6-1, A6-2, A6-3, B2-1, B2-2, B4-1, B4-2, B4-3, B5-1, B5-2, C1-1, C2-1, C2-2, C2-4, C3-1, C3-2, C3-3, C3-4, C4-1, C4-2, C5-1, C5-2, D1-1, D1-2, D1-3, D3-1, D3-2, D3-3, D4-1, D4-2, D5-1, D5-2, D6-1, D6-2, D6-3, E1-1, E1-2, E1-3, E2-1, E2-2, E3-1, E3-2, E3-3, E4-1, E4-2, E4-3, E5-1, E5-2
4. Evolution and equilibrium	A4-3, A5-2, A5-3, A6-1, A6-2, A6-3, B2-1, C3-1, C3-2, C3-3, C3-4, C4-1, C5-1, C5-2, E1-2, E1-3, E2-1, E3-1
5. Form and function	A1-1, A1-2, A1-3, A1-4, A2-1, A2-2, A2-3, A3-1, A3-2, A3-3, A4-3, A5-1, A5-2, A5-3, A6-1, A6-2, A6-3, B1-1, B1-2, B2-1, B2-2, B3-1, B3-2, B3-3, B4-1, B4-2, B4-3, B5-1, B5-2, C1-1, C1-2, C1-3, C2-1, C2-2, C2-3, C2-4, C3-2, C3-3, C3-4, C4-1, C4-2, D1-1, D1-2, D1-3, D3-1, D3-2, D3-3, D4-1, D4-2, D5-1, D5-2, D6-1, D6-2, D6-3, D7-1, D7-2, D7-3, E1-2, E1-3, E2-2, E3-1, E3-2, E3-3, E5-1, E5-2
(A) Science as Inquiry	
1. Abilities necessary to do scientific inquiry	A1-1, A1-2, A1-3, A1-4, A2-1, A2-2, A2-3, A3-1, A3-2, A3-3, A4-1, A4-2, A4-3, A5-1, A5-2, A5-3, A6-1, A6-2, A6-3, B1-1, B1-2, B2-1, B2-2, B3-1, B3-2, B3-3, B4-1, B4-2, B4-3, B5-1, B5-2, C1-1, C1-2, C1-3, C2-1, C2-2, C2-3, C2-4, C3-1, C3-2, C3-3, C3-4, C4-1, C4-2, C5-1, C5-2, D1-1, D1-2, D1-3, D2-1, D2-2, D3-1, D3-2, D3-3, D4-1, D4-2, D5-1, D5-2, D6-1, D6-2, D6-3, D7-1, D7-2, D7-3, E1-1, E1-2, E1-3, E2-1, E2-2, E3-1, E3-2, E3-3, E4-1, E4-2, E4-3, E5-1, E5-2
2. Understandings about scientific inquiry	A1-1, A1-2, A1-3, A1-4, A2-1, A2-2, A2-3, A3-1, A3-2, A3-3, A4-1, A4-3, A5-1, A5-3, A6-1, A6-2, A6-3, B1-1, B2-1, B2-2, B3-1, B4-1, B4-2, B4-3, B5-1, B5-2, C1-2, C1-3, C2-1, C2-2, C2-3, C2-4, C3-1, C3-2, C3-3, C3-4, C4-1, C4-2, C5-1, C5-2, D1-1, D1-2, D1-3, D2-1, D2-2, D3-1, D3-2, D3-3, D4-1, D4-2, D5-1, D5-2, D6-1, D6-2, D6-3, D7-1, D7-3, E5-1, E5-2, E1-1, E1-2, E1-3, E2-1, E2-2, E3-1, E3-2, E3-3, E4-1, E4-2, E4-3
(B) Physical Science	
1. Properties and changes of properties in matter	A3-1
2. Motions and forces	D1-2
3. Transfer of energy	D1-3, D2-1, D2-2, D6-3, E2-1, E2-2, E2-3
(C) Life Science	
1. Structure and function in living systems	A1-2, A1-3, A2-1, A2-2, A2-3, A3-1, A3-2, A3-3, A4-1, A4-2, A4-3, A5-1, A5-2, A5-3, A6-1, A6-3, B1-1, B2-1, B3-1, B3-2, B3-3, B4-1, B4-2, B4-3, B5-1, B5-2, C1-1, C1-2, C1-3, C2-1, C2-2, C2-3, C2-4, C3-1, C3-2, C3-3, C3-4, C4-1, D1-1, D1-2, D1-3, D2-1, D2-2, D3-1, D3-2, D3-3, D4-1, D4-2, D5-1, D5-2, D6-1, D6-2, D6-3, D7-1, D7-2, D7-3, E1-2, E1-3, E2-2, E2-3, E3-1, E3-2, E3-3, E5-1
2. Reproduction and heredity	A1-2, A2-3, A4-1, A4-2, A4-3, A5-1, A5-2, A5-3, B1-1, B2-1, B3-1, B3-2, B3-3, B4-1, B4-2, B4-3, C1-2, C2-1, C2-2, C2-3, C2-4, C3-1, C3-2, C3-3, C3-4, C4-1, C5-2, D1-1, D1-2, D1-3, D6-1, D6-2, D7-1, D7-2, D7-3, E1-1, E1-2, E1-3
3. Regulation and behavior	A1-2, A2-1, A3-2, A4-1, A4-2, A4-3, B1-1, B1-2, B2-1, B3-1, B3-2, B3-3, B5-1, B5-2, C1-1, C1-2, C1-3, C2-1, C2-2, C2-3, C2-4, C3-1, C3-2, C3-3, C3-4, C4-1, C5-1, D1-2, D1-3, D2-1, D2-2, D3-1, D3-2, D3-3, D4-1, D4-2, D5-1, D5-2, D6-1, D6-2, D6-3, D7-1, D7-2, D7-3, E1-2, E2-2, E3-1, E3-2, E3-3
4. Populations and ecosystems	A6-1, B1-1, B1-2, B2-1, B3-1, B3-2, B3-3, B4-3, C1-2, C1-3, C4-1, C5-2, E1-2, E1-3, E2-2, E3-1, E3-2, E3-3, E5-1
5. Diversity and adaptations of organisms	A1-4, A2-3, A4-2, A6-1, A6-3, B1-1, B1-2, B2-1, B3-1, B3-2, B3-3, B4-1, B4-2, B4-3, B5-2, C1-1, C1-2, C1-3, C2-2, C2-3, C2-4, C3-1, C3-2, C3-3, C3-4, C4-1, C5-1, C5-2, D1-1, D1-2, D1-3, D2-1, D2-2, D3-1, D3-2, D3-3, D5-1, D5-2, D6-3, D7-1, D7-2, D7-3, E1-2, E1-3, E2-2, E2-3, E3-3, E5-1

Content Standard	Chapter and Section
(D) Earth and Space Science	
1. Structure of the Earth system	E2-1, E2-2, E2-3
2. Earth's history	A1-3, A6-1, A6-2
3. Earth in the solar system	E2-3
(E) Science and Technology	
1. Abilities of technological design	B1-2
2. Understandings about science and technology	B1-2
(F) Science in Personal and Social Perspectives	
1. Personal health	A5-2, A6-2, B1-2, B2-2, C3-2, C4-2, D1-1, D1-2, D1-3, D2-1, D2-2, D5-1, D7-1, D7-2, D7-3
2. Populations, resources, and environments	A2-3, B4-3, E1-2, E1-3, E3-3, E4-1, E4-2, E4-3, E5-1, E5-2
3. Natural hazards	A5-3, D7-2
4. Risks and benefits	B1-2, C2-3, C3-4, D7-2, D7-3, E5-1, E5-2
5. Science and technology in society	A5-3, B1-2, C3-4, D7-2, D7-3
(G) History and Nature of Science	
1. Science as a human endeavor	A5-2, A6-1, A6-3, B3-2, D1-3, D2-2, D3-3, D7-3
2. Nature of science	A1-1, A3-3, A5-1, A6-1, A6-2, A6-3, B3-2, B4-3, C1-1, D1-3, D2-2, D3-3, D7-1, D7-3
3. History of science	A1-3, A1-4, A2-2, A2-3, A3-3, A5-1, A6-1, A6-3, B3-2, B4-3, C5-2, D4-2, D5-1

How Glencoe Science Aligns with the National Science Education Standards

The correlations at the left and above show the close alignment between Glencoe Science and the grade-appropriate standards. Glencoe Science allows students to discover concepts within each of the content standards and gives students opportunities to make connections among the science disciplines. Hands on activities and inquiry-based lessons reinforce the science processes emphasized in the standards.

How Glencoe Science Aligns with the NCTM Standards for Grades 6–8

Throughout Glencoe Science, each Applying Math activity provides students with the opportunity to practice and apply some of the mathematical concepts and applications described in the NCTM Standards. These activities serve to reinforce mathematical skills in real-life situations, thus preparing students to meet their needs in an ever-changing world.

Correlation of *Glencoe Science* to NCTM Standards

Math Standard	Page
1. Number and Operations	A-35, A-46, A-74, A-93, A-123, A-133, A-151, A-181, B-57, B-112, B-119, B-135, B-145, C-25, C-33, C-52, C-67, C-82, C-129, C-155, D-11, D-31, D-59, D-87, D-133, D-141, D-147, D-171, D-201, E-31, E-40, E-59, E-80, E-121, E-150
2. Algebra	A-35, A-46, A-93, A-181, B-57, B-112, B-119, C-25, C-52, D-133, E-40, E-80, E-121
3. Geometry	A-46, D-11
4. Measurement	A-35, A-46, A-63, D-59, E-40, E-80
5. Data Analysis and Probability	A-35, A-93, A-133, A-151, A-181, B-27, B-57, B-89, B-119, B-135, B-145, C-67, C-101, C-129, C-155, D-31, D-59, D-87, D-113, D-141, D-171, D-201, E-40, E-59, E-89, E-121, E-150
6. Problem Solving	A-46, A-93, A-123, B-27, B-89, C-25, C-33, C-101, D-59, D-133, D-201, E-31, E-40, E-59, E-150
7. Reasoning and Proof	C-82, E-31
8. Communication	A-179, D-167
9. Connections	A-35, A-46, A-63, A-74, A-93, A-123, A-133, A-151, A-181, B-27, B-57, B-112, B-119, B-135, B-145, C-25, C-33, C-52, C-67, C-129, C-155, D-59, D-87, D-133, D-141, D-147, D-171, E-40, E-59, E-80, E-121
10. Representation	A-133, E-40

Foldables™

Foldables™ are easy-to-make, three-dimensional, interactive graphic organizers that students create out of simple sheets of paper. These unique hands-on tools for studying and reviewing were created exclusively for Glencoe by education specialist Dinah Zike.

Research Behind Foldables™

According to research (Bransford, 1979; Corno, 1994), study strategies help students understand, organize, remember and apply new information presented in science textbooks. Some study strategies include concept mapping, highlighting, outlining, note taking, summarizing, and underlining (Peverly, Brobst, Graham & Shaw, 2003). Glencoe Science offers Dinah Zike's Foldables™ Study Organizers as an organizational tool and study guide for students.

Foldables™

- Build prereading skills
- Encourage active reading and writing
- Summarize content for review

FOLDABLES™ Study Organizer

Primary and Secondary Succession Make the following Foldable to help you illustrate the main ideas about succession.

STEP 1 **Fold** a vertical sheet of paper in half from top to bottom.

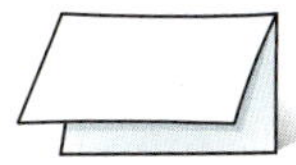

For more ideas on how to incorporate Foldables™ into your lessons consult **Dinah Zike's *Teaching Science with Foldables*™**

Educational Partnerships

NATIONAL GEOGRAPHIC

Some topics in the chapter either require or benefit from a larger, more detailed visual explanation. The National Geographic Society has created *Visualizing* features that call out an important concept from the chapter and illustrate it in a way that will inform, excite, and motivate your students.

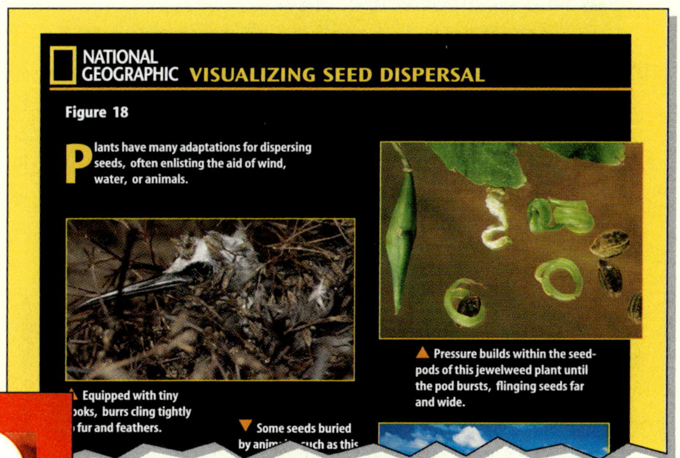

TIME

TIME magazine brings science topics and history together to further explain the chapter's main ideas and show how science relates to real life.

Safety in the Laboratory

All activities are designed to minimize dangers in the laboratory. Careful laboratory planning and management by both the instructor and the student are essential to a safe laboratory experience. **Local, state, and federal laboratory safety laws and regulations must be strictly followed.** The information provided here is one of the many resources to which you can refer for information about laboratory safety.

Classroom and Laboratory Preparation

1. Store equipment properly and securely and other thing.
   a. Clean and dry all equipment before storing.
   b. Protect electronic equipment and microscopes from dust, humidity, and extreme temperatures.
   c. Number, catalog, and organize equipment.
2. Ensure adequate work space for each student.
3. Ensure adequate classroom and storeroom ventilation.
4. Explain and post safety and evacuation guidelines along with expectations of conduct.
5. Ensure that all safety equipment is functioning properly and is clearly visible.
6. Provide hot plates as a heat source whenever possible. If gas burners are used, know where the central gas supply shutoff valve is located.
7. Ensure that each workstation has a GFCI-protected electrical source.
8. Provide safety goggles consistent with ANSI Standard Z87.1 for each student, including students who wear corrective lenses.

Before Each MiniLAB or Lab

1. Arrange the lab in such a way that equipment and supplies are clearly labeled and easily accessible.
2. Have available only equipment and supplies needed to complete the assigned investigation.
3. Review the procedure with students, emphasizing any caution statements or safety symbols that appear.
4. Be sure all students know the proper procedures to follow if an accident should occur.

After the MiniLAB or Lab

1. Be certain that students have returned all equipment and disposed of broken glassware and chemicals properly.
2. Be sure that all hot plates and electrical connections are off.

Storage of Chemicals

Be sure to store all chemicals properly. The following are guidelines commonly used. Your school, city, county, or state may have additional requirements for handling chemicals. It is the responsibility of each teacher to become informed of the rules or guidelines in effect in his or her area.

1. Separate chemicals by reaction type. Strong acids should be stored together. Likewise, strong bases should be stored together and should be separated from acids. Oxidants should be stored away from easily oxidized materials, and so on.
2. Be sure all chemicals are stored in labeled containers indicating contents, concentration, source, date purchased (or prepared), any precautions for handling and storage, and expiration date.
3. Hazardous chemicals require special storage containers and conditions. Be sure to know which chemicals those are and the accepted practices for your area. Some substances must be stored outside the building.

Disposal of Chemicals

Local, state, and federal laws regulate the proper disposal of chemicals. These laws should be consulted before chemical disposal is attempted. Although many substances encountered in the science classroom can be flushed down the drain with plenty of water, it is not safe to assume that this is always true.

DISCLAIMER

Glencoe Publishing Company makes no claims to the completeness of this discussion of laboratory safety and chemical storage. The material presented is not all-inclusive, nor does it address all of the hazards associated with handling, storing, and disposing of chemicals, or with laboratory management.

Preparation of Solutions

It is important to use safe laboratory techniques when handling all chemicals. Always check the MSDS (Material Safety Data Sheet) for each chemical before using it in the classroom. Many substances might appear harmless, but might be toxic, corrosive, or very reactive. Chemicals should never be ingested. Use proper techniques to smell any chemical, wear safety goggles and an apron in the laboratory, and observe the following precautions.

1. **Dilution of Acids and Bases** When diluting acids with water, always add the acids to the water. Never add water to acids. When sulfuric acid and sodium hydroxide are added to water, a large amount of thermal energy is released. Use extra care when handling these substances.

2. **Poisonous and Corrosive Liquids or Vapors** Use a fume hood if possible. Examples include hydrochloric acid, acetic acid, nitric acid, and ammonium hydroxide.

3. **Poisonous and Corrosive to Eyes, Lungs, and Skin** Examples include acids, bases, silver nitrate, iodine, and potassium permanganate.

Bromthymol blue: Add 0.5 g bromthymol blue powder to 500 mL distilled water to make a BTB stock solution. Dilute 40 mL BTB stock solution to 2 L with distilled water. Solution should be bright blue. If not, add one drop of NaOH at a time, swirling to mix. Check color.

Hydrochloric acid (HCL) solution: To make a 5% solution, add 13.6 mL concentrated HCl to 73 mL water while stirring. To make a 0.1*M* solution, add 1 mL concentrated hydrochloric acid to 100 mL water while stirring.

Iodine solution/Iodine stain: Dilute 1 part Lugol's solution with 15 parts water.

Lugol's solution: Dissolve 10 g potassium iodide in 100 mL distilled water. Then add and dissolve 5 g iodine. Store in dark bottle. Keeps indefinitely.

Phenolphthalein indicator: From a drug store, buy a package of any laxative that contains phenolphthalein. To make 1% solution, mash 4 tablets and pour the powder into 10 mL of rubbing alcohol. Let mixture soak for 15 minutes. Pour liquid into and store in a dropper bottle.

Potassium permanganate: For a 0.01*M* solution of potassium permanganate, dissolve 0.15 g $KMnO_4$ in 100 mL water.

Red cabbage concentrate: Put 5 leaves of red cabbage in a pot. Add 1 L of water, bring to a boil, and simmer until water turns a deep purple. Pour liquid through a strainer or piece of cheesecloth into a storage bottle. Keep refrigerated.

Salt solution: For a 3.5% salt (NaCl) solution that simulates the concentration of ocean water, dissolve 35 g of salt (NaCl) in 965 mL of water. For a 1% solution (weak), dissolve 1 g of salt (NaCl) in 99 mL of water. For a 6% solution, dissolve 6 g of salt (NaCl) in 94 mL of water.

Silver nitrate solution: To make a 10% solution, put 5 g of silver nitrate in 50 mL of distilled water.

Sugar solution: Add 1 tablespoon of sugar to 1 cup of warm water in a deep jar or flask. Stir to dissolve.

Sodium hydroxide (dilute): To make a 1% solution, dissolve 1 g NaOH in 99 mL of water.

Equipment and Materials List

Refer to the Chapter Organizer in front of each chapter for a list of equipment and materials used for each laboratory activity in the chapter.

Consumables			
Material	**Launch Lab (Chapter)**	**MiniLAB (Chapter-Section)**	**Lab (Chapter-Section)**
banana		5-1	1-3
bottles, clear, 1-L, plastic, soft-drink			4-2
boxes, cardboard			4-3
cloth or stocking		5-1	
coat hangers, wire			4-3
cookie, chocolate-chip		4-1	
cotton balls			5-1
cotton swabs			5-1
dish soap		2-1	5-1
food coloring, blue			5-1
houseplant fertilizer		2-2	
lettuce leaf			1-3
mustard seeds		5-1	
orange peel(s)			1-3
pH indicator paper		4-2	
plastic sheets, clear			4-3
pond sediment or debris		3-3	
poster board		3-2	4-3
radish seeds		1-2	2-3
sand	4		
soil	4	1-2, 2-1	2-1, 2-3, 4-2
sponge			5-1
toothpick		1	
vegetable or olive oil			5-1
vinegar		1	
water–distilled, pond or stream, rain		4-2	1-3
whole fruit		1-2	
grass clippings or green leaves			2-1
Nonconsumables			
basin, plastic or aluminum-foil baking pan	4		
binoculars			3-2
compass			3-2
cookware, black			4-3
culture containers (with tops)			1-3
die			5-2
feathers, bird (white or light-colored)			5-1
field guides			3-2
fruit fly culture kit			1-3
globe or world map	2		
jar		2-1, 5-1	2-1
lamp (100 W lightbulb)			4-2
net		3-3	
plant pots		1-2	
spray bottle	4		
toothbrush			5-1
Biological			
fruit flies			1-3
guppies			1-3
plant	3	3-3	
liver or meat, raw			1-3
moss	4		
organisms		3-3	
planarians			1-3
sod	1		

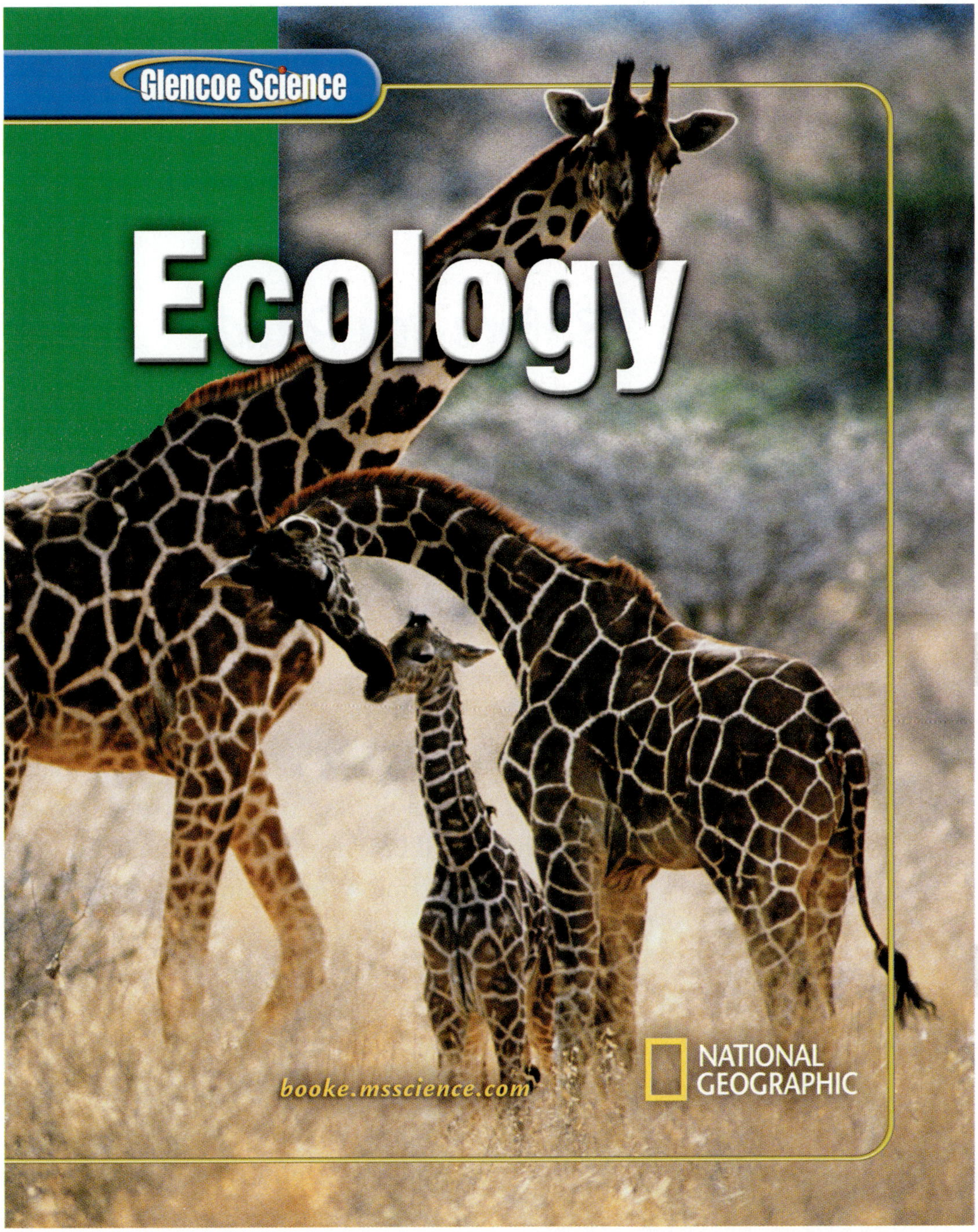

McGraw Hill Glencoe

New York, New York   Columbus, Ohio   Chicago, Illinois   Peoria, Illinois   Woodland Hills, California

HOW TO...

Use Your Science Book

Why do I need my science book?

Have you ever been in class and not understood all of what was presented? Or, you understood everything in class, but at home, got stuck on how to answer a question? Maybe you just wondered when you were ever going to use this stuff?

These next few pages are designed to help you understand everything your science book can be used for . . . besides a paperweight!

Before You Read

- **Chapter Opener** Science is occurring all around you, and the opening photo of each chapter will preview the science you will be learning about. The **Chapter Preview** will give you an idea of what you will be learning about, and you can try the **Launch Lab** to help get your brain headed in the right direction. The **Foldables** exercise is a fun way to keep you organized.
- **Section Opener** Chapters are divided into two to four sections. The **As You Read** in the margin of the first page of each section will let you know what is most important in the section. It is divided into four parts. **What You'll Learn** will tell you the major topics you will be covering. **Why It's Important** will remind you why you are studying this in the first place! The **Review Vocabulary** word is a word you already know, either from your science studies or your prior knowledge. The **New Vocabulary** words are words that you need to learn to understand this section. These words will be in **boldfaced** print and highlighted in the section. Make a note to yourself to recognize these words as you are reading the section.

As You Read

- **Headings** Each section has a title in large red letters, and is further divided into blue titles and small red titles at the beginnings of some paragraphs. To help you study, make an outline of the headings and subheadings.
- **Margins** In the margins of your text, you will find many helpful resources. The **Science Online** exercises and **Integrate** activities help you explore the topics you are studying. **MiniLabs** reinforce the science concepts you have learned.
- **Building Skills** You also will find an **Applying Math** or **Applying Science** activity in each chapter. This gives you extra practice using your new knowledge, and helps prepare you for standardized tests.
- **Student Resources** At the end of the book you will find **Student Resources** to help you throughout your studies. These include **Science, Technology,** and **Math Skill Handbooks,** an **English/Spanish Glossary,** and an **Index.** Also, use your **Foldables** as a resource. It will help you organize information, and review before a test.
- **In Class** Remember, you can always ask your teacher to explain anything you don't understand.

FOLDABLES™ Study Organizer

Science Vocabulary Make the following Foldable to help you understand the vocabulary terms in this chapter.

STEP 1 **Fold** a vertical sheet of notebook paper from side to side.

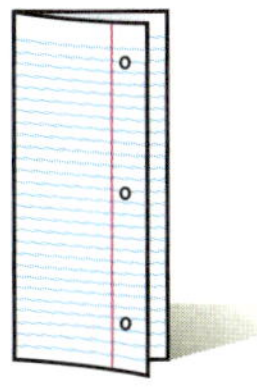

STEP 2 **Cut** along every third line of only the top layer to form tabs.

STEP 3 **Label** each tab with a vocabulary word from the chapter.

Build Vocabulary As you read the chapter, list the vocabulary words on the tabs. As you learn the definitions, write them under the tab for each vocabulary word.

In Lab

Working in the laboratory is one of the best ways to understand the concepts you are studying. Your book will be your guide through your laboratory experiences, and help you begin to think like a scientist. In it, you not only will find the steps necessary to follow the investigations, but you also will find helpful tips to make the most of your time.

- Each lab provides you with a **Real-World Question** to remind you that science is something you use every day, not just in class. This may lead to many more questions about how things happen in your world.
- Remember, experiments do not always produce the result you expect. Scientists have made many discoveries based on investigations with unexpected results. You can try the experiment again to make sure your results were accurate, or perhaps form a new hypothesis to test.
- Keeping a **Science Journal** is how scientists keep accurate records of observations and data. In your journal, you also can write any questions that may arise during your investigation. This is a great method of reminding yourself to find the answers later.

Before a Test

Admit it! You don't like to take tests! However, there *are* ways to review that make them less painful. Your book will help you be more successful taking tests if you use the resources provided to you.

- Review all of the **New Vocabulary** words and be sure you understand their definitions.
- Review the notes you've taken on your **Foldables,** in class, and in lab. Write down any question that you still need answered.
- Review the **Summaries** and **Self Check questions** at the end of each section.
- Study the concepts presented in the chapter by reading the **Study Guide** and answering the questions in the **Chapter Review.**

a or b? ? T or F?

Look For...

- **Reading Checks** and **caption questions** throughout the text.
- the **Summaries** and **Self Check questions** at the end of each section.
- the **Study Guide** and **Review** at the end of each chapter.
- the **Standardized Test Practice** after each chapter.

Scavenger HUNT

Let's Get Started

To help you find the information you need quickly, use the Scavenger Hunt below to learn where things are located in Chapter 1.

1. What is the title of this chapter?
2. What will you learn in Section 1?
3. Sometimes you may ask, "Why am I learning this?" State a reason why the concepts from Section 2 are important.
4. What is the main topic presented in Section 2?
5. How many reading checks are in Section 1?
6. What is the Web address where you can find extra information?
7. What is the main heading above the sixth paragraph in Section 2?
8. There is an integration with another subject mentioned in one of the margins of the chapter. What subject is it?
9. List the new vocabulary words presented in Section 2.
10. List the safety symbols presented in the first Lab.
11. Where would you find a Self Check to be sure you understand the section?
12. Suppose you're doing the Self Check and you have a question about concept mapping. Where could you find help?
13. On what pages are the Chapter Study Guide and Chapter Review?
14. Look in the Table of Contents to find out on which page Section 2 of the chapter begins.
15. You complete the Chapter Review to study for your chapter test. Where could you find another quiz for more practice?

Teacher Advisory Board

The Teacher Advisory Board gave the editorial staff and design team feedback on the content and design of the Student Edition. They provided valuable input in the development of the 2005 edition of ***Glencoe Science.***

John Gonzales
Challenger Middle School
Tucson, AZ

Rachel Shively
Aptakisic Jr. High School
Buffalo Grove, IL

Roger Pratt
Manistique High School
Manistique, MI

Kirtina Hile
Northmor Jr. High/High School
Galion, OH

Marie Renner
Diley Middle School
Pickerington, OH

Nelson Farrier
Hamlin Middle School
Springfield, OR

Jeff Remington
Palmyra Middle School
Palmyra, PA

Erin Peters
Williamsburg Middle School
Arlington, VA

Rubidel Peoples
Meacham Middle School
Fort Worth, TX

Kristi Ramsey
Navasota Jr. High School
Navasota, TX

Student Advisory Board

The Student Advisory Board gave the editorial staff and design team feedback on the design of the Student Edition. We thank these students for their hard work and creative suggestions in making the 2005 edition of ***Glencoe Science*** student friendly.

Jack Andrews
Reynoldsburg Jr. High School
Reynoldsburg, OH

Peter Arnold
Hastings Middle School
Upper Arlington, OH

Emily Barbe
Perry Middle School
Worthington, OH

Kirsty Bateman
Hilliard Heritage Middle School
Hilliard, OH

Andre Brown
Spanish Emersion Academy
Columbus, OH

Chris Dundon
Heritage Middle School
Westerville, OH

Ryan Manafee
Monroe Middle School
Columbus, OH

Addison Owen
Davis Middle School
Dublin, OH

Teriana Patrick
Eastmoor Middle School
Columbus, OH

Ashley Ruz
Karrer Middle School
Dublin, OH

The Glencoe middle school science Student Advisory Board taking a timeout at COSI, a science museum in Columbus, Ohio.

Contents

Nature of Science: Conservation and Native Americans—2

In each chapter, look for these opportunities for review and assessment:

- Reading Checks
- Caption Questions
- Section Review
- Chapter Study Guide
- Chapter Review
- Standardized Test Practice
- Online practice at booke.msscience.com

Contents

Contents

Student Resources

Cross-Curricular Readings/Labs

[video lab icon] available as a video lab

National Geographic VISUALIZING

TIME SCIENCE AND Society

TIME SCIENCE AND HISTORY

Science and Language Arts

Science Stats

Launch LAB

Mini LAB

Mini LAB Try at Home

Content Details

Labs/Activities

One-Page Labs

Two-Page Labs

Design Your Own Labs

Model and Invent Labs

Use the Internet Labs

Applying Math

Applying Science

INTEGRATE

Science Online

Standardized Test Practice

The NATURE OF SCIENCE

Conservation and Native Americans

Introduction

This feature introduces students to the concepts of conservation and resource management. Students will also read that all species on Earth are interconnected and about how each interacts with and affects the environment. The ideas of conservation and resource management are presented through the example of the traditional practices of Native North Americans. The feature also stresses the concept that ideas in science change as humans gain new knowledge, but also that some ideas from history are still valid.

1 Motivate

Why is conservation of resources important? Possible answer: Some resources are of limited supply. Have students make a list of what resources they use everyday. **Which of these resources may need to be conserved?** Answers will vary. **What resources do you conserve?** Possible answers: water, paper (wood), fossil fuels (electricity), plastics

The NATURE OF SCIENCE

History of Science

Conservation and Native Americans

Figure 1 Salmon were the major food source of the Pacific Northwest Native Americans.

A cool breeze blows through the trees lining the Columbia River in the state of Washington as shiny salmon dart through the water. Every year these fish make the difficult journey upstream to lay their eggs in the same waters where they began their lives.

Many years ago, before Europeans came to North America, Native Americans of the northwest depended on salmon. At the first sign of autumn, the salmon run would start. Native American oral histories tell of rivers being so full of the fish that you almost could walk across them to the other side. The native people harvested and dried enough fish to sustain them for the entire year. They also had a deep respect for the salmon. The belief that the fish had a spirit was shown by the tradition of thanking the salmon's spirit for its sacrifice before it was eaten.

Figure 2 Native Americans did not waste any part of nature. This Blackfoot shirt was made with hair, porcupine quills, and feathers.

Respect for Life

The Native American tradition is to respect the bodies and spirits of all animals that gave their lives to feed and clothe the people. On the Great Plains, the Cheyenne and other native peoples had strict rules against killing more bison than they needed. They believed in using every part of the animal. They used as much of the animal as possible for food. The animal's fat was used in cooking. Tools were made from bones. Clothing, shoes, and blankets were made from hides. Even the bisons' stomachs were used as water pouches.

Respect for life extended to plants and crops for Native Americans. The Iroquois of the northeastern United States celebrated festivals in honor of the "three sisters"—corn, squash,

Curriculum Connection

History Have students research the history of Native American fishing in the Pacific Northwest. **Which fish were an important source of food? What methods were used to catch the fish? Were there other uses for the fish besides food?** Possible answers: Herring, salmon, cod, halibut, and flounder; fish were caught with spears, harpoons, trawls, weirs, nets, and hooks on lines. In one case, the skeleton of a fish was used in a ceremony. The oil of some fish was used for flavoring foods.

and beans—their essential foods. The Maya who once lived in what is now southern Mexico and Central America, felt that if someone cut down a tree unnecessarily, that person's life would be shortened.

Every year, in the late spring or early summer, native peoples living on the Plains, including the Cree, Kiowa, Shoshone, and others, celebrated the cycle of life with the Sun Dance. This ceremony, which is performed by many tribes today, stresses the regeneration of life and humans' connection to Earth. Native Americans expressed thankfulness for Earth's gifts—the return of flowers and crops in the spring and summer, and the sacrifice made by animal spirits as they give their bodies to sustain the people. Their traditions recognize that people must cooperate with nature so that revival and rebirth can continue.

Figure 3 Agriculture students experiment with the Native American practice of growing several types of plants in one area.

Respect for Earth

Native Americans felt a deep connection to Earth. While some European settlers viewed the New World as a wilderness ready to be tamed, the native peoples believed in living in harmony with Earth. They didn't understand the European desire to own and develop land. To Native Americans, everyone shared the land. You could no more own the land than you could own the air. The Lakota chief Black Elk once spoke of each human's responsibility to Earth in this way: "Every step that we take upon You should be done in a sacred manner; every step should be taken as a prayer."

Figure 4 Native American artist Helen Hardin titled this piece *Father Sky Embracing Mother Earth.*

2 Teach

Content Background

Resources can be classified as renewable, those that are constantly recycled or replaced by nature, or nonrenewable, those that are used up more quickly than they can be replaced by natural processes. Fossil fuels such as coal, oil, and natural gas are examples of nonrenewable resources. Renewable energy resources include solar power, wind, or water.

Other resources such as groundwater or fisheries can be depleted if they are overexploited. When the removal rate of adult fish from an area exceeds their rate of reproduction, this can result in increasingly smaller numbers in the population. If overharvesting of a particular species continues for an extended period of time, the fish can disappear from an area. During the nineteenth century, halibut was overfished so drastically from the northwestern Atlantic that even today the numbers of halibut caught in this area are very low.

Discussion

How are all species on Earth interconnected? Possible answer: Species are connected by food webs and by the global environment. Pollutants in the air and ocean can have far reaching effects. Issues such as the depletion of the ozone layer, the greenhouse effect, and global warming can affect all species on Earth.

Science Journal

Resource Poems Have students write a poem or song about a resource they use everyday. Students could include why the resource is important and ways it could be managed or conserved. Students can record their writings in their Science Journals. LS **Linguistic**

Content Background

The Iroquois planted the "three sisters" (corn, squash, and beans) together, taking advantage of the natural characteristics of each plant to help the others grow. The corn plants provided shade for the bean plants. The bean and squash vines grew up and around the tall stalk of the corn. The bean plants provided a usable form of nitrogen to surrounding plants. The squash plants also grew close to the ground around the base of the corn and beans, which helped to reduce soil erosion. Also, the prickly foliage of squash vines deter some animals that eat corn.

Activity

Have students each choose one resource and devise a plan to reduce their use of the resource by ten percent. For example, they could choose a fossil fuel such as natural gas, which may be the fuel used to heat their home, heat water, provide electricity, or provide energy for the stove; water; or paper (wood). Suggest students examine a gas or electric bill, count how many pieces of paper are used, or estimate water usage by examining statistics of average use in order to determine how much of the resource their families use in a week. Conservation strategies may include turning off lights or televisions when not in the room or turning off water while brushing teeth. Have students write out their plans and follow them for a week, recording their results.

The NATURE OF SCIENCE

Figure 5 Early industrial development led to environmental pollution.

Need for Conservation

When Europeans began to settle throughout America, it seemed that their way of life would completely replace Native American conservation practices. Many farming techniques used by the new settlers eroded fertile topsoil. Some of the wildlife, which had been treasured and respected by the native peoples, was driven to either extinction or near extinction.

New technology greatly affected the environment. The steam and diesel engines, commercial oil drilling, burning of coal, and industrial development led to waste and pollution.

In recent years, Americans have become more aware of the need for conservation. In 1970, the Environmental Protection Agency was created with the goal of safeguarding the environment. Paper, plastic, glass, and metal are widely recycled. Many items that were once considered waste are now being reused. Although pollution is still a problem, Americans are returning to practices that are more like traditional Native American ways.

Science

The practice of modifying human behavior to preserve Earth is known as conservation. Conservation involves managing the use of natural resources found in many different environments. In this book, you will learn about how all species on Earth are connected and how they can affect their environments.

History of Science

History is a record of people's achievements and their mistakes. Science history teaches that scientific ideas are not limited to scientists. Many people of different ethnic backgrounds, professions, and ages—male and female—have contributed to today's conservation movement. Studying the history of science reminds people that even though accepted views may change, humans constantly gain a better understanding of nature.

Figure 6 Recycling and using nonpolluting types of energy, such as solar energy, are two of the ways Americans are becoming more environmentally conscious.

4 ◆ E

Visual Learning

Figure 7 Unlike petroleum based packing beads, some new packing beads are being manufactured from wheat, corn, or potato starch. Packing beads made from starch are biodegradable. They can also be recycled or composted. **What other items could be used in packing that are recyclable or biodegradable?** Possible answers: paper, newspaper, wood shavings, crumpled plastic grocery bags

Connections

Throughout their history, Native Americans have understood the importance of living in harmony with nature. They practiced conservation as a way of life, not just a passing fad. The people of the northwest understood that fishing for more salmon than they needed would endanger the next year's supply. Today, people are learning this same lesson as some traditional fishing grounds are closed because of overharvesting.

Recall the Mayan belief about how unnecessarily cutting down a tree shortened one's life. It is now known that plants provide oxygen and remove carbon dioxide from the air. Scientists warn that cutting down rain forest trees might lead to a buildup of carbon dioxide in the atmosphere. This could add to the trend in rising, global average temperatures known as global warming.

Figure 7 Many items that you use every day are made of recycled material. These packing beads are made from a wheat product.

Native Americans on the Plains realized the importance of using every part of the bison, learning to live with their minimum needs, and limiting their use of natural resources. This model is being followed today, in some ways, through the effort to reduce, reuse, and recycle.

As new knowledge replaces old, it's tempting to think that only new ideas are worthwhile. Native American traditions of conservation show that this is not always true. These traditions once were disregarded but the value of their conservation is now known.

Figure 8 The amount of land covered by rain forests is decreasing daily.

You Do It

Forests originally covered about 25 percent of Earth's land areas. Today, only about 13 percent of Earth's land areas are covered with forests. Research to find out the history of this change. Explain how following Native American conservation practices might have avoided the problems that face forests today.

Extension

Have students analyze their results from the activity. Were they able to reduce their usage of a resource by at least ten percent? Was the percentage of reduction less? Greater? How could they improve their plans? Have students make a conservation poster that shows the goal, the plan, and the results that might encourage other people to try it.

Use Science Words

Word Origin Have students use a dictionary to determine the origin of the word *ecology*. The word is derived from the Greek *ökologie*, the root of which, *oikos*, means "house"; *–logie* means "science."

Quick Demo

Present students with a list of resources and have them classify the resources as renewable or nonrenewable. Discuss with students the reasons why they choose to classify the resources as they did. Possible answers: nonrenewable: coal, oil, natural gas, minerals, metals in Earth's crust; renewable: Sun (solar power), water, air, crops, or wind

3 Assess

Use the question in You Do It to assess understanding. Answers should relate to what is known about the removal of rain forests and should be as specific as possible.

You Do It

Rain forests are being cut down for several reasons including as a source of wood, for space needed as populations expand, and for farmland. Some of the problems that have resulted from deforestation include habitat destruction for many types of plants and animals that live in the forests. It is the plants in the forest itself that holds most of the nutrients. Once the plants have been removed, the soil is nutrient poor and can be easily eroded.

chapter 1 Organizer

Section/Objectives	Standards		Labs/Features
	National	State/Local	
Chapter Opener	See pp. 16T–17T for a Key to Standards.		**Launch Lab:** How do lawn organisms survive? p. 7 **Foldables,** p. 7
Section 1 Living Earth 2 sessions 1 block 1. **Identify** places where life is found on Earth. 2. **Define** ecology. 3. **Observe** how the environment influences life.	National Content Standards: UCP.1, UCP.2, UCP.3, A.1, A.2, C.2		**Science Online,** p. 10
Section 2 Populations 2 sessions 1 block 4. **Identify** methods for estimating population sizes. 5. **Explain** how competition limits population growth. 6. **List** factors that influence changes in population size.	National Content Standards: UCP.1– UCP.5, A.1, A.2, C.1–C.5, F.2		**MiniLAB:** Observe Seedling Competition, p. 13 **Applying Science:** Do you have too many crickets?, p. 15 **Science Online,** p. 16 **MiniLAB:** Comparing Biotic Potential, p. 17 **Visualizing Population Growth,** p. 18
Section 3 Interactions Within Communities 3 sessions 1.5 blocks 7. **Describe** how organisms obtain energy for life. 8. **Explain** how organisms interact. 9. **Recognize** that every organism occupies a niche.	National Content Standards: UCP.1–UCP.5, A.1, A.2, C.1, C.2, C.4, C.5, F.2		**Integrate Chemistry,** p. 21 **Integrate History,** p. 23 **Lab:** Feeding Habits of Planaria, p. 25 **Lab:** Population Growth in Fruit Flies, p. 26 **Science and History:** The Census measures a human population, p. 28

Glencoe Exclusive! TeacherWorks™ All-In-One Planner and Resource Center

Lab Materials	Reproducible Resources	Section Assessment	Technology
Launch Lab: a section of sod, magnifying lens	**Chapter *FAST FILE* Resources** Foldables Worksheet, p. 17 Directed Reading Overview, p. 19 Note-taking Worksheets, pp. 33–35	Glencoe's Assessment Advantage	TeacherWorks includes: • Interactive Teacher Edition • Lesson Planner with calendar • Access to all program blacklines • Correlations to standards • Web links
Need materials? Contact Science Kit at 1-800-828-7777 or www.sciencekit.com on the Internet.	**Chapter *FAST FILE* Resources** Transparency Activity, p. 44 Enrichment, p. 30 Reinforcement, p. 27 Directed Reading, p. 20	**Portfolio** Activity, p. 10 **Performance** Applying Skills, p. 11 **Content** Challenge, p. 10 Section Review, p. 11	Section Focus Transparency Virtual Labs CD-ROM Guided Reading Audio Program Interactive Chalkboard CD-ROM
MiniLAB: 2 pots of plants, moist potting soil, radish seeds, watering can, basin, metric ruler **MiniLAB:** whole fruit, plastic knife, paper towels	**Chapter *FAST FILE* Resources** Transparency Activity, p. 45 MiniLAB, pp. 3, 4 Enrichment, p. 31 Reinforcement, p. 28 Directed Reading, p. 20 Transparency Activity, pp. 47–48 **Mathematics Skill Activities,** p. 5	**Portfolio** Differentiated Instruction, p. 15 **Performance** MiniLAB, p. 13 Applying Science, p. 15 MiniLAB, p. 17 Applying Skills, p. 19 **Content** Challenge, p. 15 Section Review, p. 19	Section Focus Transparency Teaching Transparency Virtual Labs CD-ROM Guided Reading Audio Program Interactive Chalkboard CD-ROM Video Lab
Lab: small bowl, several planarians, lettuce leaf, raw liver or meat, several guppies, pond or stream water, magnifying lens **Lab:** fruit flies; fruit fly culture kit; food items (banana, orange peel, or other fruit); water; culture containers; cloth, plastic, or other tops for containers; magnifying lens; heating or cooling source	**Chapter *FAST FILE* Resources** Transparency Activity, p. 46 Lab Worksheets, pp. 5–6, 7–8 Enrichment, p. 32 Reinforcement, p. 29 Directed Reading, pp. 21, 22 Lab Activities, pp. 9–11, 13–16 **Home and Community Involvement,** p. 47 **Lab Management and Safety,** p. 71	**Portfolio** Science Journal, p. 23 **Performance** Applying Skills, p. 24 **Content** Challenge, p. 22 Section Review, p. 24	Section Focus Transparency Virtual Labs CD-ROM Guided Reading Audio Program Interactive Chalkboard CD-ROM

End of Chapter Assessment

Glencoe's Assessment Advantage

Blackline Masters	Technology	Professional Series
Chapter *FAST FILE* Resources Chapter Review, pp. 37–38 Chapter Tests, pp. 39–42 **Standardized Test Practice,** pp. 7–10	MindJogger Videoquiz Virtual Labs CD-ROM Exam*View*® Pro Testmaker TeacherWorks CD-ROM Interactive Chalkboard CD-ROM	**Performance Assessment in the Science Classroom (PASC)**

Transparencies

Section Focus

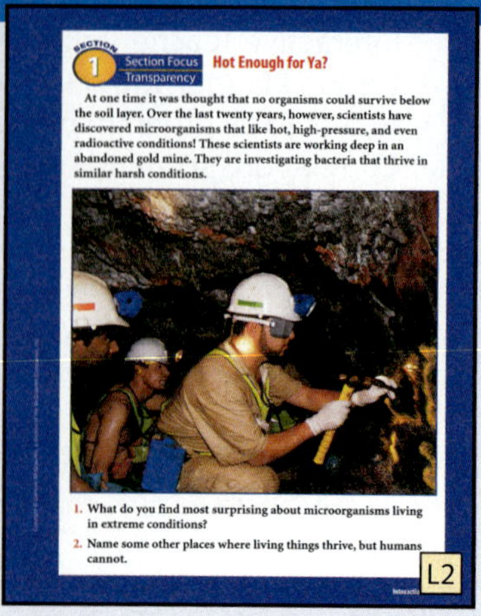

This is a representation of key blackline masters available in the Teacher Classroom Resources. See Resource Manager boxes within the chapter for additional information.

Assessment

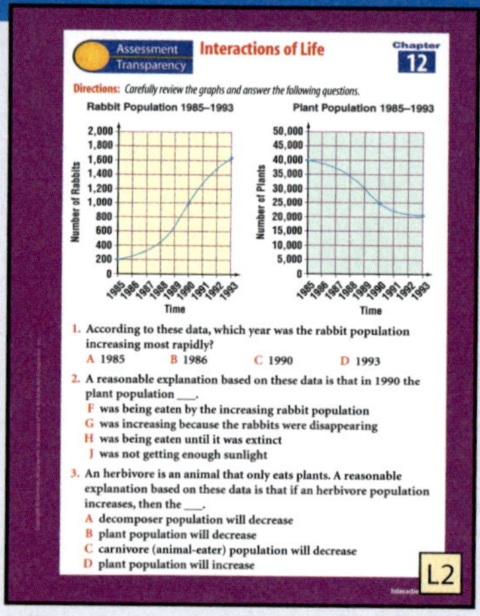

Teaching

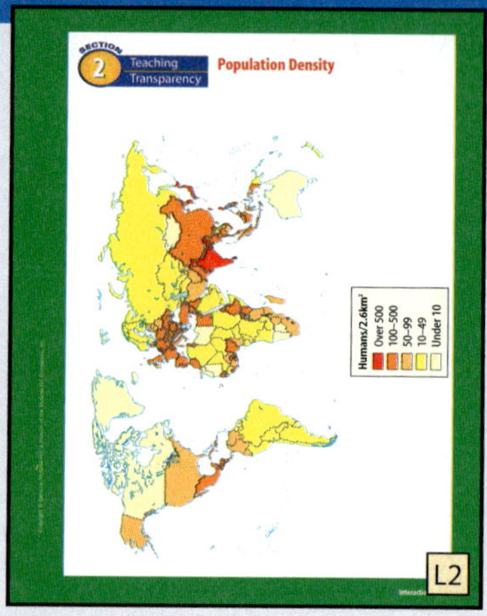

Key to Teaching Strategies

The following designations will help you decide which activities are appropriate for your students.

- L1 Level 1 activities should be appropriate for students with learning difficulties.
- L2 Level 2 activities should be within the ability range of all students.
- L3 Level 3 activities are designed for above-average students.
- ELL ELL activities should be within the ability range of English-Language Learners.
- COOP LEARN Cooperative Learning activities are designed for small group work.
- LS Multiple Learning Styles logos, as described on page 12T, are used throughout to indicate strategies that address different learning styles.
- P These strategies represent student products that can be placed into a best-work portfolio.
- PBL Problem-Based Learning activities apply real-world situations to learning.

Hands-on Activities

Student Text Lab Worksheet

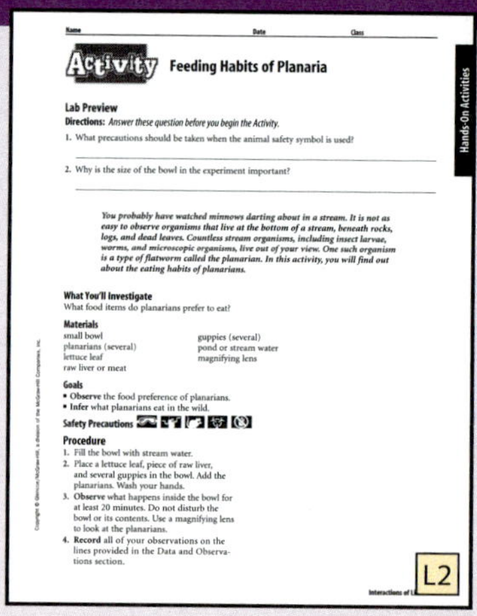

Laboratory Activities

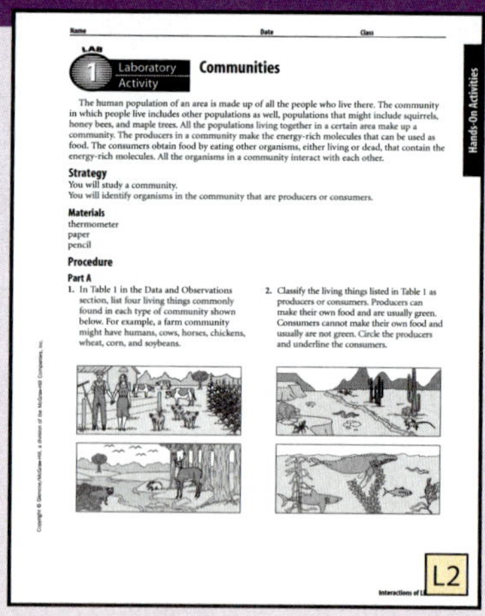

Meeting Different Ability Levels

Content Outline

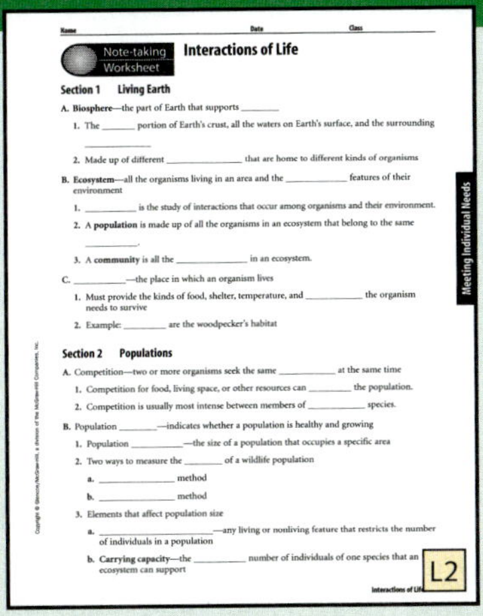

Reinforcement

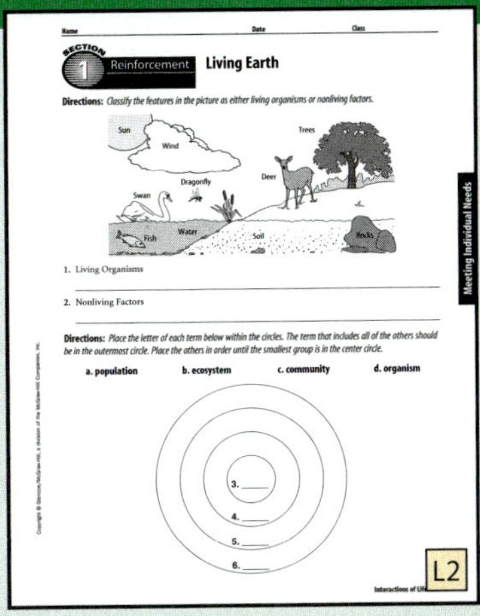

Enrichment

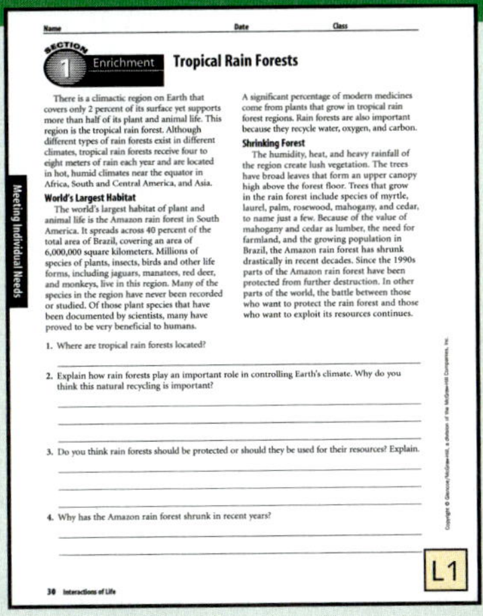

Directed Reading (English/Spanish)

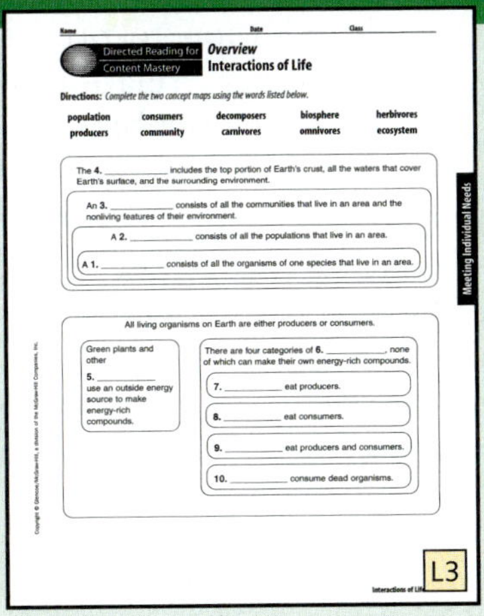

Study Guide

Reading Essentials

Assessment

Test Practice Workbook

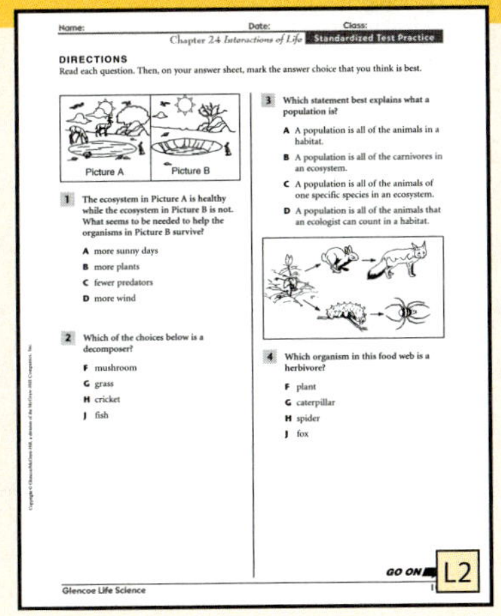

Chapter Review

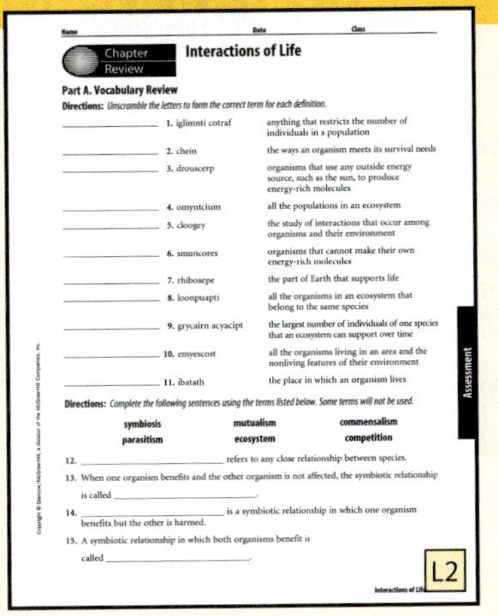

Chapter Tests

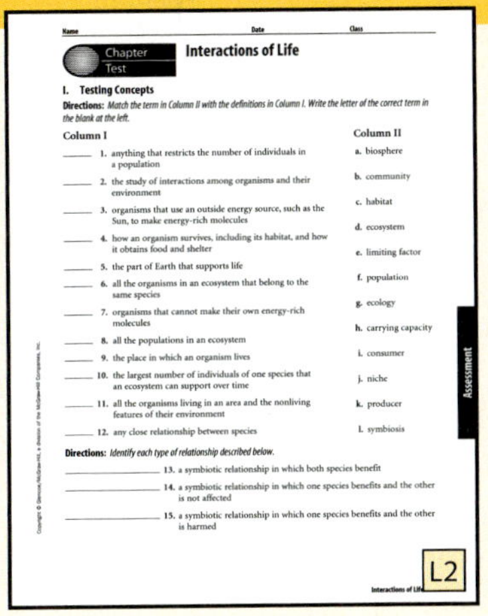

Science Content Background

section 1 Living Earth

Understanding Ecosystems

The biosphere is the total part of Earth where life can exist and includes a great variety of conditions. An ecosystem is a smaller area consisting of the organisms and nonliving features that interact in the system. Sir Arthur George Tansley, a British plant ecologist, coined the word *ecosystem* in 1935.

section 2 Populations

Population Size

Population size is an important characteristic, but it can be difficult to measure. Animal population size is influenced by the amount of food and space available. Natural populations cannot increase forever. Population density measures how crowded a population is. It is always expressed as the number of individuals per unit area or volume.

Carrying Capacity

When a population's size is no longer increasing, it has reached the carrying capacity of its ecosystem. The carrying capacity is the greatest number of individuals that a given environment is capable of supporting under a given set of conditions. If a population is at carrying capacity, the number of organisms born in a given period of time is balanced by the number of organisms that die during that same time.

section 3 Interactions Within Communities

Types of Interactions

Organisms within a community interact in many ways. Plants and other photosynthetic organisms produce sugars that they use as food. Animals, fungi, and other nonphotosynthetic organisms must consume other organisms for food. In addition to predator/prey or producer/consumer relationships, organisms may participate in symbiotic relationships, such as mutualism, commensalism, and parasitism, or they may compete with other species with similar needs for resources. A single species interacts directly or indirectly with nearly all the other species in its community. For example, a squirrel in a forest interacts not only with the plants it eats or with the predators that eat it but also with the plants it uses for cover or shelter, with insects that share its nest, with other animals that use its abandoned nest for shelter, with the bacteria that live on its skin, and so on.

chapter content resources

Internet Resources

For additional content background, visit booke.msscience.com to:

- access your book online
- find references to related articles in popular science magazines
- access Web links with related content background
- access current events with science journal topics

Print Resources

Ecology, A Pocket Guide, by Ernest Callenbach, University of California Press, 1998

Basics of Environmental Science, by Michael Allaby, Routledge, 2000

The Ages of Gaia, by James Lovelock, W.W. Norton, 1988

IDENTIFYING Misconceptions

Find Out What Students Think

Students may think that . . .

Plants do not depend on other organisms.

Organisms do not depend on one another.

Interdependence is primarily predator/prey relationships.

While most students can draw a picture of a natural community that includes producers, herbivores, and predators, few can explain their selections in terms of ecological relationships. Many students understand that the organisms they have depicted belong in the environment, but they don't understand how the organisms interact with one another and with the nonliving environment. While students may be able to identify predator/prey relationships, or plant/herbivore relationships, they seldom understand other types of relationships, nor do they see how interconnected one organism may be with many other organisms in its environment.

Activity

Have students draw a terrestrial community, using the local environment as a model. Their drawings should include several different types of organisms. Ask students to indicate as many interactions between organisms as they can. Then have them describe, aloud or in essay form, how each organism in the picture interacts with other organisms. Note the types and complexity of interactions that students describe.

Promote Understanding

Activity

After students read Section 3, organize the class into teams of three or four students. Assign each team a color. Have each team make 10 or so flags from paper corresponding to their team's color and toothpicks.

- Take the class outside to a grassy or weedy area of the school grounds.
- Have teams hunt for examples of organisms interacting in various ways, including consumption, competition, and symbiosis, and mark with a flag each area where an interaction is occurring. On each flag, team members should identify the type of interaction and write a short description in their Science Journals of what they found. Encourage students to include drawings along with their verbal descriptions.
- Give students time to examine the findings of other teams before gathering up the flags and returning to the classroom.
- Have representatives from each team report their findings to the class. Encourage teams to include their drawings of the interactions they observed.

When all teams have reported, discuss the relationships discovered by the class. Focus the discussion on a few organisms identified by most teams. Challenge students to brainstorm a list of ways in which those organisms interact with other organisms. L2

Assess

After completing the chapter, see *Identifying Misconceptions* in the Study Guide at the end of the chapter.

chapter 1

Chapter Vocabulary

biosphere, p. 8
ecosystem, p. 9
ecology, p. 9
population, p. 10
community, p. 10
habitat, p. 11
limiting factor, p. 14
carrying capacity, p. 15
producer, p. 20
consumer, p. 21
symbiosis, p. 22
mutualism, p. 22
commensalism, p. 22
parasitism, p. 22
niche, p. 23

Science Journal Student responses will vary, but may include feeding relationships, such as birds eating fruit or nuts from trees, or other relationships, such as trees providing nesting sites for birds or insects.

INTERACTIVE CHALKBOARD with Image Bank

PowerPoint® Presentations

This CD-ROM is an editable Microsoft® PowerPoint® presentation that includes:

- a pre-made presentation for every chapter
- interactive graphics
- animations
- audio clips
- image bank
- all new section and chapter questions
- Standardized Test Practice
- transparencies
- pre-lab questions for all labs
- Foldables directions
- links to booke.msscience.com

chapter 1

Interactions of Life

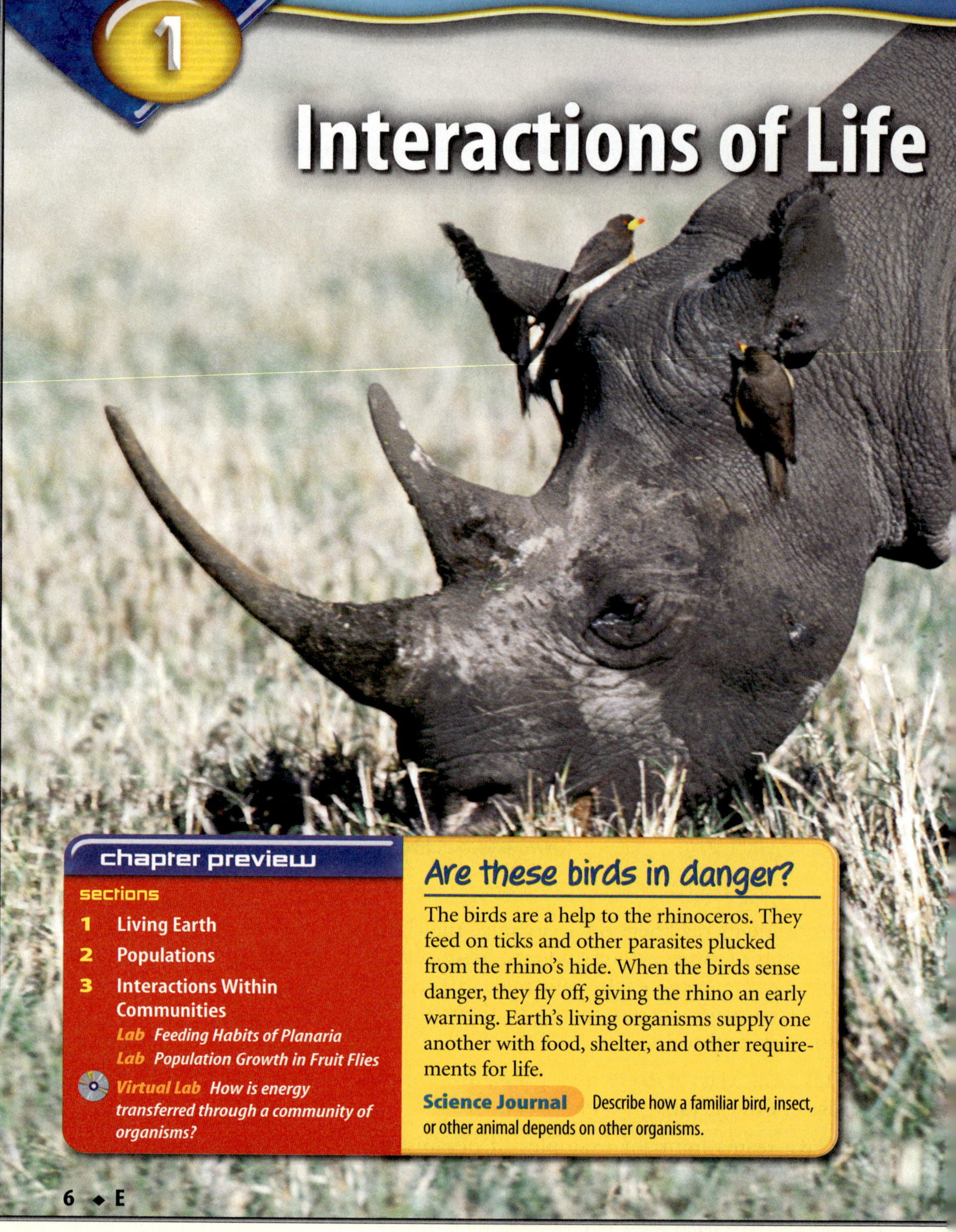

chapter preview

sections

1 Living Earth
2 Populations
3 Interactions Within Communities
Lab *Feeding Habits of Planaria*
Lab *Population Growth in Fruit Flies*
Virtual Lab *How is energy transferred through a community of organisms?*

Are these birds in danger?

The birds are a help to the rhinoceros. They feed on ticks and other parasites plucked from the rhino's hide. When the birds sense danger, they fly off, giving the rhino an early warning. Earth's living organisms supply one another with food, shelter, and other requirements for life.

Science Journal Describe how a familiar bird, insect, or other animal depends on other organisms.

Theme Connection

Systems and Interactions A discussion of ecosystems, habitats, and communities shows how organisms interact. These interactions within the system affect population sizes and organisms' abilities to survive in their environments.

About the Photo

Mutualism The tick bird, or oxpecker, has a mutualistic relationship with the rhinoceros and with other mammals. The bird helps keep mammals free from ticks and other external parasites. The legs and sharp claws of the oxpecker are well adapted for clinging to fur or hide. Each bird obtains almost all of its food from the mammal host and can consume thousands of tick larvae or close to 100 adult ticks daily.

Start-Up Activities

Launch LAB

How do lawn organisms survive?

You probably have taken thousands of footsteps on grassy lawns or playing fields. If you look closely at the grass, you'll see that each blade is attached to roots in the soil. How do grass plants obtain everything they need to live and grow? What other kinds of organisms live in the grass? The following lab will give you a chance to take a closer look at the life in a lawn.

1. Examine a section of sod from a lawn.
2. How do the roots of the grass plants hold the soil?
3. Do you see signs of other living things besides grass?
4. **Think Critically** In your Science Journal, answer the above questions and describe any organisms that are present in your section of sod. Explain how these organisms might affect the growth of grass plants. Draw a picture of your section of sod.

FOLDABLES™ Study Organizer

Ecology Make the following Foldable to help organize information about one of your favorite wild animals and its role in an ecosystem.

STEP 1 **Fold** a vertical sheet of paper from side to side. Make the front edge 1.25 cm shorter than the back edge.

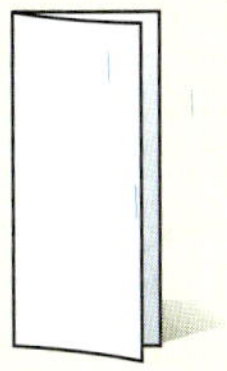

STEP 2 **Turn** lengthwise and **fold** into thirds.

STEP 3 **Unfold and cut** only the top layer along both folds to make three tabs. **Label** each tab.

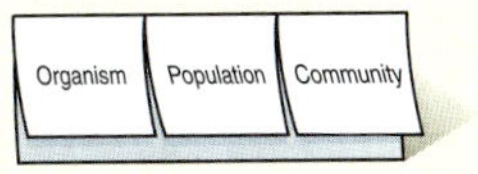

Identify Questions Before you read the chapter, write what you already know about your favorite animal under the left tab of your Foldable. As you read the chapter, write how the animal is part of a population and a community under the appropriate tabs.

Science Online
Preview this chapter's content and activities at booke.msscience.com

Launch LAB

Purpose Use the Launch Lab to introduce students to characteristics of ecosystems and communities. Explain that they will be learning about interactions among organisms of the same or different species. L1 ELL LS **Kinesthetic**

Preparation Purchase sod from a garden center, or dig up a square of sod, including the soil that surrounds the roots. Temporarily fill the area with soil so you don't leave a divot. Replace the sod when you are finished. Keep the sod in a cool, moist area so that it stays fresh and retains small animal life.

Materials small piece of sod for each group, magnifying lens (optional)

Teaching Strategy After a few minutes of examination, instruct students to tear the sod in half. This may reveal animals that have moved to the interior of the sod.

Think Critically

Organisms may include one or more grasses, clover, ants, grasshopper nymphs, earthworms, fungi, or bacteria. The effects of the organisms will vary. A grasshopper might slow the growth of grass, but an earthworm might accelerate it.

Assessment

Process Have students show each other an organism that they observed. Then have them infer the effect of the organism on the grass. Use **Performance Assessment in the Science Classroom,** p. 89. L2

FOLDABLES™ Study Organizer **Dinah Zike Study Fold**

Student preparation materials for this Foldable are available in the Chapter *FAST FILE* Resources.

section 1 Living Earth

1 Motivate

Bellringer

Section Focus Transparencies also are available on the Interactive Chalkboard CD-ROM. L2 ELL

Section Focus Transparency 1 — Hot Enough for Ya?

At one time it was thought that no organisms could survive below the soil layer. Over the last twenty years, however, scientists have discovered microorganisms that like hot, high-pressure, and even radioactive conditions! These scientists are working deep in an abandoned gold mine. They are investigating bacteria that thrive in similar harsh conditions.

1. What do you find most surprising about microorganisms living in extreme conditions?
2. Name some other places where living things thrive, but humans cannot.

Tie to Prior Knowledge

Habitats Ask students to identify several habitats—the places where plants and animals live. Have them describe living and nonliving things in those habitats. L2

Reading Check

Answer the top portion of the crust, surface water, and Earth's atmosphere

as you read

What You'll Learn

- **Identify** places where life is found on Earth.
- **Define** ecology.
- **Observe** how the environment influences life.

Why It's Important

All living things on Earth depend on each other for survival.

Review Vocabulary

adaptation: any variation that makes an organism better suited to its environment

New Vocabulary

- biosphere
- ecosystem
- ecology
- population
- community
- habitat

The Biosphere

What makes Earth different from other planets in the solar system? One difference is Earth's abundance of living organisms. The part of Earth that supports life is the **biosphere** (BI uh sfihr). The biosphere includes the top portion of Earth's crust, all the waters that cover Earth's surface, and the atmosphere that surrounds Earth.

Reading Check *What three things make up the biosphere?*

As **Figure 1** shows, the biosphere is made up of different environments that are home to different kinds of organisms. For example, desert environments receive little rain. Cactus plants, coyotes, and lizards are included in the life of the desert. Tropical rain forest environments receive plenty of rain and warm weather. Parrots, monkeys, and tens of thousands of other organisms live in the rain forest. Coral reefs form in warm, shallow ocean waters. Arctic regions near the north pole are covered with ice and snow. Polar bears, seals, and walruses live in the arctic.

Figure 1 Earth's biosphere consists of many environments, including ocean waters, polar regions, and deserts.

Section 1 Resource Manager

Chapter *FAST FILE* Resources

- Transparency Activity, p. 44
- Reinforcement, p. 27
- Directed Reading for Content Mastery, pp. 19, 20
- Enrichment, p. 30
- Note-taking Worksheets, pp. 33–35

INTEGRATE Astronomy

Life on Earth In our solar system, Earth is the third planet from the Sun. The amount of energy that reaches Earth from the Sun helps make the temperature just right for life. Mercury, the planet closest to the Sun, is too hot during the day and too cold at night to make life possible there. Venus, the second planet from the Sun, has a thick, carbon dioxide atmosphere and high temperatures. It is unlikely that life could survive there. Mars, the fourth planet, is much colder than Earth because it is farther from the Sun and has a thinner atmosphere. It might support microscopic life, but none has been found. The planets beyond Mars probably do not receive enough heat and light from the Sun to have the right conditions for life.

Ecosystems

On a visit to Yellowstone National Park in Wyoming, you might see a prairie scene like the one shown in **Figure 2.** Bison graze on prairie grass. Cowbirds follow the bison, catching grasshoppers that jump away from the bisons' hooves. This scene is part of an ecosystem. An **ecosystem** consists of all the organisms living in an area, as well as the nonliving parts of that environment. Bison, grass, birds, and insects are living organisms of this prairie ecosystem. Water, temperature, sunlight, soil, and air are nonliving features of this prairie ecosystem. **Ecology** is the study of interactions that occur among organisms and their environments. Ecologists are scientists who study these interactions.

Reading Check *What is an ecosystem?*

Figure 2 Ecosystems are made up of living organisms and the nonliving factors of their environment. In this prairie ecosystem, cowbirds eat insects and bison graze on grass.
List *other kinds of organisms that might live in this ecosystem.*

2 Teach

Caption Answer

Figure 2 Possible answers: rabbits, prairie dogs, antelope, small shrubs, hawks

Reading Check

Answer all living organisms and nonliving features of an area

Fun Fact

Both penguins and polar bears live in ice-covered regions, so why is it impossible to see a penguin and a polar bear in the wild at the same time? Polar bears live only in the Arctic, near the North Pole. Penguins live only in the southern hemisphere and in the Antarctic, near the South Pole.

Quick Demo

Locating Ecosystems

Materials globe or large world map

Estimated Time 10 minutes

Procedure Use the globe to show students the location of Earth's tropical regions (area between the Tropic of Cancer and the Tropic of Capricorn). Point out that tropical rain forests and coral reefs are two ecosystems that occur only in tropical areas. Also show students the Arctic region, (area above the Arctic circle) and point out that this icy region is where polar bears, walruses, harp seals, and other cold-tolerant organisms live.

Teacher FYI

Ecosystems are the basic unit of study of ecology. They contain living (biotic) and nonliving (abiotic) components. The terms *biotic* and *abiotic* are introduced later in this unit.

Cultural Diversity

Varieties of Ecosystems Have students who are originally from different cities, states, or countries describe to the class ecosystems from where they used to live. Encourage these students to explain why the ecosystem is important to the people in that area. Show the locations of these areas on a map after students describe them. L2

SECTION 1 Living Earth **E** ◆ 9

IDENTIFYING Misconceptions

Populations Uses of the term *population* may lead to confusion. When discussing human population size it is common to simply say population. For example, "What is the population of San Diego?" In ecology, a population refers to all the organisms of a species in an area. A population has many characteristics, only one of which is size.

Activity

Communities Ask students to observe a natural setting and draw or photograph a community that contains visible plant and animal life. Have students place their drawings or photographs, along with a list of the populations observed, in their Science Journals. L2 ELL LS **Naturalist** P

Visual Learning

Figure 3 What communities are present in the ecosystem shown in this figure? deer, birds, rabbits, grasses, trees

Science Online

Topic: Human Population Data

Visit booke.msscience.com for Web links to information about the estimated human population size for the world today.

Activity Create a graph that shows how the human population has changed between the year 2000 and this year.

Populations

Suppose you meet an ecologist who studies how a herd of bison moves from place to place and how the female bison in the herd care for their young. This ecologist is studying the members of a population. A **population** is made up of all organisms of the same species that live in an area at the same time. For example, all the bison in a prairie ecosystem are one population. All the cowbirds in this ecosystem make up a different population. The grasshoppers make up yet another population.

Ecologists often study how populations interact. For example, an ecologist might try to answer questions about several prairie species. How does grazing by bison affect the growth of prairie grass? How does grazing influence the insects that live in the grass and the birds that eat those insects? This ecologist is studying a community. A **community** is all the populations of all species living in an ecosystem. The prairie community is made of populations of bison, grasshoppers, cowbirds, and all other species in the prairie ecosystem. An arctic community might include populations of fish, seals that eat fish, and polar bears that hunt and eat seals. **Figure 3** shows how organisms, populations, communities, and ecosystems are related.

Figure 3 The living world is arranged in several levels of organization.

Differentiated Instruction

Learning Disabled Give student pairs a one-inch column of newspaper text. Ask them to find, circle, and count different populations. For example, have them find the population of the letter *P* by circling each *P* and counting to determine population size. Explain that one letter is an organism, each type of letter is a population, and all the types of letters are a community. L1

Challenge Have students research the natural habitat of an organism and then tell the class five interesting things about this habitat. L3 LS **Naturalist**

Figure 4 The trees of the forest provide a habitat for woodpeckers and other birds. This salamander's habitat is the moist forest floor.

Habitats

Each organism in an ecosystem needs a place to live. The place in which an organism lives is called its **habitat.** The animals shown in **Figure 4** live in a forest ecosystem. Trees are the woodpecker's habitat. These birds use their strong beaks to pry insects from tree bark or break open acorns and nuts. Woodpeckers usually nest in holes in dead trees. The salamander's habitat is the forest floor, beneath fallen leaves and twigs. Salamanders avoid sunlight and seek damp, dark places. This animal eats small worms, insects, and slugs. An organism's habitat provides the kinds of food and shelter, the temperature, and the amount of moisture the organism needs to survive.

section 1 review

Summary

The Biosphere

- The biosphere is the portion of Earth that supports life.

Ecosystems

- An ecosystem is made up of the living organisms and nonliving parts of an area.

Populations

- A population is made up of all members of a species that live in the same ecosystem.
- A community consists of all the populations in an ecosystem.

Habitats

- A habitat is where an organism lives.

Self Check

1. **List** three parts of the Earth included in the biosphere.
2. **Define** the term *ecology*.
3. **Compare and contrast** the terms *habitat* and *biosphere*.
4. **Identify** the major difference between a community and a population, and give one example of each.
5. **Think Critically** Does the amount of rain that falls in an area determine which kinds of organisms can live there? Why or why not?

Applying Skills

6. **Form a hypothesis** about how a population of dandelion plants might be affected by a population of rabbits.

Science Online booke.msscience.com/self_check_quiz

Discussion

Habitats Point out that a habitat may describe a very small location, such as a nest in a tree, or a much larger area, such as a freshwater lake. What is your habitat? house or apartment What is the habitat of a humpback whale? the ocean

3 Assess

DAILY INTERVENTION

Check for Understanding

Linguistic Have students write a paragraph in their Science Journals explaining the relationships among a population, a community, and an ecosystem. They should include that populations interact to form a community and that communities and nonliving things together make up the ecosystem. L3 LS

Reteach

Ecosystem Display Place a photograph or drawing of a tree on the wall or bulletin board, and identify it as a single organism. Add more trees of the same species and identify the grouping as a population. Add other species of trees as well as birds, mammals, and other organisms, and identify the grouping as a community. Finally, add abiotic factors, such as water, soil, etc., and identify the entire display as an ecosystem.

Assessment

Content Provide students with photographs from discarded magazines. Have them identify living and nonliving things in each photograph and explain how the nonliving things might affect living things. Use **Performance Assessment in the Science Classroom,** p. 89. L1

section 1 review

1. Earth's crust, surface waters, atmosphere
2. the study of the interactions between organisms and between organisms and their environment
3. Habitat—place where organisms live; the biosphere includes all Earth's habitats.
4. A population is a single species living in an area. Example: all the maple trees of the Catskill Mountains. A community is all the populations that live in an area. Example: all the species of the Catskill Mountains.
5. Yes; organisms that need a lot of water cannot exist in areas with little rainfall.
6. Answers will vary. If students are aware that rabbits eat dandelions, their hypotheses should indicate that a rabbit population would limit the size of a dandelion population in the same community.

1 Motivate

INTERACTIVE CHALKBOARD PowerPoint® Presentations

Bellringer

Section Focus Transparencies also are available on the Interactive Chalkboard CD-ROM. L2 ELL

Tie to Prior Knowledge

Requirements for Life Have students recall the needs of living things. food, water, oxygen, shelter, living space Knowing what living things need to grow and reproduce will be helpful in understanding limiting factors. L1

section 2 Populations

as you read

What You'll Learn

- **Identify** methods for estimating population sizes.
- **Explain** how competition limits population growth.
- **List** factors that influence changes in population size.

Why It's Important

Competition caused by population growth reduces the amount of food, living space, and other resources available to organisms, including humans.

Review Vocabulary
natural selection: hypothesis that states organisms with traits best suited to their environment are more likely to survive and reproduce

New Vocabulary
- limiting factor
- carrying capacity

Competition

Wild crickets feed on plant material at night. They hide under leaves or in dark damp places during the day. In some pet shops, crickets are raised in cages and fed to pet reptiles. Crickets require plenty of food, water, and hiding places. As a population of caged crickets grows, extra food and more hiding places are needed. To avoid crowding, some crickets might have to be moved to other cages.

Food and Space Organisms living in the wild do not always have enough food or living space. The Gila woodpecker, shown in **Figure 5,** lives in the Sonoran Desert of Arizona and Mexico. This woodpecker makes its nest by drilling a hole in a saguaro (suh GWAR oh) cactus. Woodpeckers must compete with each other for nesting spots. Competition occurs when two or more organisms seek the same resource at the same time.

Growth Limits Competition limits population size. If available nesting spaces are limited, some woodpeckers will not be able to raise young. Gila woodpeckers eat cactus fruit, berries, and insects. If food becomes scarce, some woodpeckers might not survive to reproduce. Competition for food, living space, or other resources can limit population growth.

In nature, the most intense competition is usually among individuals of the same species, because they need the same kinds of food and shelter. Competition also takes place among different species. For example, after a Gila woodpecker has abandoned its nest, owls, flycatchers, snakes, and lizards might compete for the shelter of the empty hole.

Figure 5 Gila woodpeckers make nesting holes in the saguaro cactus. Many animals compete for the shelter these holes provide.

Section 2 Resource Manager

Chapter FAST FILE Resources
- Transparency Activity, pp. 45, 47–48
- Directed Reading for Content Mastery, p. 20
- MiniLAB, pp. 3, 4
- Enrichment, p. 31
- Reinforcement, p. 28

Reading and Writing Skill Activities, p. 1
Mathematics Skill Activities, p. 5
Life Science Critical Thinking/Problem Solving, pp. 12, 13

12 ◆ E CHAPTER 1 Interactions of Life

Population Size

Ecologists often need to measure the size of a population. This information can indicate whether or not a population is healthy and growing. Population counts can help identify populations that could be in danger of disappearing.

Some populations are easy to measure. If you were raising crickets, you could measure the size of your cricket population simply by counting all the crickets in the container. What if you wanted to compare the cricket populations in two different containers? You would calculate the number of crickets per square meter (m^2) of your container. The number of individuals of one species per a specific area is called population density. **Figure 6** shows Earth's human population density.

✓ Reading Check *What is population density?*

Measuring Populations Counting crickets can be tricky. They look alike, move a lot, and hide. The same cricket could be counted more than once, and others could be completely missed. Ecologists have similar problems when measuring wildlife populations. One of the methods they use is called trap-mark-release. Suppose you want to count wild rabbits. Rabbits live underground and come out at dawn and dusk to eat. Ecologists set traps that capture rabbits without injuring them. Each captured rabbit is marked and released. Later, another sample of rabbits is captured. Some of these rabbits will have marks, but many will not. By comparing the number of marked and unmarked rabbits in the second sample, ecologists can estimate the population size.

Mini LAB

Observing Seedling Competition

Procedure

1. Fill **two plant pots** with **moist potting soil.**
2. Plant **radish seeds** in one pot, following the spacing instructions on the seed packet. Label this pot *Recommended Spacing*.
3. Plant **radish seeds** in the second pot, spaced half the recommended distance apart. Label this pot *Densely Populated*. Wash your hands.
4. Keep the soil moist. When the seeds sprout, move them to a well-lit area.
5. Measure and record in your **Science Journal** the height of the seedlings every two days for two weeks.

Analysis

1. Which plants grew faster?
2. Which plants looked healthiest after two weeks?
3. How did competition influence the plants?

Try at Home

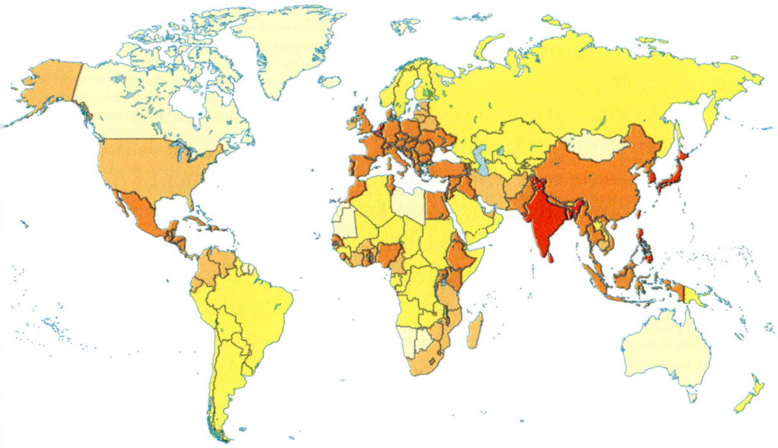

Figure 6 This map shows human population density.
Interpret Illustrations *Which countries have the highest population density?*

2 Teach

✓ Reading Check

Answer the size of a population that occupies a specific area

Mini LAB

Purpose Students determine if population density affects the growth of certain plants. L2 LS **Logical-Mathematical**

Materials 2 plant pots, potting soil, radish seeds, watering can, large basin, metric ruler

Teaching Strategies

- Have students premeasure water before adding.

Troubleshooting Follow seed packet instructions as to the number of seeds to plant per hole. Thin plants so the density of seedlings in one pot is half the density in the other.

Analysis

1. In many cases the densely planted seedlings grew faster.
2. The plants with more space will usually look healthier. They may not be as tall, but they look greener, fuller, and more robust.
3. Competition probably caused the less densely planted seedlings to grow better because they were able to get more light.

Assessment

Oral Have students explain why more densely planted seedlings might grow faster. Have them predict which plants would be healthier if the experiment continued for four weeks. Use **PASC**, p. 99. L2

Try at Home

Caption Answer

Figure 6 India, Japan, the Philippines, and South Korea

Discussion

Population Size Why would people want to know the population size of organisms such as deer? Ecologists may want the data for research. Game and fish departments may want the information to decide how many hunting permits to issue.

Caption Answer

Figure 7 Answers will vary. Students may suggest that they can count the number of wildebeests in the enlarged square, and then multiply by 20, the number of squares in the entire photograph.

Inquiry Lab

Measuring Populations

Purpose To explore methods for estimating population size, have students study a natural population.

Possible Materials magnifying lens, tape measure, field guide, stakes and string, trowel

Teaching Strategies

- Students can use sample counts to estimate the population of a grass or wildflower species in a field. They could use stakes and string to mark out sample plots for counting.
- Students could also count earthworms or other organisms from a soil sample and use this information to estimate population density in a given area.
- Allow students to explore other questions that arise. L2

For additional inquiry activities, see *Science Inquiry Labs*.

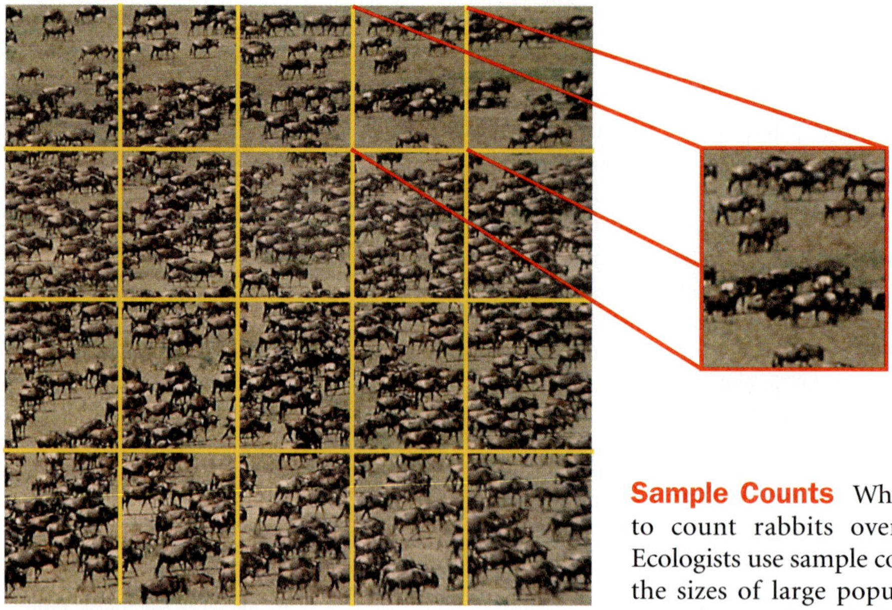

Figure 7 Ecologists can estimate population size by making a sample count. Wildebeests graze on the grassy plains of Africa.
Draw Conclusions *How could you use the enlarged square to estimate the number of wildebeests in the entire photograph?*

Sample Counts What if you wanted to count rabbits over a large area? Ecologists use sample counts to estimate the sizes of large populations. To estimate the number of rabbits in an area of 100 acres, for example, you could count the rabbits in one acre and multiply by 100 to estimate the population size. **Figure 7** shows another approach to sample counting.

Limiting Factors One grass plant can produce hundreds of seeds. Imagine those seeds drifting onto a vacant field. Many of the seeds sprout and grow into grass plants that produce hundreds more seeds. Soon the field is covered with grass. Can this grass population keep growing forever? Suppose the seeds of wildflowers or trees drift onto the field. If those seeds sprout, trees and flowers would compete with grasses for sunlight, soil, and water. Even if the grasses did not have to compete with other plants, they might eventually use up all the space in the field. When no more living space is available, the population cannot grow.

In any ecosystem, the availability of food, water, living space, mates, nesting sites, and other resources is often limited. A **limiting factor** is anything that restricts the number of individuals in a population. Limiting factors include living and nonliving features of the ecosystem.

A limiting factor can affect more than one population in a community. Suppose a lack of rain limits plant growth in a meadow. Fewer plants produce fewer seeds. For seed-eating mice, this reduction in the food supply could become a limiting factor. A smaller mouse population could, in turn, become a limiting factor for the hawks and owls that feed on mice.

LAB DEMONSTRATION

Purpose to estimate population size

Materials overhead projector (OHP), 4 sheets of paper, transparency, marker, ruler, cardboard to screen the OHP surface, hole punch

Preparation Punch 100 discs. Draw a grid on the transparency consisting of twenty-five 1.25″ × 1.25″ squares.

Procedure Display the grid. Erect the screen. Evenly sprinkle 25 discs. Position papers to reveal one square. Turn on the OHP; count the discs. Repeat twice. Estimate and then count population size. Repeat for 50 and 100 discs.

Expected Outcome Estimates will approximate actual population size.

Assessment

How could a better estimate be made? Collect data from a greater number of cells.

Carrying Capacity A population of robins lives in a grove of trees in a park. Over several years, the number of robins increases and nesting space becomes scarce. Nesting space is a limiting factor that prevents the robin population from getting any larger. This ecosystem has reached its carrying capacity for robins. **Carrying capacity** is the largest number of individuals of one species that an ecosystem can support over time. If a population begins to exceed the environment's carrying capacity, some individuals will not have enough resources. They could die or be forced to move elsewhere, like the deer shown in **Figure 8.**

Figure 8 These deer might have moved into a residential area because a nearby forest's carrying capacity for deer has been reached.

Reading Check *How are limiting factors related to carrying capacity?*

Applying Science

Do you have too many crickets?

You've decided to raise crickets to sell to pet stores. A friend says you should not allow the cricket population density to go over 210 crickets/m^2. Use what you've learned in this section to measure the population density in your cricket tanks.

Identifying the Problem

The table on the right lists the areas and populations of your three cricket tanks. How can you determine if too many crickets are in one tank? If a tank contains too many crickets, what could you do? Explain why too many crickets in a tank might be a problem.

Cricket Population

Tank	Area (m^2)	Number of Crickets
1	0.80	200
2	0.80	150
3	1.5	315

Solving the Problem

1. Do any of the tanks contain too many crickets? Could you make the population density of the three tanks equal by moving crickets from one tank to another? If so, which tank would you move crickets into?
2. Wild crickets living in a field have a population density of 2.4 crickets/m^2. If the field's area is 250 m^2, what is the approximate size of the cricket population? Why would the population density of crickets in a field be lower than the population density of crickets in a tank?

Activity

Carrying Capacity Have the classroom represent an ecosystem. Set the carrying capacity of the room at 80 percent of the number of students present. Have the remaining 20 percent of the students stand to the side. Have these "extra" individuals enter the ecosystem through birth or immigration and have others leave through death or emigration so that the size of the population remains steady. L2 LS **Kinesthetic**

Reading Check

Answer Carrying capacity exists because of limiting factors. A shortage of one or more things limits population size.

Applying Science

Teaching Strategy

Remind students to calculate the population density of each tank listed in the table before determining which tank(s) are overpopulated.

Answers

1. The population densities of the three tanks are 1: 250 crickets/m^2 2: 187.5 crickets/m^2 3: 210 crickets/m^2. Tank 1 has too many crickets. You might try putting some of the extra crickets into tank 2, because it has a lower population density. Tank 3 has the maximum population density allowed.
2. Multiply the population density by the area: 250 m^2 $\times$ 2.4 crickets/m^2 = 600 crickets. The field's population density would be less than the tank's because the crickets have more space and less competition.

Differentiated Instruction

Challenge Crabgrass is an annual plant. Assume one crabgrass plant produces 100 seeds every season, and each of those sprouts into a plant next season that will also produce 100 seeds. If you start with one grass plant, how many will you have after three years? year 1–1 plant; year 2–100 plants; year 3–10,000 plants L3

Learning Disabled Have students research populations that live in areas in which limiting factors severely restrict the kinds of organisms that can live there. Have them make a collage that shows several different populations and environments, such as penguins in Antarctica and cacti in a hot, dry desert. L2 P

Visual Learning

Table 1 Which country has the fastest population growth? Zimbabwe Which country has the slowest growth? Germany What does a negative growth rate mean? Death rate is greater than birthrate.

Quick Demo

Observing Seed Distribution

Materials seeds that travel, such as dandelion, maple, and foxtails or burr clover

Estimated Time 10 minutes

Safety Precautions Be sure students wear appropriate outdoor clothing and use insect protection, if necessary.

Procedure Blow on the dandelion seeds to show how easily they move through the air. Drop a maple seed to show how its "wings" make the seed move. Demonstrate how foxtails and burr clover cling to fur or clothing. Discuss with students the advantages of a seed being able to move away from the parent plant. reduces competition for space and nutrients between offspring and parent

Fun Fact

Why are the ants that sometimes invade people's kitchens so hard to get rid of? One reason could be population size. A typical colony of Argentine ants numbers about 1 million individuals. Scientists have discovered a "supercolony" of Argentine ants that extends from Italy to Spain and contains billions of individuals.

Science Online

Topic: Birthrates and Death Rates

Visit booke.msscience.com for Web links to information about birthrates and death rates for the human population.

Activity Find out whether the human population worldwide is increasing because of rising birthrates or declining death rates.

Biotic Potential What would happen if no limiting factors restricted the growth of a population? Think about a population that has an unlimited supply of food, water, and living space. The climate is favorable. Population growth is not limited by diseases, predators, or competition with other species. Under ideal conditions like these, the population would continue to grow.

The highest rate of reproduction under ideal conditions is a population's biotic potential. The larger the number of offspring that are produced by parent organisms, the higher the biotic potential of the species will be. Compare an avocado tree to a tangerine tree. Assume that each tree produces the same number of fruits. Each avocado fruit contains one large seed. Each tangerine fruit contains a dozen seeds or more. Because the tangerine tree produces more seeds per fruit, it has a higher biotic potential than the avocado tree.

Changes in Populations

Birthrates and death rates also influence the size of a population and its rate of growth. A population gets larger when the number of individuals born is greater than the number of individuals that die. When the number of deaths is greater than the number of births, populations get smaller. Take the squirrels living in New York City's Central Park as an example. In one year, if 900 squirrels are born and 800 die, the population increases by 100. If 400 squirrels are born and 500 die, the population decreases by 100.

The same is true for human populations. **Table 1** shows birthrates, death rates, and population changes for several countries around the world. In countries with faster population growth, birthrates are much higher than death rates. In countries with slower population growth, birthrates are only slightly higher than death rates. In Germany, where the population is getting smaller, the birthrate is lower than the death rate.

Table 1 Population Growth

	Birthrate*	Death Rate*	Population Increase (percent)
Rapid-Growth Countries			
Jordan	38.8	5.5	3.3
Uganda	50.8	21.8	2.9
Zimbabwe	34.3	9.4	5.2
Slow-Growth Countries			
Germany	9.4	10.8	−1.5
Sweden	10.8	10.6	0.1
United States	14.8	8.8	0.6

*Number per 1,000 people

Differentiated Instruction

Hearing Impaired Have students use index cards to make flipbooks showing variations in population growth. On the first cards, have them show a few organisms followed by exponential growth. Next, have them show how competition for resources affects the population. Finally, have them show a constant population size. Suggest they include a line graph that indicates change in population size in each scene. L2

Teacher FYI

Offspring Some organisms produce few offspring and some produce many. Fewer offspring benefit because they can be supplied with more of the parents' resources. Having many offspring increases the odds that one or more will survive to reproduce.

Figure 9 Mangrove seeds sprout while they are still attached to the parent tree. Some sprouted seeds drop into the mud below the parent tree and continue to grow. Others drop into the water and can be carried away by tides and ocean currents. When they wash ashore, they might start a new population of mangroves or add to an existing mangrove population.

Moving Around Most animals can move easily from place to place, and these movements can affect population size. For example, a male mountain sheep might wander many miles in search of a mate. After he finds a mate, their offspring might establish a completely new population of mountain sheep far from the male's original population.

Many bird species move from one place to another during their annual migrations. During the summer, populations of Baltimore orioles are found throughout eastern North America. During the winter, these populations disappear because the birds migrate to Central America. They spend the winter there, where the climate is mild and food supplies are plentiful. When summer approaches, the orioles migrate back to North America.

Even plants and microscopic organisms can move from place to place, carried by wind, water, or animals. The tiny spores of mushrooms, mosses, and ferns float through the air. The seeds of dandelions, maple trees, and other plants have feathery or winglike growths that allow them to be carried by wind. Spine-covered seeds hitch rides by clinging to animal fur or people's clothing. Many kinds of seeds can be transported by river and ocean currents. Mangrove trees growing along Florida's Gulf Coast, shown in **Figure 9,** provide an example of how water moves seeds.

Mini LAB

Comparing Biotic Potential

Procedure

1. Remove all the seeds from a **whole fruit.** Do not put fruit or seeds in your mouth.
2. Count the total number of seeds in the fruit. Wash your hands, then record these data in your **Science Journal.**
3. Compare your seed totals with those of classmates who examined other types of fruit.

Analysis

1. Which type of fruit had the most seeds? Which had the fewest seeds?
2. What is an advantage of producing many seeds? Can you think of a possible disadvantage?
3. To estimate the total number of seeds produced by a tomato plant, what would you need to know?

Mini LAB

Purpose Students observe and compare the number of seeds in different types of fruit. L2 ELL

LS **Logical-Mathematical**

Materials plastic knives, paper towels, assorted fruits

Teaching Strategies

- Ask students to bring in fruits from home.
- Explain that many of our vegetables are actually fruits, including tomatoes, eggplants, and squashes.

Safety Precautions Instruct students on the proper way to handle and use the plastic knives.

Analysis

1. Tomatoes and strawberries have many seeds; avocados and peaches have one seed.
2. Advantage: There is a greater chance that a seed will find an optimal place to grow. Disadvantage: The seeds are small and carry few nutrients.
3. how many tomatoes the plant produces and about how many seeds are produced by each tomato

Assessment

Oral Based on the number of seeds contained in each fruit, which has a greater biotic potential—a nectarine or a watermelon? Explain. A watermelon—it contains many seeds, while a nectarine contains only one. Use **Performance Assessment in the Science Classroom,** p. 89.

Active Reading

Four-Corner Discussion This strategy encourages debate. Make four signs: Strongly Agree, Agree, Disagree, Strongly Disagree. Place one sign in each corner of the room. Write on the board a statement that will elicit reactions from students. Have students respond on paper. After several minutes, direct them to move to the corner with the sign that most closely reflects their opinions. In the corners, students share responses. Each group selects a spokesperson to report the group's opinions. After all reports, open the floor for debate. Allow students who change their opinions to change corners. Use this strategy to discuss the causes of change in populations. L2

NATIONAL GEOGRAPHIC

Visualizing Population Growth

Have students examine the pictures and read the captions. Then ask the following question.

What happens when a population exceeds carrying capacity? Possible answer: There may not be enough resources to support all members of the population. It is likely that members will begin to die from lack of resources such as food and water.

Activity

Data Table Have students graph the following data on predator-prey cycles between lynx and snowshoe hares.

Years	# of lynx	# of hares
1845	30,000	18,000
1850	10,000	40,000
1855	30,000	78,000
1860	8,000	21,000
1865	67,000	71,000
1870	7,000	12,000
1875	40,000	99,000
1880	12,000	9,000

What is the pattern between the population numbers of the lynx and the hares? As the number of hares increases, the number of lynx increases. As the larger lynx population feeds on hares, the number of hares decreases, followed by a decrease in the number of lynx. L2

NATIONAL GEOGRAPHIC **VISUALIZING POPULATION GROWTH**

Figure 10

When a species enters an ecosystem that has abundant food, water, and other resources, its population can flourish. Beginning with a few organisms, the population increases until the number of organisms and available resources are in balance. At that point, population growth slows or stops. A graph of these changes over time produces an S-curve, as shown here for coyotes.

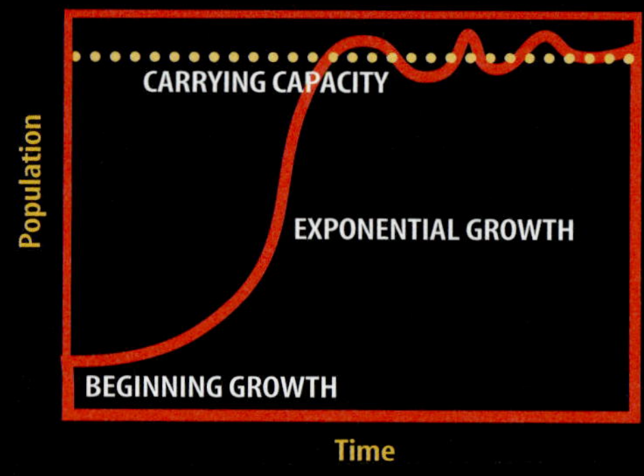

BEGINNING GROWTH During the first few years, population growth is slow, because there are few adults to produce young. As the population grows, so does the number of breeding adults.

EXPONENTIAL GROWTH As the number of adults in the population grows, so does the number of births. The coyote population undergoes exponential growth, quickly increasing in size.

CARRYING CAPACITY As resources become less plentiful, the birthrate declines and the death rate may rise. Population growth slows. The coyote population has reached the environmental carrying capacity—the maximum number of coyotes that the environment can sustain.

Curriculum Connection

Math The bacterium *E. coli* can double in population size every 20 minutes under suitable conditions. Starting with a single bacterium, have students calculate the length of time it would take for the population to reach 1 million. The population would reach 1 million in only 6 hours and 40 minutes. L2

LS **Logical-Mathematical**

Exponential Growth When a species moves into a new area with plenty of food, living space, and other resources, the population grows quickly, in a pattern called exponential growth. Exponential growth means that the larger a population gets, the faster it grows. Over time, the population will reach the ecosystem's carrying capacity for that species. **Figure 10** shows each stage in this pattern of population growth.

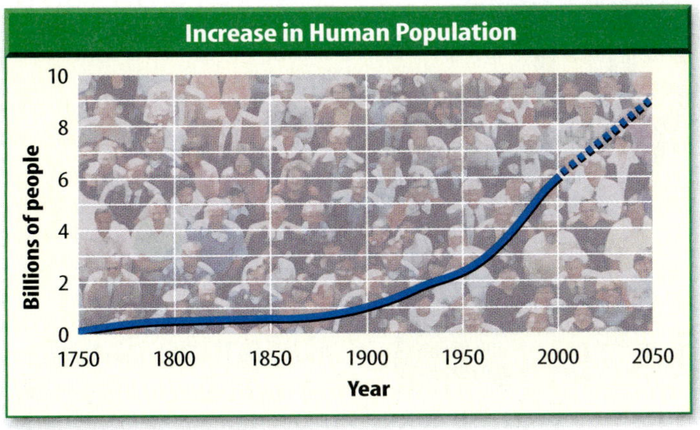

Figure 11 The size of the human population is increasing by about 1.6 percent per year.
Identify *the factors that affect human population growth.*

As a population approaches its ecosystem's carrying capacity, competition for living space and other resources increases. As you can see in **Figure 11,** Earth's human population shows exponential growth. By the year 2050, the population could reach 9 billion. You probably have read about or experienced some of the competition associated with human population growth, such as freeway traffic jams, crowded subways and buses, or housing shortages. As population density increases, people are forced to live closer to one another. Infectious diseases can spread easily when people are crowded together.

section 2 review

Summary

Competition

- When more than one organism needs the same resource, competition occurs.
- Competition limits population size.

Population Size

- Population density is the number of individuals per unit area.
- Limiting factors are resources that restrict population size.
- An ecosystem's carrying capacity is the largest population it can support.
- Biotic potential is the highest possible rate of growth for a population.

Changes in Populations

- Birthrates, death rates, and movement from place to place affect population size.

Self Check

1. **Describe** three ways in which ecologists can estimate the size of a population.
2. **Explain** how birthrates and death rates influence the size of a population.
3. **Explain** how carrying capacity influences the number of organisms in an ecosystem.
4. **Think Critically** Why are food and water the limiting factors that usually have the greatest effect on population size?

Applying Skills

5. **Make and use a table** on changes in the size of a deer population in Arizona. Use the following data. In 1910 there were 6 deer; in 1915, 36 deer; in 1920, 143 deer; in 1925, 86 deer; and in 1935, 26 deer. Explain what might have caused these changes.

Science Online booke.msscience.com/self_check_quiz

section 2 review

1. Count all individuals. Trap-mark-release: trap and mark two samples of the population, then compare the number of marked and unmarked individuals in the second sample. Sample counts: count the number of individuals in one part of an area, then multiply by the number of portions making up the entire area.
2. population increase, birthrate > death rate; population decrease, death rate > birthrate; population unchanged, birthrate = death rate
3. places a limit on the number of organisms that can survive
4. All living things require food and water, so their supply can affect population size.
5. Possible answer: The population grew when predators were eliminated. When the deer exceeded carrying capacity, their numbers were again reduced.

Caption Answer

Figure 11 Accept all reasonable answers. Sample responses: availability of resources (e.g., food, water, and space), birthrate, death rate.

3 Assess

DAILY INTERVENTION

Check for Understanding

Coop Learn Ask students to work in groups to (a) guess the size of your school's seventh-grade population, (b) devise a way to more accurately estimate that population size, and (c) give reasons for seventh-grade immigration and emigration. L1

Reteach

Ecosystem Interactions Display photographs of organisms common to an ecosystem, such as a coniferous forest or a desert. Work with students to create a list of species present in the ecosystem. Discuss interactions that take place among the species, and how those interactions relate to competition, limiting factors, and carrying capacity. L1

Assessment

Performance Have students research the population trends of an endangered species. Ask students to graph the data, showing numbers of individuals on the vertical axis and year on the horizontal axis. Have students describe the growth curve. Use **Performance Assessment in the Science Classroom,** p. 111. L2

1 Motivate

INTERACTIVE CHALKBOARD
PowerPoint® Presentations

Bellringer

Section Focus Transparencies also are available on the Interactive Chalkboard CD-ROM. L2 ELL

Tie to Prior Knowledge

Community Interactions Have students consider a familiar ecosystem. Ask them to name organisms that make up the community and describe interactions among these organisms. L1

Virtual Labs

Energy Transfer *How is energy transferred through a community of organisms?*

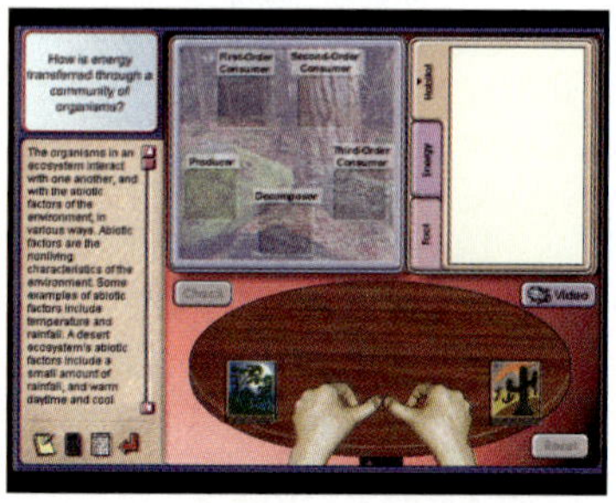

section 3 Interactions Within Communities

as you read

What You'll Learn

- **Describe** how organisms obtain energy for life.
- **Explain** how organisms interact.
- **Recognize** that every organism occupies a niche.

Why It's Important

Obtaining food, shelter, and other needs is crucial to the survival of all living organisms, including you.

Review Vocabulary
social behavior: interactions among members of the same species

New Vocabulary
- producer
- consumer
- symbiosis
- mutualism
- commensalism
- parasitism
- niche

Obtaining Energy

Just as a car engine needs a constant supply of gasoline, living organisms need a constant supply of energy. The energy that fuels most life on Earth comes from the Sun. Some organisms use the Sun's energy to create energy-rich molecules through the process of photosynthesis. The energy-rich molecules, usually sugars, serve as food. They are made up of different combinations of carbon, hydrogen, and oxygen atoms. Energy is stored in the chemical bonds that hold the atoms of these molecules together. When the molecules break apart—for example, during digestion—the energy in the chemical bonds is released to fuel life processes.

Producers Organisms that use an outside energy source like the Sun to make energy-rich molecules are called **producers.** Most producers contain chlorophyll (KLOR uh fihl), a chemical that is required for photosynthesis. As shown in **Figure 12,** green plants are producers. Some producers do not contain chlorophyll and do not use energy from the Sun. Instead, they make energy-rich molecules through a process called chemosynthesis (kee moh SIHN thuh sus). These organisms can be found near volcanic vents on the ocean floor. Inorganic molecules in the water provide the energy source for chemosynthesis.

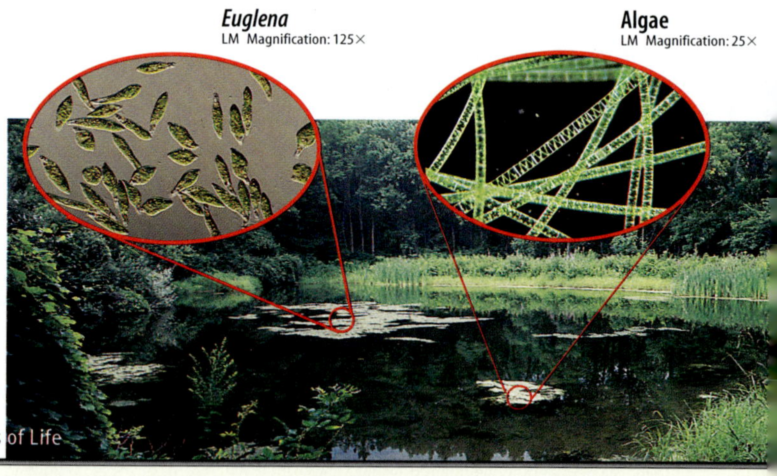

Figure 12 Green plants, including the grasses that surround this pond, are producers. The pond water also contains producers, including microscopic organisms like *Euglena* and algae.

Section 3 Resource Manager

Chapter *Fast File* Resources
- Transparency Activity, p. 46
- Enrichment, p. 32
- Directed Reading for Content Mastery, pp. 21, 22
- Reinforcement, p. 29
- Lab Worksheets, pp. 5–6, 7–8
- Lab Activities, pp. 9–11, 13–16

Reading and Writing Skill Activities, p. 49
Cultural Diversity, p. 13
Home and Community Involvement, p. 47
Lab Management and Safety, p. 71

Figure 13 Four categories of consumers are shown.
Identify *the consumer category that would apply to a bear. What about a mushroom?*

Consumers Organisms that cannot make their own energy-rich molecules are called **consumers.** Consumers obtain energy by eating other organisms. **Figure 13** shows the four general categories of consumers. Herbivores are the vegetarians of the world. They include rabbits, deer, and other plant eaters. Carnivores are animals that eat other animals. Frogs and spiders are carnivores that eat insects. Omnivores, including pigs and humans, eat mostly plants and animals. Decomposers, including fungi, bacteria, and earthworms, consume wastes and dead organisms. Decomposers help recycle once-living matter by breaking it down into simple, energy-rich substances. These substances might serve as food for decomposers, be absorbed by plant roots, or be consumed by other organisms.

Reading Check *How are producers different from consumers?*

Food Chains Ecology includes the study of how organisms depend on each other for food. A food chain is a simple model of the feeding relationships in an ecosystem. For example, shrubs are food for deer, and deer are food for mountain lions, as illustrated in **Figure 14.** What food chain would include you?

INTEGRATE Chemistry

Glucose The nutrient molecule produced during photosynthesis is glucose. Look up the chemical structure of glucose and draw it in your Science Journal.

Figure 14 Food chains illustrate how consumers obtain energy from other organisms in an ecosystem.

2 Teach

Caption Answer

Figure 13 bear: omnivore; mushroom: decomposer.

INTEGRATE Chemistry

Glucose Energy for life is stored in the chemical bonds of organic molecules such as glucose. A molecule of glucose is made of 6 carbons, 12 hydrogens, and 6 oxygen atoms ($C_6H_{12}O_6$). The structure is as follows.

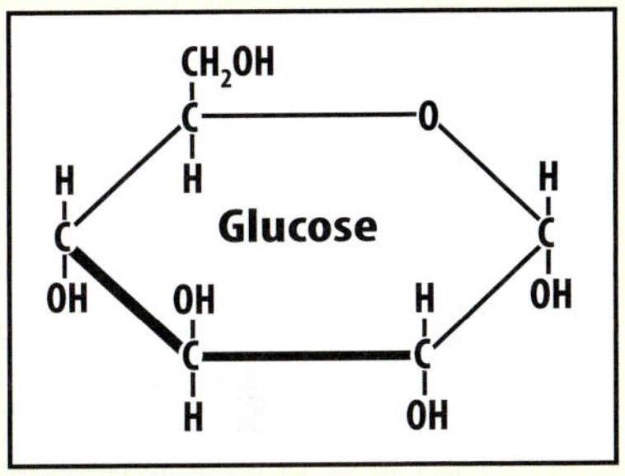

Research Have students look up the chemical structure of sucrose and compare it with glucose. Ask students to identify which sugar stores more energy. Sucrose ($C_{12}H_{22}O_{11}$) stores more energy because it contains a larger number of chemical bonds.

Reading Check

Answer producers can make food; consumers must obtain food from other organisms

Use Science Words

Word Origin Types of consumers are described by words derived from Latin. *Vorare* means "to devour," *herba* means "grass," *caro* means "flesh," and *omnis* means "all." Present this information to students and have them use it to explain the terms *herbivore*, *carnivore*, and *omnivore*.

Text Question Answer

Accept any answer that shows a human eating food derived from a plant or animal. For example, in Figure 14, humans might replace the mountain lion, because humans eat venison.

Activity

Lichens Bring in stones or pieces of tree bark with lichens growing on them. Have students look at the lichens using hand lenses. Then have them prepare wet-mount slides of lichen pieces and observe them using microscopes. How do the algae and fungus benefit each other? Algae make food for the fungus, and the fungus provides a habitat and moisture for the algae. L2 LS **Visual-Spatial**

Use an Analogy

Commensalism Challenge students to describe a human interaction that is analogous to commensalism. Possible example: A person may go through another's trash and find something useful. This is analogous to commensalism because one person benefits, while the other is neither helped nor harmed. L3

Fun Fact

Some people might say that animals are solar powered. Going back in the food chain, all energy can be traced back to producers, which make carbohydrates using the Sun's energy.

Make a Model

Food Web Have students create a display or diagram that models the food web of an ecosystem of their choice. Challenge students to include symbiotic relationships in their model. Possible answer: Fungi are fed by algae in the body of a lichen; caribou feed on lichens; human hunters eat caribou. L2

Figure 15 Many examples of symbiotic relationships exist in nature.

Symbiotic Relationships

Not all relationships among organisms involve food. Many organisms live together and share resources in other ways. Any close relationship between species is called **symbiosis.**

Lichens are a result of mutualism.

Clown fish and sea anemones have a commensal relationship.

LM Magnification: 128×

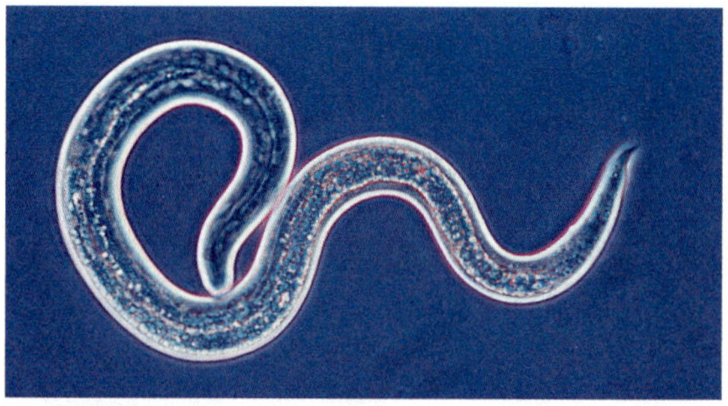

Some roundworms are parasites that rob nutrients from their hosts.

Mutualism You may have noticed crusty lichens growing on fences, trees, or rocks. Lichens, like those shown in **Figure 15,** are made up of an alga or a cyanobacterium that lives within the tissues of a fungus. Through photosynthesis, the cyanobacterium or alga supplies energy to itself and the fungus. The fungus provides a protected space in which the cyanobacterium or alga can live. Both organisms benefit from this association. A symbiotic relationship in which both species benefit is called **mutualism** (MYEW chuh wuh lih zum).

Commensalism If you've ever visited a marine aquarium, you might have seen the ocean organisms shown in **Figure 15.** The creature with gently waving, tubelike tentacles is a sea anemone. The tentacles contain a mild poison. Anemones use their tentacles to capture shrimp, fish, and other small animals to eat. The striped clown fish can swim among the tentacles without being harmed. The anemone's tentacles protect the clown fish from predators. In this relationship, the clown fish benefits but the sea anemone is not helped or hurt. A symbiotic relationship in which one organism benefits and the other is not affected is called **commensalism** (kuh MEN suh lih zum).

Parasitism Pet cats or dogs sometimes have to be treated for worms. Roundworms, like the one shown in **Figure 15,** are common in puppies. This roundworm attaches itself to the inside of the puppy's intestine and feeds on nutrients in the puppy's blood. The puppy may have abdominal pain, bloating, and diarrhea. If the infection is severe, the puppy might die. A symbiotic relationship in which one organism benefits but the other is harmed is called **parasitism** (PER uh suh tih zum).

Differentiated Instruction

Challenge Have students research an organism that has a symbiotic relationship with another organism. Ask students to write a report or create a display that describes the niches of both the organism and its symbiotic partner. Possible topics: ants and aphids; termites and bacteria that live in the termite's digestive system; honeybees and flowers; ticks and deer L3

Niches

One habitat might contain hundreds or even thousands of species. Look at the rotting log habitat shown in **Figure 16.** A rotting log in a forest can be home to many species of insects, including termites that eat decaying wood and ants that feed on the termites. Other species that live on or under the rotting log include millipedes, centipedes, spiders, and worms. You might think that competition for resources would make it impossible for so many species to live in the same habitat. However, each species has different requirements for its survival. As a result, each species has its own niche (NICH). An organism's **niche** is its role in its environment—how it obtains food and shelter, finds a mate, cares for its young, and avoids danger.

Reading Check *Why does each species have its own niche?*

Special adaptations that improve survival are often part of an organism's niche. Milkweed plants contain a poison that prevents many insects from feeding on them. Monarch butterfly caterpillars have an adaptation that allows them to eat milkweed. Monarchs can take advantage of a food resource that other species cannot use. Milkweed poison also helps protect monarchs from predators. When the caterpillars eat milkweed, they become slightly poisonous. Birds avoid eating monarchs because they learn that the caterpillars and adult butterflies have an awful taste and can make them sick.

INTEGRATE History

Plant Poisons The poison in milkweed is similar to the drug digitalis. Small amounts of digitalis are used to treat heart ailments in humans, but it is poisonous in large doses. Research the history of digitalis as a medicine. In your Science Journal, list diseases for which it was used but is no longer used.

Figure 16 Different adaptations enable each species living in this rotting log to have its own niche. Termites eat wood. They make tunnels inside the log. Millipedes feed on plant matter and find shelter beneath the log. Wolf spiders capture insects living in and around the log.

Discussion

Niches The term *niche* is sometimes described as an activity that one person is good at, but others find difficult. How is this similar to the biological definition? Organisms with a particular niche often have adaptations that give them an advantage in their environment.

INTEGRATE History

Plant Poisons According to *Neal's Medical Digest* published in 1877, digitalis cured over 32 illnesses including dropsy, adenitis, bronchitis, tuberculosis, and typhoid.

Reading Check

Answer Each species has different requirements for its survival.

IDENTIFYING Misconceptions

Plants Students may think that plants do not depend on other organisms. Refer to page F at the beginning of this chapter for teaching strategies that address this misconception.

Use Science Words

Word Origin Use word origins to emphasize the differences among the three types of symbiosis. Mutualism comes from the word *mutual*, which means "shared." Parasitism comes from a Greek word that means "feeding from the table of another." Commensalism means "eating from the same table." Symbiosis means "living together."

Visual Learning

Figure 16 Work with students to produce a food web diagram of the rotting log ecosystem. Have students describe each organism's niche as they add it to the web. Fungi and termites eat wood; insects and millipedes eat leaves; earthworms eat dead material beneath the log; spiders eat insects; birds eat spiders and millipedes. L2

Science Journal

Local Niches Have students identify organisms in local habitats that have different niches. Encourage them to research one organism and to write descriptive paragraphs in their Science Journals about how the organism survives, obtains food and shelter, finds a mate, cares for young, and avoids danger. L3 P

SECTION 3 Interactions Within Communities E ◆ 23

Quick Demo

Habitat and Niche

Materials aquarium with snails, algae

Estimated Time 10 minutes

Procedure Ask students to describe the habitat and niche of these two aquarium organisms. Habitat for both organisms: freshwater; snail niche: feeds on algae, shelters under rocks or in aquarium plants; algae niche: manufactures food through photosynthesis, obtains needed nutrients from water

3 Assess

DAILY INTERVENTION

Check for Understanding

Interpersonal Have pairs of students brainstorm together to answer the following question: What might happen if all insects were eliminated? Possible responses: Other animals that eat insects for food would be in danger of dying if they could not substitute another food source; insects would no longer decompose wastes or once-living materials, allowing wastes to accumulate. Bees and other insects would not pollinate plants, so they wouldn't form fruits and seeds. This would result in less food for people and other animals. L2 LS **Naturalist**

Reteach

Niche Describe to students the niche of a bird or other animal species common to an ecosystem near your school. Ask students to suggest reasons why the animal is able to survive there. L1

Assessment

Process Have students classify the following organisms based on how each obtains nutrition: spider (carnivore); pig (omnivore); goat (herbivore); fungus (decomposer); frog (carnivore). Use **Performance Assessment in the Science Classroom**, p. 121. L2

Figure 17 The alligator is a predator. The turtle is its prey.

Predator and Prey When you think of survival in the wild, you might imagine an antelope running away from a lion. An organism's niche includes how it avoids being eaten and how it finds or captures its food. Predators, like the one shown in **Figure 17,** are consumers that capture and eat other consumers. The prey is the organism that is captured by the predator. The presence of predators usually increases the number of different species that can live in an ecosystem. Predators limit the size of prey populations. As a result, food and other resources are less likely to become scarce, and competition between species is reduced.

Cooperation Individual organisms often cooperate in ways that improve survival. For example, a white-tailed deer that detects the presence of wolves or coyotes will alert the other deer in the herd. Many insects, such as ants and honeybees, live in social groups. Different individuals perform different tasks required for the survival of the entire nest. Soldier ants protect workers that go out of the nest to gather food. Worker ants feed and care for ant larvae that hatch from eggs laid by the queen. These cooperative actions improve survival and are a part of the specie's niche.

section 3 review

Summary

Obtaining Energy

- All life requires a constant supply of energy.
- Most producers make food by photosynthesis using light energy.
- Consumers cannot make food. They obtain energy by eating producers or other consumers.
- A food chain models the feeding relationships between species.

Symbiotic Relationships

- Symbiosis is any close relationship between species.
- Mutualism, commensalism, and parasitism are types of symbiosis.
- An organism's niche describes the ways in which the organism obtains food, avoids danger, and finds shelter.

Self Check

1. **Explain** why all consumers depend on producers for food.
2. **Describe** a mutualistic relationship between two imaginary organisms. Name the organisms and explain how each benefits.
3. **Compare and contrast** the terms *habitat* and *niche*.
4. **Think Critically** A parasite can obtain food only from a host organism. Explain why most parasites weaken, but do not kill, their hosts.

Applying Skills

5. **Design an experiment** to classify the symbiotic relationship that exists between two hypothetical organisms. Animal A definitely benefits from its relationship with Plant B, but it is not clear whether Plant B benefits, is harmed, or is unaffected.

Science Online booke.msscience.com/self_check_quiz

section 3 review

1. Producers can make their own carbohydrates from nonliving materials; other types of organisms cannot do this and must feed on producers, either directly or indirectly.
2. Answers will vary but both organisms should benefit from the relationship.
3. The place an organism lives is its habitat. How an organism survives is its niche.
4. If a parasite kills its host, the parasite also will die.
5. Answers will vary but should include a sample of Animal A grown with Plant B. This will be compared to a sample of Plant B grown alone.

LAB

BENCH TESTED

Feeding Habits of Planaria

You probably have watched minnows darting about in a stream. It is not as easy to observe organisms that live at the bottom of a stream, beneath rocks, logs, and dead leaves. Countless stream organisms, including insect larvae, worms, and microscopic organisms, live out of your view. One such organism is a type of flatworm called a planarian. In this lab, you will find out about the eating habits of planarians.

Real-World Question

What food items do planarians prefer to eat?

Goals

- **Observe** the food preference of planarians.
- **Infer** what planarians eat in the wild.

Materials

small bowl
planarians (several)
lettuce leaf
raw liver or meat
guppies (several)
pond or stream water
magnifying lens

Safety Precautions

Magnification: Unknown

Procedure

1. Fill the bowl with stream water.
2. Place a lettuce leaf, piece of raw liver, and several guppies in the bowl. Add the planarians. Wash your hands.
3. **Observe** what happens inside the bowl for at least 20 minutes. Do not disturb the bowl or its contents. Use a magnifying lens to look at the planarians.
4. **Record** all of your observations in your Science Journal.

Conclude and Apply

1. **Name** the food the planarians preferred.
2. **Infer** what planarians might eat when in their natural environment.
3. **Describe,** based on your observations during this lab, a planarian's niche in a stream ecosystem.
4. **Predict** where in a stream you might find planarians. Use references to find out whether your prediction is correct.

Communicating Your Data

Share your results with other students in your class. Plan an adult-supervised trip with several classmates to a local stream to search for planarians in their native habitat. **For more help, refer to the Science Skill Handbook.**

Real-World Question

Purpose Students observe the eating habits of planarians and infer what wild planarians eat. L2 LS **Visual-Spatial**

Process Skills observe, predict, infer

Time Required 30 minutes

Procedure

Safety Precautions Instruct students to handle planarians with care and to wash their hands after the lab.

Teaching Strategy To house the planarians for an extended period of time, change the water of their bowls every other day and keep a small piece of liver in the bowl.

Troubleshooting

- Set up a bowl of planarians to serve as a reference.
- The piece of liver should be about the size of a quarter.
- Tell students to cover their dishes for a few minutes; planaria do not like light.

Conclude and Apply

1. liver
2. Most planaria are carnivorous night feeders. They eat protozoans, tiny snails, worms, and dead animals.
3. Although planarians prey on slow-moving animals, they primarily scavenge dead organisms from the bottom of streams.
4. Planarians live in slow-moving or still portions of a stream where dead organisms settle to the stream bottom. They are found beneath rotting leaves or rocks.

Communicating Your Data

Have students write a paragraph describing their results. They can exchange papers with a classmate to see their results. L2

Assessment

Performance Ask students to infer how their results might have differed had they used tap water instead of stream water. Tap water contains chemicals toxic to soft bodied animals and could kill the planarians. Use **Performance Assessment in the Science Classroom,** p. 89. L2

BENCH TESTED

LAB Design Your Own

POPULATION GROWTH IN FRUIT FLIES

Goals

- **Identify** the environmental factors needed by a population of fruit flies.
- **Design** an experiment to investigate how a change in one environmental factor affects in any way the size of a fruit fly population.
- **Observe** and **measure** changes in population size.

Possible Materials

fruit flies
standard fruit fly culture kit
food items (banana, orange peel, or other fruit)
water
heating or cooling source
culture containers
cloth, plastic, or other tops for culture containers
magnifying lens

Safety Precautions

Real-World Question

Populations can grow at an exponential rate only if the environment provides the right amount of food, shelter, air, moisture, heat, living space, and other factors. You probably have seen fruit flies hovering near ripe bananas or other fruit. Fruit flies are fast-growing organisms often raised in science laboratories. The flies are kept in culture tubes and fed a diet of specially prepared food flakes. Can you improve on this standard growing method to achieve faster population growth? Will a change in one environmental factor affect the growth of a fruit fly population?

Form a Hypothesis

Based on your reading about fruit flies, state a hypothesis about how changing one environmental factor will affect the rate of growth of a fruit fly population.

Test Your Hypothesis

Make a Plan

1. As a group, decide on one environmental factor to investigate. Agree on a hypothesis about how a change in this factor will affect population growth. Decide how you will test your hypothesis, and identify the experimental results that would support your hypothesis.
2. **List** the steps you will need to take to test your hypothesis. Describe exactly what you will do. List your materials.
3. **Determine** the method you will use to measure changes in the size of your fruit fly populations.

26 ◆ E CHAPTER 1 Interactions of Life

Real-World Question

Purpose Students will test the effect of a change in one environmental factor on the rate of growth of a fruit fly population. L2

Process Skills identify a question, form hypotheses, test a hypothesis, identify and manipulate variables, collect data, make and use tables, record data, record observations, analyze results, form operational definitions, evaluate other's data and conclusions, communicate, make and use graphs

Time Required two 45-minute periods and once weekly observations for one to two months

Possible Materials Standard fruit fly culture kits are available from biological supply houses. Possible culture containers include mayonnaise jars or other empty, sanitized food jars.

Safety Precautions

- Students should take care not to release their fruit flies within the school building.
- Used fruit fly culture containers can contain bacteria and mold, and should not be reused for other purposes.

Data Table:

Fruit Fly Population				
Culture number	Number of Flies			
	Date	Date	Date	Date
1				
2				
3				

Form a Hypothesis

Possible Hypothesis Students' hypotheses should reflect a change in one environmental factor, such as type of food, amount of water, or size of container, and its possible effect on the rate of growth of the fruit fly population. For example: A larger living space will result in a faster-growing fruit fly population.

Using Scientific Methods

4. Prepare a data table in your Science Journal to record weekly measurements of your fruit fly populations.
5. Read the entire experiment and make sure all of the steps are in a logical order.
6. **Research** the standard method used to raise fruit flies in the laboratory. Use this method as the control in your experiment.
7. **Identify** all constants, variables, and controls in your experiment.

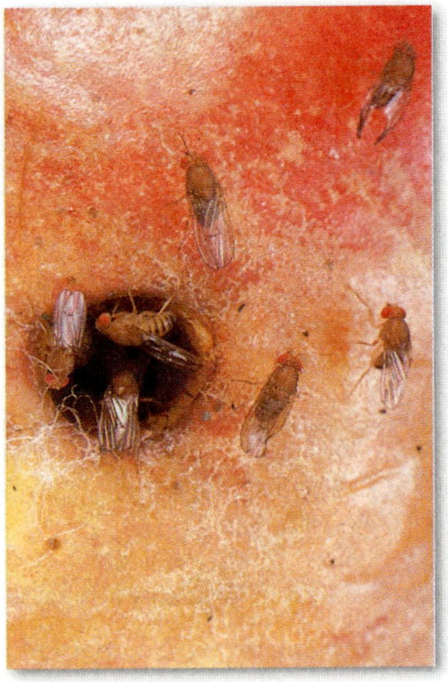

Follow Your Plan

1. Make sure your teacher approves your plan before you start.
2. Carry out your experiment.
3. **Measure** the growth of your fruit fly populations weekly and record the data in your data table.

Analyze Your Data

1. **Identify** the constants and the variables in your experiment.
2. **Compare** changes in the size of your control population with changes in your experimental population. Which population grew faster?
3. **Make and Use Graphs** Using the information in your data table, make a line graph that shows how the sizes of your two fruit fly populations changed over time. Use a different colored pencil for each population's line on the graph.

Conclude and Apply

1. **Explain** whether or not the results support your hypothesis.
2. **Compare** the growth of your control and experimental populations. Did either population reach exponential growth?
How do you know?

Communicating Your Data

Compare the results of your experiment with those of other students in your class.
For more help, refer to the Science Skill Handbook.

Test Your Hypothesis

Possible Procedures Provide two equal-sized populations of fruit flies with the same food, moisture, and temperature, but two different-sized containers to test the effect of living area on fruit fly population growth. Make observations over the next one to two months. The fruit fly's life cycle is about two weeks, so students should see rapid changes in population size.

Teaching Strategy Most students are familiar with fruit flies "appearing" near ripe fruit. Have students consider the source of these flies. Sometimes eggs were already present; other times flies have used their strong sense of smell to locate the ripe fruit.

Expected Outcome Most results will show that any extreme change from the standard method of maintaining fruit flies will result in a decreased rate of population growth.

Analyze Your Data

Answers to Questions

1. Answers will vary.
2. Answers will vary.
3. Student graphs will vary. Check students' work.

Error Analysis Have students compare their results and their hypotheses and explain why differences occurred. L2

Conclude and Apply

1. Answers will be determined by student's hypotheses.
2. A population shows exponential growth if it increases at a fixed percentage per time period. Student answers will depend on results.

Communicating Your Data

Students can make an electronic spreadsheet that allows comparison of data about different environmental factors that affect the rate of growth of the fruit fly population. L3

Assessment

Oral Have students explain the relationship between environmental conditions and population growth in fruit flies. Flies achieve the fastest rate of growth when all the conditions in their environment are at optimal levels. Any change in these conditions will result in a decrease in population growth rates. Use **Performance Assessment in the Science Classroom**, p. 89. L1

TIME

Content Background

The U.S. Congress and the Electoral College are made up of representatives from each state. The number of representatives and electors from each state is determined by the state's population and updated by the census every 10 years. To make sure each Congress member represents an equal number of people, congressional districts are reapportioned according to the census.

Discussion

Age Why do you think it is important for the census to include information about the ages of U.S residents? Possible answers: Knowing the ages of children indicates the number of schools needed in the near future. The number of middle-aged adults indicates how many seniors might enter government programs such as Social Security and Medicare in coming years.

Activity

2000 Census Results Have students research results of the 2000 Census for their state, their region of the country, and the U.S. as a whole. Ask them to write a paragraph in their Science Journal comparing changes in the population of their state or region with changes that took place nationwide. Nationwide, the population rose by 32.7 million, or 13.2 percent, between 1990 and 2000. The western region grew faster than other regions of the country. L2

TIME **SCIENCE AND HISTORY** SCIENCE CAN CHANGE THE COURSE OF HISTORY!

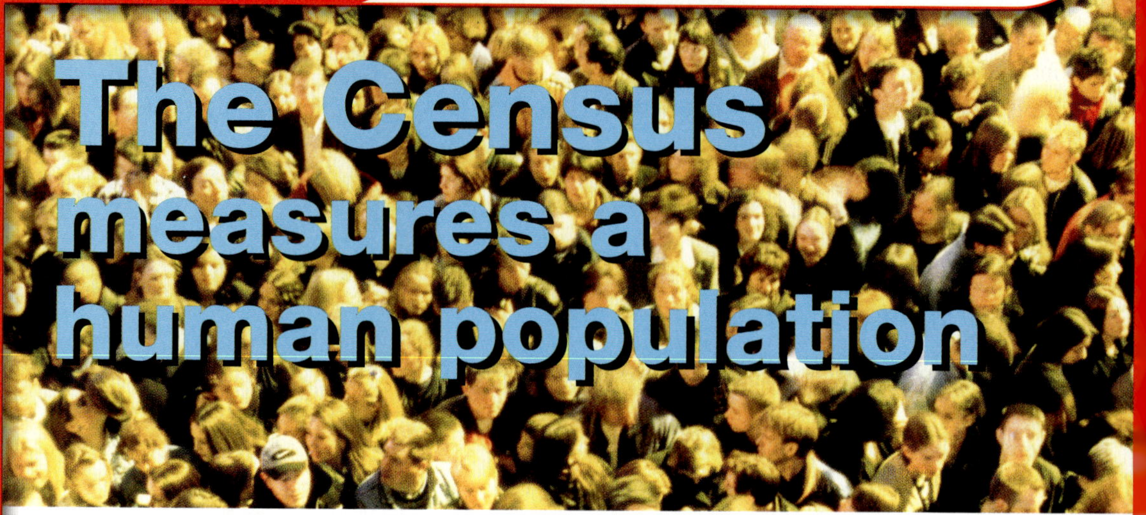

The Census measures a human population

Counting people is important to the United States and to many other countries around the world. It helps governments determine the distribution of people in the various regions of a nation. To obtain this information, the government takes a census—a count of how many people are living in their country on a particular day at a particular time, and in a particular place. A census is a snapshot of a country's population.

Counting on the Count

When the United States government was formed, its founders set up the House of Representatives based on population. Areas with more people had more government representatives, and areas with fewer people had fewer representatives. In 1787, the requirement for a census became part of the U.S. Constitution. A census must be taken every ten years so the proper number of representatives for each state can be calculated.

The Short Form

Before 1970, United States census data was collected by field workers. They went door to door to count the number of people living in each household. Since then, the census has been done mostly by mail. Census data are important in deciding how to distribute government services and funding.

The 2000 Snapshot

One of the findings of the 2000 Census is that the U.S. population is becoming more equally spread out across age groups. Census officials estimate that by 2020 the population of children, middle-aged people, and senior citizens will be about equal. It's predicted also that there will be more people who are over 100 years old than ever before. Federal, state, and local governments will be using the results of the 2000 Census for years to come as they plan our future.

Census **Develop a school census. What questions will you ask? (Don't ask questions that are too personal.) Who will ask them? How will you make sure you counted everyone? Using the results, can you make any predictions about your school's future or its current students?**

Science Online
For more information, visit booke.msscience.com/time

Census Today, the entire census is taken using a mail-in form. Have students develop a short mail-in form to take a school census. Have groups write questions to include on the forms. Challenge students to design the form so that it is understandable. After the census is taken, have students discuss challenges they encountered. L2

Resources for Teachers and Students

Who Counts? The Politics of Census-Taking in Contemporary America, by Margo J. Anderson and Stephen E. Fienberg, Russell Sage Foundation, New York. 1999

Census Brief 2000: Population Change and Distribution 1990–2000, by Marc J. Perry and Paul J. Mackun, U.S. Census Bureau, U.S. Department of Commerce, 2001

Reviewing Main Ideas

Section 1 Living Earth

1. Ecology is the study of interactions that take place in the biosphere.
2. A population is made up of all organisms of one species living in an area at the same time.
3. A community is made up of all the populations living in one ecosystem.
4. Living and nonliving factors affect an organism's ability to survive in its habitat.

Section 2 Populations

1. Population size can be estimated by counting a sample of a total population.
2. Competition for limiting factors can restrict the size of a population.
3. Population growth is affected by birthrate, death rate, and the movement of individuals into or out of a community.
4. Exponential population growth can occur in environments that provide a species with plenty of food, shelter, and other resources.

Section 3 Interactions Within Communities

1. All life requires energy.
2. Most producers use light to make food in the form of energy-rich molecules. Consumers obtain energy by eating other organisms.
3. Mutualism, commensalism, and parasitism are the three kinds of symbiosis.
4. Every species has its own niche, which includes adaptations for survival.

Visualizing Main Ideas

Copy and complete the following concept map on communities.

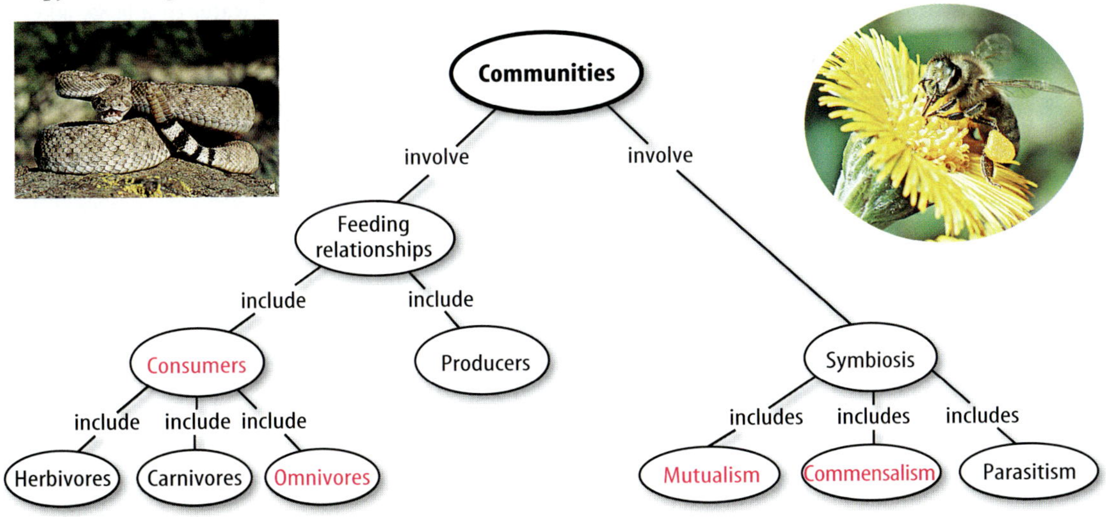

Science Online booke.msscience.com/interactive_tutor

Identifying Misconceptions

Assess

Use the assessment as follow-up to page F at the beginning of the chapter.

Activity After completing the activity in which students find and label interactions outdoors, have students again draw a picture of a natural community. Have students describe the interactions in the picture. L1

Expected Outcome In this second drawing, students should be able to identify more relationships among organisms. The interactions they depict should include not only predation, but also various types of symbiosis. L3

Reviewing Main Ideas

Summary statements can be used by students to review the major concepts of the chapter.

Visualizing Main Ideas

See student page.

Science Online

Visit **booke.msscience.com**
- /self_check_quiz
- /interactive_tutor
- /vocabulary_puzzlemaker
- /chapter_review
- /standardized_test
- /field_guide

Assessment Transparency

For additional assessment questions, use the *Assessment Transparency* located in the transparency book.

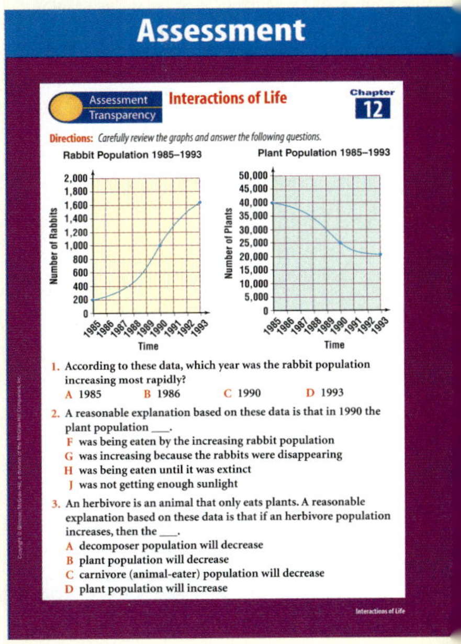

Assessment

Assessment Transparency — Interactions of Life — Chapter 12

Directions: *Carefully review the graphs and answer the following questions.*

1. According to these data, which year was the rabbit population increasing most rapidly?
   A 1985 B 1986 C 1990 D 1993
2. A reasonable explanation based on these data is that in 1990 the plant population ____.
   F was being eaten by the increasing rabbit population
   G was increasing because the rabbits were disappearing
   H was being eaten until it was extinct
   J was not getting enough sunlight
3. An herbivore is an animal that only eats plants. A reasonable explanation based on these data is that if an herbivore population increases, then the ____.
   A decomposer population will decrease
   B plant population will decrease
   C carnivore (animal-eater) population will decrease
   D plant population will increase

Interactions of Life

chapter 1 Review

Using Vocabulary

1. Habitat is the place an organism lives. Niche refers to adaptations that help an organism survive.
2. Both describe symbiotic relationships. In mutualism both organisms benefit. In commensalism, one organism benefits, and the other neither benefits nor is harmed.
3. Carrying capacity is the largest number of individuals of one species that an ecosystem can support. Limiting factors restrict carrying capacity.
4. An ecosystem is all the living and nonliving components of an area. The biosphere is the part of Earth that supports life. It is made up of many ecosystems.
5. Producers make their own food. Consumers eat producers or other consumers for food.
6. An ecosystem is all the living and nonliving components of an area. A population is all the individuals of one species living in an area.
7. A community is all the different species in an area. A population refers to one species in an area.
8. Symbiosis describes a close relationship between species. Parasitism is a type of symbiosis in which one species benefits and the other species is harmed.
9. An ecosystem is all the living and nonliving components of an area. Ecology is the study of interactions between these components.
10. Both are types of symbiosis. In parasitism, one species benefits, and the other is harmed. In commensalism, one organism benefits, and the other neither benefits nor is harmed.

Checking Concepts

11. A	14. D	17. C
12. C	15. B	18. C
13. D	16. A	19. A

chapter 1 Review

Using Vocabulary

biosphere p. 684	limiting factor p. 690
carrying capacity p. 691	mutualism p. 698
commensalism p. 698	niche p. 699
community p. 686	parasitism p. 698
consumer p. 697	population p. 686
ecology p. 685	producer p. 696
ecosystem p. 685	symbiosis p. 698
habitat p. 687	

Explain the difference between the vocabulary words in each of the following sets.

1. niche—habitat
2. mutualism—commensalism
3. limiting factor—carrying capacity
4. biosphere—ecosystem
5. producer—consumer
6. population—ecosystem
7. community—population
8. parasitism—symbiosis
9. ecosystem—ecology
10. parasitism—commensalism

Checking Concepts

Choose the word or phrase that best answers the question.

11. Which of the following is a living factor in the environment?
    A) animals **C)** sunlight
    B) air **D)** soil
12. What is made up of all the populations in an area?
    A) niches **C)** community
    B) habitats **D)** ecosystem
13. What does the number of individuals in a population that occupies an area of a specific size describe?
    A) clumping **C)** spacing
    B) size **D)** density
14. Which of the following animals is an example of an herbivore?
    A) wolf **C)** tree
    B) moss **D)** rabbit
15. What term best describes a symbiotic relationship in which one species is helped and the other is harmed?
    A) mutualism **C)** commensalism
    B) parasitism **D)** consumerism
16. Which of the following conditions tends to increase the size of a population?
    A) births exceed deaths
    B) population size exceeds the carrying capacity
    C) movements out of an area exceed movements into the area
    D) severe drought
17. Which of the following is most likely to be a limiting factor in a population of fish living in the shallow water of a large lake?
    A) sunlight **C)** food
    B) water **D)** soil
18. In which of the following categories does the pictured organism belong?

    A) herbivore
    B) carnivore
    C) producer
    D) consumer
19. Which pair of words is incorrect?
    A) black bear—carnivore
    B) grasshopper—herbivore
    C) pig—omnivore
    D) lion—carnivore

Science Online booke.msscience.com/vocabulary_puzzlemaker

Use the ExamView® Pro Testmaker CD-ROM to:
- create multiple versions of tests
- create modified tests with one mouse click for inclusion students
- edit existing questions and add your own questions
- build tests aligned with state standards using built-in State Curriculum Tags
- change English tests to Spanish with one mouse click and vice versa

Thinking Critically

20. **Infer** why a parasite has a harmful effect on the organism it infects.

21. **Explain** what factors affect carrying capacity.

22. **Describe** your own habitat and niche.

23. **Make and Use Tables** Copy and complete the following table.

Types of Symbiosis

Organism A	Organism B	Relationship
Gains	Doesn't gain or lose	Commensalism
Gains	Gains	Mutualism
Gains	Loses	Parasitism

24. **Explain** how several different niches can exist in the same habitat.

25. **Make a model** of a food chain using the following organisms: grass, snake, mouse, and hawk.

26. **Predict** Dandelion seeds can float great distances on the wind with the help of white, featherlike attachments. Predict how a dandelion seed's ability to be carried on the wind helps reduce competition among dandelion plants.

27. **Classify** the following relationships as parasitism, commensalism, or mutualism: a shark and a remora fish that cleans and eats parasites from the shark's gills; head lice and a human; a spiny sea urchin and a tiny fish that hides from predators by floating among the sea urchin's spines.

28. **Compare and contrast** the diets of omnivores and herbivores. Give examples of each.

29. **List** three ways exponential growth in the human population affects people's lives.

Performance Activities

30. **Poster** Use photographs from old magazines to create a poster that shows at least three different food chains. Illustrate energy pathways from organism to organism and from organisms to the environment. Display your poster for your classmates.

Applying Math

31. **Measuring Populations** An ecologist wants to know the size of a population of wild daisy plants growing in a meadow that measures 1,000 m^2. The ecologist counts 30 daisy plants in a sample area of 100 m^2. What is the estimated population of daisies in the entire meadow?

Use the table below to answer question 32.

Lynx and Hare Pelts Sold to the Hudson's Bay Company*

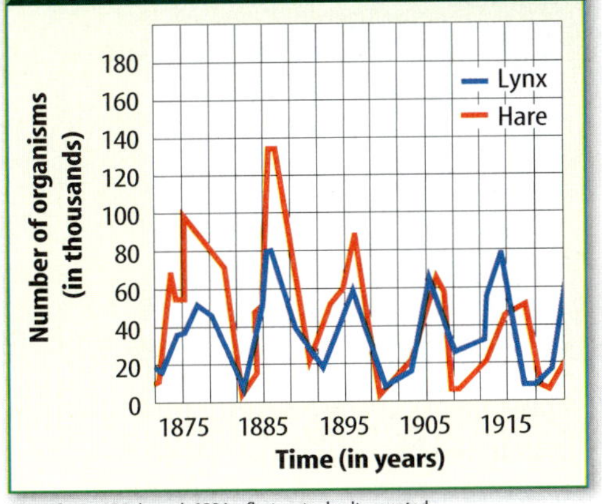

* Data from 1875 through 1904 reflects actual pelts counted. Data from 1905 through 1915 is based on answers to questionnaire.

32. **Changes in Populations** The graph above shows changes over time in the sizes of lynx and rabbit populations in an ecosystem. What does the graph tell you about the relationship between these two species? Explain how they influence each other's population size.

Assessment Resources

Reproducible Masters

Chapter *Fast File* Resources
- Chapter Review, pp. 37–38
- Chapter Tests, pp. 39–42
- Assessment Transparency Activity, p. 49

Glencoe Science Web site
- Chapter Review Test
- Standardized Test Practice

Glencoe Technology

- Assessment Transparency
- Exam*View*® Pro Testmaker
- MindJogger Videoquiz
- Interactive Chalkboard

chapter 1 Review

Thinking Critically

20. It may absorb nutrients from the organism, as does a tapeworm living inside a human's intestine, or spread disease from one organism to another, as mosquitos spread malaria.

21. Possible answers: food, water, shelter, space, birthrates and death rates, immigration, emigration

22. Description of habitat should include where the student lives. Description of niche should include adaptations and behaviors that enhance the student's survival.

23. See student page.

24. Organisms are adapted to feed on different foods, hunt at different times, and nest in different places. Therefore many niches can exist in the same habitat.

25. grass, mouse, snake, hawk

26. Individual seeds are able to move far from the parent plant, which helps reduce population density.

27. mutualism; parasitism; commensalism

28. Both eat plants. Omnivores, such as pigs and humans, eat both plants and animals. Herbivores, such as goats, eat only plants.

29. traffic, lack of housing, more rapid spread of disease, food shortages

Performance Activities

30. Food chains should move from producers to consumers. Use **PASC**, p. 145.

Applying Math

National Math Standards

1, 6, 7

31. 300

32. Shows a predator-prey relationship: lynxes eat hares. As the hare population rises, the lynx population also rises, because more food is available. As the hare population decreases, there is less food for lynxes, so the lynx population declines.

chapter 1 Standardized Test Practice

FAST FILE

Answer Sheet A practice answer sheet can be found at booke.msscience.com/answer_sheet.

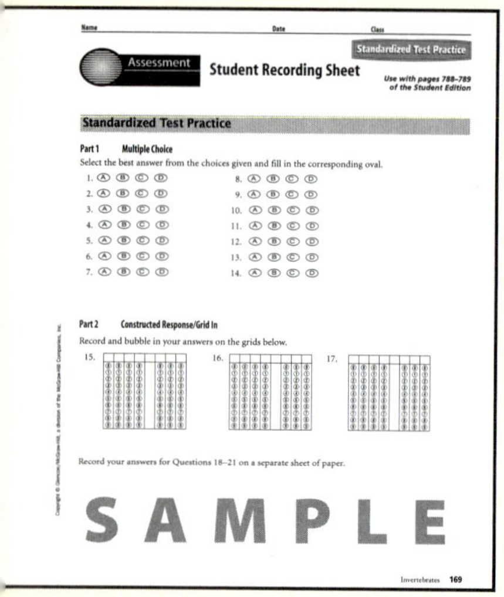
Name Date Class

Assessment **Student Recording Sheet** Standardized Test Practice

Use with pages 788–789 of the Student Edition

Standardized Test Practice

Part 1 Multiple Choice

Select the best answer from the choices given and fill in the corresponding oval.

1. Ⓐ Ⓑ Ⓒ Ⓓ
2. Ⓐ Ⓑ Ⓒ Ⓓ
3. Ⓐ Ⓑ Ⓒ Ⓓ
4. Ⓐ Ⓑ Ⓒ Ⓓ
5. Ⓐ Ⓑ Ⓒ Ⓓ
6. Ⓐ Ⓑ Ⓒ Ⓓ
7. Ⓐ Ⓑ Ⓒ Ⓓ
8. Ⓐ Ⓑ Ⓒ Ⓓ
9. Ⓐ Ⓑ Ⓒ Ⓓ
10. Ⓐ Ⓑ Ⓒ Ⓓ
11. Ⓐ Ⓑ Ⓒ Ⓓ
12. Ⓐ Ⓑ Ⓒ Ⓓ
13. Ⓐ Ⓑ Ⓒ Ⓓ
14. Ⓐ Ⓑ Ⓒ Ⓓ

Part 2 Constructed Response/Grid In

Record and bubble in your answers on the grids below.

15. 16. 17.

Record your answers for Questions 18–21 on a separate sheet of paper.

SAMPLE

Invertebrates 169

Part 1 Multiple Choice

1. C
2. C
3. B
4. D
5. A
6. A
7. A
8. B
9. D
10. C

Part 1 Multiple Choice

Record your answers on the answer sheet provided by your teacher or on a sheet of paper.

1. Which of the following terms is defined in part by nonliving factors?

A. population **C.** ecosystem
B. community **D.** niche

2. Which of the follow terms would include all places where organisms live on Earth?

A. ecosystem **C.** biosphere
B. habitat **D.** community

3. Which of the following is not a method of measuring populations?

A. total count **C.** sample count
B. trap-release **D.** trap-mark-release

Use the photo below to answer questions 4 and 5.

4. Dead plants at the bottom of this pond are consumed by

A. omnivores. **C.** carnivores.
B. herbivores. **D.** decomposers.

5. If the pond shrinks in size, what effect will this have on the population density of the pond's minnow species?

A. It will increase.
B. It will decrease.
C. It will stay the same.
D. No effect; it is not a limiting factor.

6. Which of the following includes organisms that can directly convert energy from the Sun into food?

A. producers **C.** omnivores
B. decomposers **D.** consumers

7. You have a symbiotic relationship with bacteria in your digestive system. These bacteria break down food you ingest, and you get vital nutrients from them. Which type of symbiosis is this?

A. mutualism **C.** commensalism
B. barbarism **D.** parasitism

Use the photo below to answer questions 8 and 9.

8. An eastern screech owl might compete with which organism most intensely for resources?

A. mouse **C.** mountain lion
B. hawk **D.** wren

9. Which of the following organisms might compete with the mouse for seeds?

A. hawk **C.** fox
B. lion **D.** sparrow

10. Which of the following is an example of a community?

A. all the white-tailed deer in a forest
B. all the trees, soil, and water in a forest
C. all the plants and animals in a wetland
D. all the cattails in a wetland

Part 2 Short Response/Grid In

11. Hawks are a limiting factor for population A but not population B. Limiting factors for population B could be another predator, food, water, habitat space, or another competitive species for these resources.

12. Diagram should depict energy flow arrows going from the Sun (and possibly inorganic chemicals) to producers, from producers to consumers, from producers and consumers to decomposers.

Part 2 Short Response/Grid In

Record your answers on the answer sheet provided by your teacher or on a sheet of paper.

Use the graph below to answer question 11.

Mouse Population Exposed to Predators

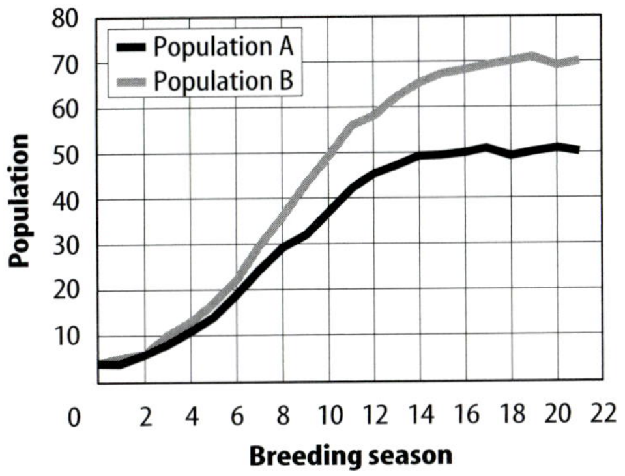

11. The graph depicts the growth of two white-footed mice populations, one exposed to hawks (population A) and one without hawks (population B). Are hawks a limiting factor for either mouse population? If not, then what other factor could be a limiting factor for that population?

12. Diagram the flow of energy through an ecosystem. Include the sources of energy, producers, consumers, and decomposers in the ecosystem.

Test-Taking Tip

Understand the Question Be sure you understand the question before you read the answer choices. Make special note of words like NOT or EXCEPT. Read and consider choices before you mark your answer sheet.

Question 11 Make sure you understand which mouse population is subject to predation by hawks and which mouse population do hawks not affect.

Part 3 Open Ended

Record your answers on a sheet of paper.

13. The colors and patterns of the viceroy butterfly are similar to the monarch butterfly, however, the viceroy caterpillars don't feed on milkweed. How does the viceroy butterfly benefit from this adaptation of its appearance? Under what circumstance would this adaptation not benefit the viceroy? Why?

Use the illustration below to answer question 14.

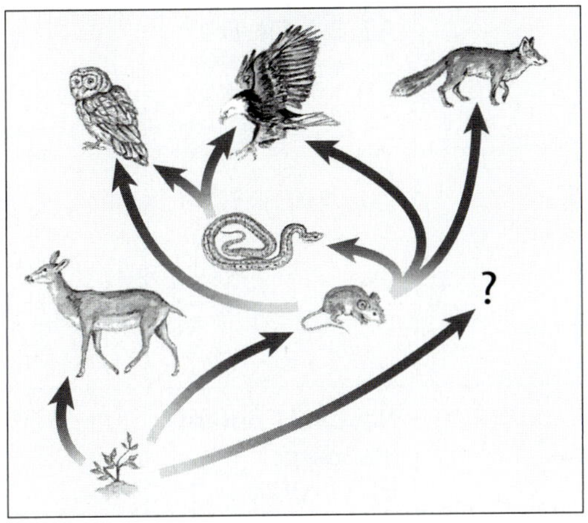

14. The illustration depicts a food web for a particular ecosystem. If the "?" is another mouse species population that is introduced into the ecosystem, explain what impact this would have on the species populations in the ecosystem.

15. Identify and explain possible limiting factors that would control the size of an ant colony.

16. How would you measure the size of a population of gray squirrels in a woodland? Explain which method you would choose and why.

Part 3 Open Ended

13. The viceroy butterfly benefits because predators avoid eating them due to their similar appearance to the monarch butterfly. This adaptation would not be beneficial if viceroys moved into an area where there are no monarchs. In this case, predators would not have learned to avoid monarch butterflies and would eat the viceroy butterflies.

14. The existing mouse population would have the most intense competition for resources with the introduced mouse population. This might lead to a reduction in the size of the original mouse population. However, the new mouse population can also serve as a prey species for the predators, owls, eagles, foxes and snakes, which may reduce predation pressure on the original mouse population. If the introduced mouse population eats the same vegetation as the deer population, then the deer population may have to compete for food resources. The diagram should be modified to include energy flow arrows from the introduced mouse population to the four predator populations.

15. Possible limiting factors: temperature, water, food, space, predators. Answer should include a reasonable explanation as to why the factor is limiting.

16. Trap-mark-release would ensure individuals were not counted more than once. A nest count would give an estimate of the number of breeding pairs in the woodland. A complete population count would be difficult, since squirrels in the wild can be wary of humans and a means of identifying each individual would be needed.

Rubrics

For more help evaluating open-ended assessment questions, see the rubric on p. 10T.

chapter 2 Organizer

Section/Objectives	Standards: National	Standards: State/Local	Labs/Features
Chapter Opener	See pp. 16T–17T for a Key to Standards.		**Launch Lab:** Earth Has Many Ecosystems, p. 35 **Foldables,** p. 35
Section 1 Abiotic Factors 2 sessions 1 block 1. **Identify** common abiotic factors in most ecosystems. 2. **List** the components of air that are needed for life. 3. **Explain** how climate influences life in an ecosystem.	National Content Standards: UCP.1–UCP.4 A.1, A.2, B.3, C.4, D.1		**MiniLAB:** Determining Soil Makeup, p. 38 **Applying Math:** Temperature Changes, p. 40 **Integrate Career,** p. 41 **Science Online,** p. 41 **Lab:** Humus Farm, p. 43
Section 2 Cycles in Nature 2 sessions 1 block 4. **Explain** the importance of Earth's water cycle. 5. **Diagram** the carbon cycle. 6. **Recognize** the role of nitrogen in life on Earth.	National Content Standards: UCP.1–UCP.3, UCP.5, A.1, A.2, B.3, C.1, C.3–C.5, D.1		**MiniLAB:** Comparing Fertilizers, p. 47 **Visualizing the Carbon Cycle,** p. 48 **Science Online,** p. 49
Section 3 Energy Flow 3 sessions 1.5 blocks 7. **Explain** how organisms produce energy-rich compounds. 8. **Describe** how energy flows through ecosystems. 9. **Recognize** how much energy is available at different levels in a food chain.	National Content Standards: B.3, C.1, C.5, D.1		**Integrate Earth Science,** p. 51 **Lab:** Where does the mass of a plant come from?, pp. 54–55 **Science Stats:** Extreme Climates, p. 56

Glencoe Exclusive! TeacherWorks™ All-In-One Planner and Resource Center

Lab Materials	Reproducible Resources	Section Assessment	Technology
Launch Lab: globe or world map, weather references	**Chapter *Fast File* Resources** Foldables Worksheet, p. 17 Directed Reading Overview, p. 19 Note-taking Worksheets, pp. 33–34	Glencoe's Assessment Advantage	TeacherWorks includes: • Interactive Teacher Edition • Lesson Planner with calendar • Access to all program blacklines • Correlations to standards • Web links
MiniLAB: soil (2 cups), quart jar with lid, water, dishwashing liquid (1 teaspoon), watch or clock, metric ruler **Lab:** widemouthed jar, soil, grass clippings or green leaves, water, marker, metric ruler, graduated cylinder	**Chapter *Fast File* Resources** Transparency Activity, p. 44 MiniLAB, p. 3 Enrichment, p. 30 Reinforcement, p. 27 Directed Reading, p. 20 Lab Activity, pp. 9–12 Lab Worksheet, pp. 5–6 **Cultural Diversity,** p. 33 **Reading and Writing Skill Activities,** p. 3	**Portfolio** Assessment, p. 42 **Performance** MiniLAB, p. 38 Applying Math, p. 40 Applying Math, p. 42 **Content** Section Review, p. 42	Section Focus Transparency Virtual Labs CD-ROM Guided Reading Audio Program Interactive Chalkboard CD-ROM Video Lab
MiniLAB: labels (and prices) of 3 brands of houseplant fertilizer	**Chapter *Fast File* Resources** Transparency Activity, p. 45 MiniLAB, p. 4 Enrichment, p. 31 Reinforcement, p. 28 Directed Reading, p. 21 Lab Activity, pp. 13–16	**Portfolio** Activity, p. 48 **Performance** MiniLAB, p. 47 Applying Skills, p. 49 **Content** Section Review, p. 49	Section Focus Transparency Virtual Labs CD-ROM Guided Reading Audio Program Interactive Chalkboard CD-ROM
Lab: 8 oz. plastic or paper cup, potting soil to fill cup, scale or balance, radish seeds (4), water, paper towels *Need materials?* Contact Science Kit at 1-800-828-7777 or www.sciencekit.com on the Internet.	**Chapter *Fast File* Resources** Transparency Activity, p. 46 Enrichment, p. 32 Reinforcement, p. 29 Directed Reading, pp. 21, 22 Transparency Activity, pp. 47–48 Lab Worksheet, pp. 7–8 **Lab Management and Safety,** p. 38	**Portfolio** Activity, p. 51 Visual Learning, p. 52 **Performance** Applying Math, p. 53 **Content** Section Review, p. 53	Section Focus Transparency Teaching Transparency Virtual Labs CD-ROM Guided Reading Audio Program Interactive Chalkboard CD-ROM

End of Chapter Assessment

Blackline Masters	Technology	Professional Series
Chapter *Fast File* Resources Chapter Review, pp. 37–38 Chapter Tests, pp. 39–42 **Standardized Test Practice,** pp. 11–14	MindJogger Videoquiz Virtual Labs CD-ROM Exam*View*® Pro Testmaker TeacherWorks CD-ROM Interactive Chalkboard CD-ROM	**Performance Assessment in the Science Classroom (PASC)**

chapter 2 The Nonliving Environment

Transparencies

Section Focus

Section 1 Section Focus Transparency **All Aglow**

What are the hanging lights in this picture? They're a species of glowworm that lives in New Zealand. The worms secrete a sticky thread that glows. This light attracts other insects that become stuck to the thread and are soon eaten! The glowing threads also attract mates. If disturbed, the glowworms can douse their lights.

1. What is the main factor of cave life that makes glowing an advantage for the glowworm?
2. When might glowing not be an advantage?
3. What environmental factors might a glowworm need in order to stay alive?

L2

Section 2 Section Focus Transparency **A Vital Job**

All living things generate waste products. Luckily, some living things decompose and recycle that waste. Fungi, a group of plants that includes mushrooms, are great decomposers. The decomposing actions of fungi release nutrients back into the environment.

1. What are these mushrooms living on?
2. If there were no decomposers, what might happen?
3. Describe the role of fungi in recycling nutrients.

L2

Section 3 Section Focus Transparency **What's for dinner?**

Living things need energy to survive. Plants can get the energy they need directly from the Sun. Other organisms feed on plants or animals in order to get energy. This family is having a nutritious dinner. How many different kinds of food can you see in this picture?

1. How do living things use energy?
2. How do apple trees get energy? How do fish get energy?
3. Some organisms are said to be at the top of the food chain. What does this expression mean? Does a food chain ever end? Explain.

L2

This is a representation of key blackline masters available in the Teacher Classroom Resources. See Resource Manager boxes within the chapter for additional information.

Assessment

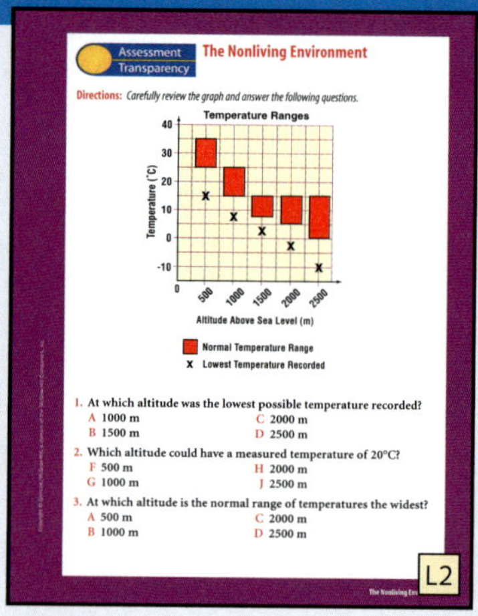

Assessment Transparency **The Nonliving Environment**

Directions: *Carefully review the graph and answer the following questions.*

1. At which altitude was the lowest possible temperature recorded?
   A 1000 m C 2000 m
   B 1500 m D 2500 m
2. Which altitude could have a measured temperature of 20°C?
   F 500 m H 2000 m
   G 1000 m J 2500 m
3. At which altitude is the normal range of temperatures the widest?
   A 500 m C 2000 m
   B 1000 m D 2500 m

L2

Teaching

Section 3 Teaching Transparency **Food Web**

L2

Key to Teaching Strategies

The following designations will help you decide which activities are appropriate for your students.

L1 Level 1 activities should be appropriate for students with learning difficulties.

L2 Level 2 activities should be within the ability range of all students.

L3 Level 3 activities are designed for above-average students.

ELL ELL activities should be within the ability range of English-Language Learners.

COOP LEARN Cooperative Learning activities are designed for small group work.

LS Multiple Learning Styles logos, as described on page 12T, are used throughout to indicate strategies that address different learning styles.

P These strategies represent student products that can be placed into a best-work portfolio.

PBL Problem-Based Learning activities apply real-world situations to learning.

Hands-on Activities

Student Text Lab Worksheet

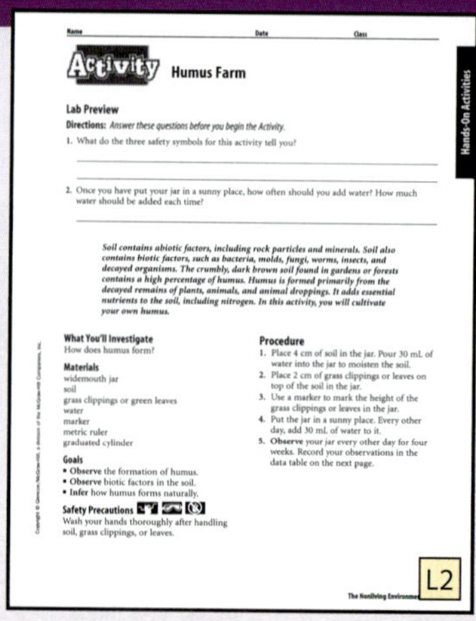

Name Date Class

Activity **Humus Farm**

Lab Preview

Directions: *Answer these questions before you begin the Activity.*

1. What do the three safety symbols for this activity tell you?
2. Once you have put your jar in a sunny place, how often should you add water? How much water should be added each time?

Soil contains abiotic factors, including rock particles and minerals. Soil also contains biotic factors, such as bacteria, molds, fungi, worms, insects, and decayed organisms. The crumbly, dark brown soil found in gardens or forests contains a high percentage of humus. Humus is formed primarily from the decayed remains of plants, animals, and animal droppings. It adds essential nutrients to the soil, including nitrogen. In this activity, you will cultivate your own humus.

What You'll Investigate

How does humus form?

Materials

widemouth jar
soil
grass clippings or green leaves
water
marker
metric ruler
graduated cylinder

Goals

- **Observe** the formation of humus.
- **Observe** biotic factors in the soil.
- **Infer** how humus forms naturally.

Safety Precautions

Wash your hands thoroughly after handling soil, grass clippings, or leaves.

Procedure

1. Place 4 cm of soil in the jar. Pour 30 mL of water into the jar to moisten the soil.
2. Place 2 cm of grass clippings or leaves on top of the soil in the jar.
3. Use a marker to mark the height of the grass clippings or leaves in the jar.
4. Put the jar in a sunny place. Every other day, add 30 mL of water to it.
5. **Observe** your jar every other day for four weeks. Record your observations in the data table on the next page.

Hands-On Activities

L2

Laboratory Activities

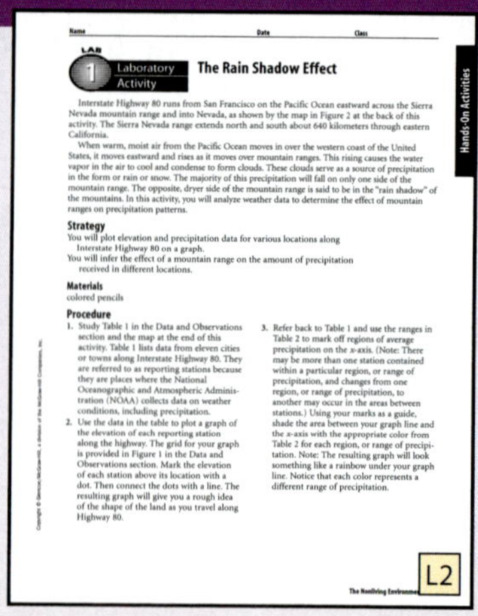

Name Date Class

Lab 1 Laboratory Activity **The Rain Shadow Effect**

Interstate Highway 80 runs from San Francisco on the Pacific Ocean eastward across the Sierra Nevada mountain range and into Nevada, as shown by the map in Figure 2 at the back of this activity. The Sierra Nevada range extends north and south about 640 kilometers through eastern California.

When warm, moist air from the Pacific Ocean moves in over the western coast of the United States, it moves eastward and rises as it moves over mountain ranges. This rising causes the water vapor in the air to cool and condense to form clouds. These clouds serve as a source of precipitation in the form or rain or snow. The majority of this precipitation will fall on only one side of the mountain range. The opposite, dryer side of the mountain range is said to be in the "rain shadow" of the mountains. In this activity, you will analyze weather data to determine the effect of mountain ranges on precipitation patterns.

Strategy

You will plot elevation and precipitation data for various locations along Interstate Highway 80 on a graph.

You will infer the effect of a mountain range on the amount of precipitation received in different locations.

Materials

colored pencils

Procedure

1. Study Table 1 in the Data and Observations section and the map at the end of this activity. Table 1 lists data from eleven cities or towns along Interstate Highway 80. They are referred to as reporting stations because they are places where the National Oceanographic and Atmospheric Administration (NOAA) collects data on weather conditions, including precipitation.
2. Use the data in the table to plot a graph of the elevation of each reporting station along the highway. The grid for your graph is provided in Figure 1 in the Data and Observations section. Mark the elevation of each station above its location with a dot. Then connect the dots with a line. The resulting graph will give you a rough idea of the shape of the land as you travel along Highway 80.
3. Refer back to Table 1 and use the ranges in Table 2 to mark off regions of average precipitation on the x-axis. (Note: There may be more than one station contained within a particular region, or range of precipitation, and changes from one region, or range of precipitation, to another may occur in the areas between stations.) Using your marks as a guide, shade the area between your graph line and the x-axis with the appropriate color from Table 2 for each region, or range of precipitation. Note: The resulting graph will look something like a rainbow under your graph line. Notice that each color represents a different range of precipitation.

Hands-On Activities

L2

Meeting Different Ability Levels

Content Outline

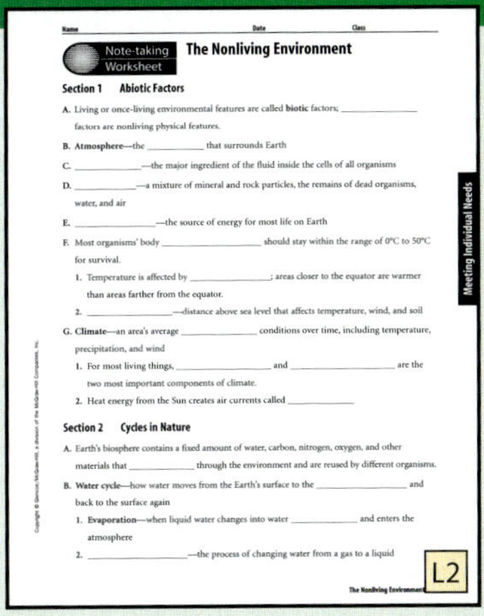
Note-taking Worksheet — The Nonliving Environment

L2

Reinforcement

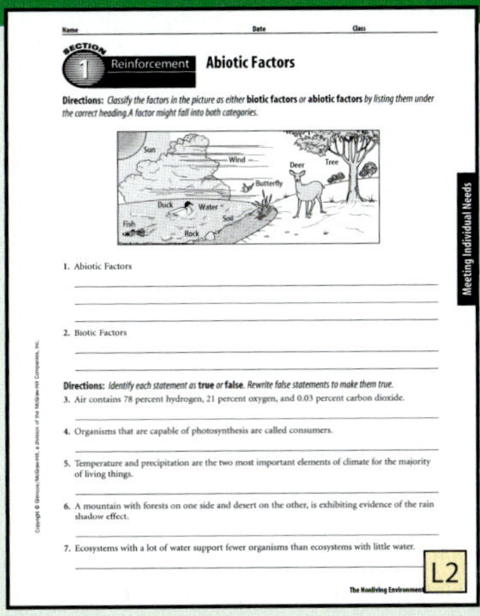
Reinforcement — Abiotic Factors

L2

Enrichment

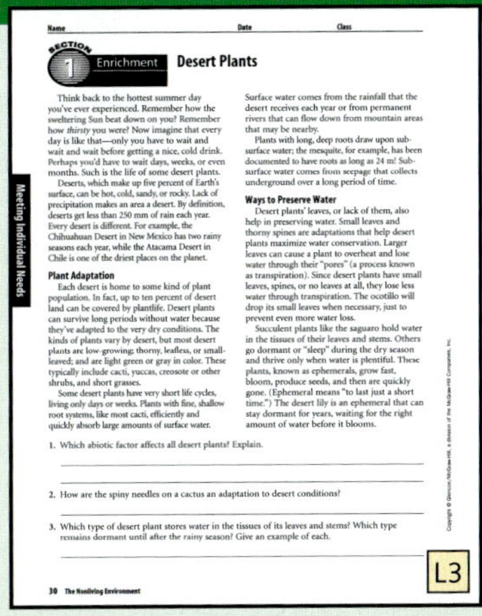
Enrichment — Desert Plants

L3

Directed Reading (English/Spanish)

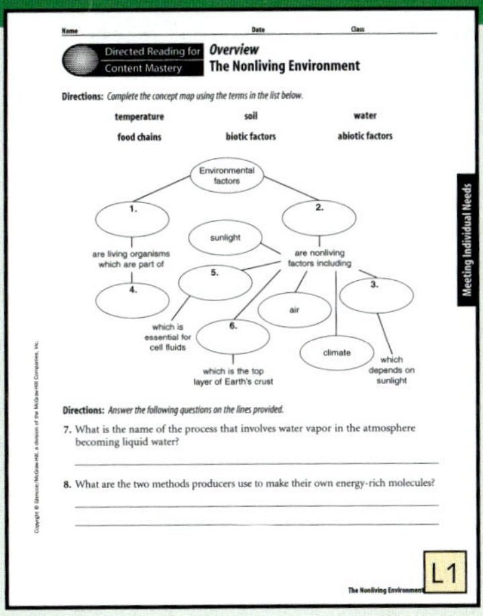
Directed Reading for Content Mastery — Overview — The Nonliving Environment

L1

Study Guide

L2

Reading Essentials

L1

Assessment

Test Practice Workbook

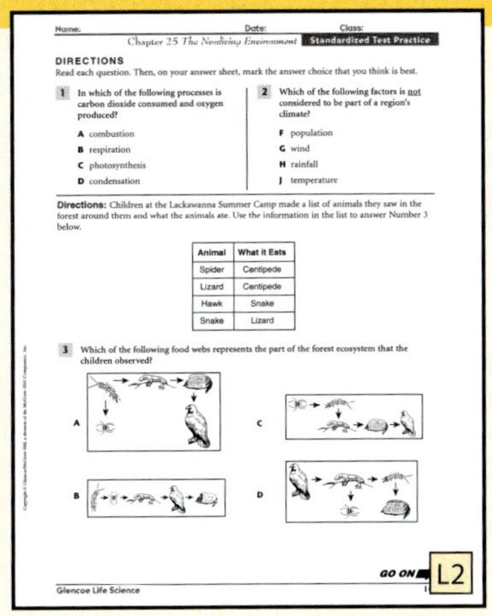
Chapter 25 The Nonliving Environment — Standardized Test Practice

L2

Chapter Review

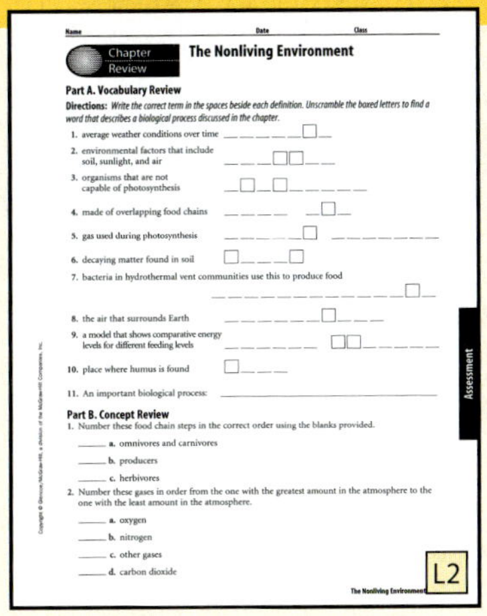
Chapter Review — The Nonliving Environment

L2

Chapter Tests

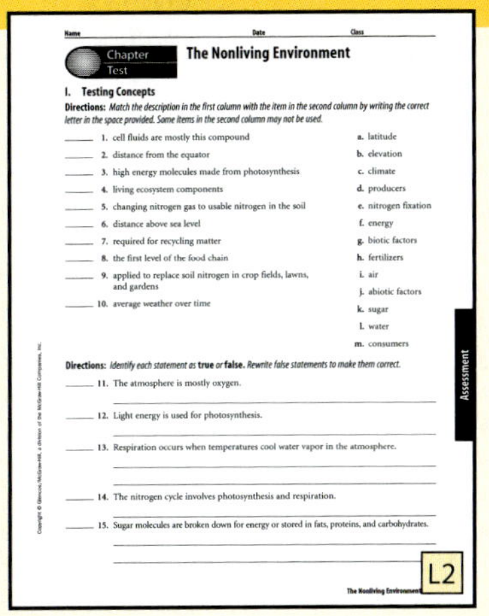
Chapter Test — The Nonliving Environment

L2

Science Content Background

section 1 Abiotic Factors

Importance to Life

Important abiotic factors include air, water, soil, sunlight, temperature, and climate. These factors are interrelated and can vary from environment to environment and over time. Consider temperature, which changes from hour to hour, day to day, season to season, and year to year. Abiotic and biotic factors are not independent. Lack of rainfall can cause a drought in a grassland. The animals that depend on plants for food would find it hard to survive. Many of the factors are predictable and cyclic, but may have extreme variations. Some can be influenced by other factors such as pollution in air, water, and soil.

section 2 Cycles in Nature

Natural Recycling

Matter, in the form of nutrients, moves through the organisms at each trophic level in an ecosystem. But matter cannot be replenished like the energy from sunlight. The atoms of carbon, nitrogen, and other elements that make up the bodies of organisms alive today are the same atoms that have been on Earth since life began. Matter is constantly being recycled.

Water Cycle

The amount of water on Earth remains fairly constant. While some of the water on Earth cycles, much of it is held in oceans or as ice and does not enter the cycle. At a given time, the atmosphere holds about 12,000 cubic kilometers of water, while all the freshwater rivers and lakes hold 120,000 cubic kilometers. Every day about 1,200 cubic kilometers of water evaporates from the ground or transpires from plants and about the same amount falls back to the surface as rain.

Nitrogen Cycle

Nitrogen gas makes up about 79 percent of the atmosphere, but cannot be utilized by plants in this form. In addition to fixation in the soil by bacteria, the high energy of lightning and cosmic radiation combines nitrogen with oxygen into usable nitrates that are carried to Earth with precipitation. The nitrogen in fertilizers is artificially fixed in processing plants. Nitrogen, which is taken up directly by plants and incorporated into plant tissue, enters the food web when plants are consumed.

Carbon Cycle

In 1994, it was reported that the total amount of carbon available on Earth was about 47,000 gigatons (1 gigaton = 1 billion metric tons). About 83 percent of the available carbon was in the oceans, 22 percent in fossil fuels, and four percent in dead organic matter, living organisism, and soils. Only two percent was in the atmosphere. Many scientist believe that the accelerated greenhouse effect is caused by an increased amount of carbon dioxide in the atmosphere, due largely to the burning of fossil fuels.

chapter content resources

Internet Resources

For additional content background, visit booke.msscience.com to:

- access your book online
- find references to related articles in popular science magazines
- access Web links with related content background
- access current events with science journal topics

Print Resources

The Weather Book (2nd edition), by Jack Williams, Vintage Books, 1997

The Water Cycle, by Theresa Greenaway, Raintree/Steck Vaughn, 2000

Food Webs, by Stuart L. Pimm, University of Chicago Press, 2002

section 3 Energy Flow

Food Chains

The law of conservation of energy states that energy cannot be created or destroyed, but can change from one form to another. The energy in an ecosystem is passed from one organism to another through a series of interactions called a food chain. Food chains, food webs and ecological pyramids show how energy moves in only one direction through the trophic levels of an ecosystem and how energy is lost at each transition, from one trophic level to the next. The energy is lost to the environment in the form of heat generated by the body processes of organisms. Although the heat is ultimately lost to the environment, it also serves the purpose of maintaining the body temperature of the organism. Sunlight is the source of most energy, so energy is constantly being replaced.

Food Webs

Some food webs are terrestrial, some are aquatic, and some are combinations. Two types of food webs are grazing and detrital. Detritivores are organisms such as crabs and earthworms that consume dead or decomposing organic matter. The largest amount of energy passes through detrital food webs.

Fritz Polking/Peter Arnold, Inc.

chapter 2

Chapter Vocabulary

biotic, p. 36
abiotic, p. 36
atmosphere, p. 37
soil, p. 38
climate, p. 41
evaporation, p. 44
condensation, p. 45
water cycle, p. 45
nitrogen fixation, p. 46
nitrogen cycle, p. 46
carbon cycle, p. 49
chemosynthesis, p. 51
food web, p. 52
energy pyramid, p. 53

Science Journal Answers may include water, sand, rocks, and sunlight. Order of importance and explanations will vary.

INTERACTIVE CHALKBOARD with Image Bank

PowerPoint® Presentations

This CD-ROM is an editable Microsoft® PowerPoint® presentation that includes:

- a pre-made presentation for every chapter
- interactive graphics
- animations
- audio clips
- image bank
- all new section and chapter questions
- Standardized Test Practice
- transparencies
- pre-lab questions for all labs
- Foldables directions
- links to booke.msscience.com

chapter 2

The Nonliving Environment

chapter preview

sections

1 Abiotic Factors
Lab *Humus Farm*
2 Cycles in Nature
3 Energy Flow
Lab *Where does the mass of a plant come from?*
Virtual Lab *How do organisms react to changes in abiotic factors?*

Sun, Surf, and Sand

Living things on this coast directly or indirectly depend on nonliving things, such as sunlight, water, and rocks, for energy and raw materials needed for their life processes. In this chapter, you will read how these and other nonliving things affect life on Earth.

Science Journal List all the nonliving things that you can see in this picture in order of importance. Explain your reasoning for the order you chose.

34 ◆ E

Theme Connection

Systems and Interactions The ingredients for life are a part of the nonliving environment. Organisms can't exist without interaction with nonliving systems.

About the Photo

Nonliving Factors Everywhere on Earth nonliving factors can be observed, including along the coast of Oregon, as shown in this picture. These nonliving factors provide every living organism with the essential elements of life.

Start-Up Activities

Launch LAB

Earth Has Many Ecosystems

Do you live in a dry, sandy region covered with cactus plants or desert scrub? Is your home in the mountains? Does snow fall during the winter? In this chapter, you'll learn why the nonliving factors in each ecosystem are different. The following lab will get you started.

1. Locate your city or town on a globe or world map. Find your latitude. Latitude shows your distance from the equator and is expressed in degrees, minutes, and seconds.
2. Locate another city with the same latitude as your city but on a different continent.
3. Locate a third city with latitude close to the equator.
4. Using references, compare average annual precipitation and average high and low temperatures for all three cities.
5. **Think Critically** Hypothesize how latitude affects average temperatures and rainfall.

Science Online | Preview this chapter's content and activities at booke.msscience.com

FOLDABLES Study Organizer

Nonliving Factors Make the following Foldable to help you understand the cause and effect relationships within the nonliving environment.

STEP 1 **Fold** two vertical sheets of paper in half from top to bottom. **Cut** the papers in half along the folds.

STEP 2 Discard one piece and **fold** the three vertical pieces in half from top to bottom.

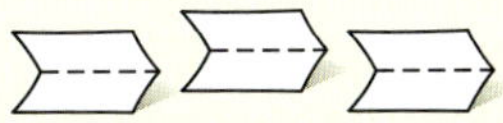

STEP 3 **Turn** the papers horizontally. **Tape** the short ends of the pieces together (overlapping the edges slightly).

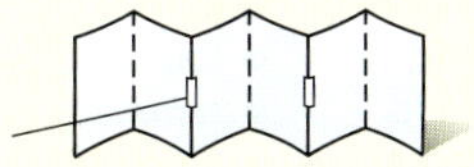

STEP 4 On one side, **label** the folds: *Nonliving, Water, Soil, Wind, Temperature,* and *Elevation.* **Draw** a picture of a familiar ecosystem on the other side.

Sequence As you read the chapter, write on the folds how each nonliving factor affects the environment that you draw.

Launch LAB

Purpose Use the Launch Lab to show students how latitude affects climate. L2 **Visual-Spatial**

Preparation Obtain reference information that provides climate data for various cities.

Materials world maps or globes, climate reference information

Teaching Strategy Have students compare their findings with other students' findings. Discuss differences and similarities.

Think Critically

In general, the lower the latitude, the higher the temperature. Rainfall is influenced by many variables, so there was probably no consistent relationship found between latitude and rainfall.

Assessment

Performance Give students a map of the United States and ask them to put Phoenix, San Francisco, and Los Angeles in order from lowest average temperature to highest. Use **Performance Assessment in the Science Classroom,** p. 89. L2

FOLDABLES Study Organizer — Dinah Zike Study Fold

Student preparation materials for this Foldable are available in the Chapter *FAST FILE* Resources.

1 Motivate

INTERACTIVE CHALKBOARD
PowerPoint® Presentations

Bellringer

Section Focus Transparencies also are available on the Interactive Chalkboard CD-ROM. L2 ELL

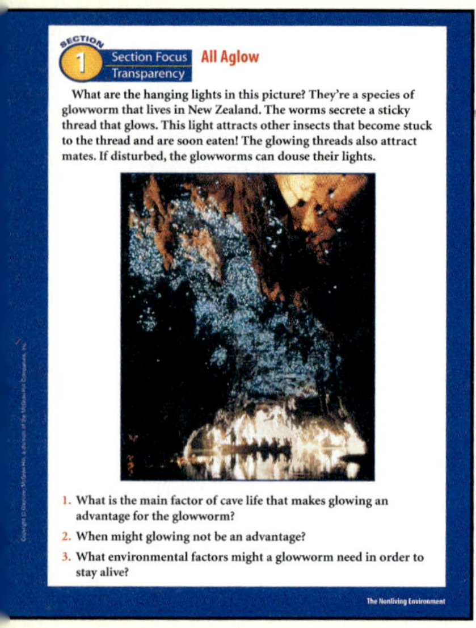

SECTION 1 Section Focus Transparency **All Aglow**

What are the hanging lights in this picture? They're a species of glowworm that lives in New Zealand. The worms secrete a sticky thread that glows. This light attracts other insects that become stuck to the thread and are soon eaten! The glowing threads also attract mates. If disturbed, the glowworms can douse their lights.

1. What is the main factor of cave life that makes glowing an advantage for the glowworm?
2. When might glowing not be an advantage?
3. What environmental factors might a glowworm need in order to stay alive?

The Nonliving Environment

Tie to Prior Knowledge

Survival of Living Things Have students examine a fish tank or a scene from nature. Ask them to identify nonliving things that are important for living things to survive. L2

section 1

Abiotic Factors

as you read

What You'll Learn

- **Identify** common abiotic factors in most ecosystems.
- **List** the components of air that are needed for life.
- **Explain** how climate influences life in an ecosystem.

Why It's Important

Knowing how organisms depend on the nonliving world can help humans maintain a healthy environment.

Review Vocabulary
environment: everything, such as climate, soil, and living things, that surrounds and affects an organism

New Vocabulary
- biotic
- abiotic
- atmosphere
- soil
- climate

Environmental Factors

Living organisms depend on one another for food and shelter. The leaves of plants provide food and a home for grasshoppers, caterpillars, and other insects. Many birds depend on insects for food. Dead plants and animals decay and become part of the soil. The features of the environment that are alive, or were once alive, are called **biotic** (bi AH tihk) factors. The term *biotic* means "living."

Biotic factors are not the only things in an environment that are important to life. Most plants cannot grow without sunlight, air, water, and soil. Animals cannot survive without air, water, or the warmth that sunlight provides. The nonliving, physical features of the environment are called **abiotic** (ay bi AH tihk) factors. The prefix *a* means "not." The term *abiotic* means "not living." Abiotic factors include air, water, soil, sunlight, temperature, and climate. The abiotic factors in an environment often determine which kinds of organisms can live there. For example, water is an important abiotic factor in the environment, as shown in **Figure 1.**

Figure 1 Abiotic factors—air, water, soil, sunlight, temperature, and climate—influence all life on Earth.

Section 1 Resource Manager

Chapter *FAST FILE* Resources
- Transparency Activity, p. 44
- Directed Reading for Content Mastery, pp. 19, 20
- Note-taking Worksheets, pp. 33–34
- MiniLAB, p. 3
- Lab, pp. 9–12
- Enrichment, p. 30
- Reinforcement, p. 27
- Lab Worksheet, pp. 5–6

Cultural Diversity, p. 33
Reading and Writing Skill Activities, p. 3
Performance Assessment in the Science Classroom, p. 48

Air

Air is invisible and plentiful, so it is easily overlooked as an abiotic factor of the environment. The air that surrounds Earth is called the **atmosphere.** Air contains 78 percent nitrogen, 21 percent oxygen, 0.94 percent argon, 0.03 percent carbon dioxide, and trace amounts of other gases. Some of these gases provide substances that support life.

Carbon dioxide (CO_2) is required for photosynthesis. Photosynthesis—a series of chemical reactions—uses CO_2, water, and energy from sunlight to produce sugar molecules. Organisms, like plants, that can use photosynthesis are called producers because they produce their own food. During photosynthesis, oxygen is released into the atmosphere.

When a candle burns, oxygen from the air chemically combines with the molecules of candle wax. Chemical energy stored in the wax is converted and released as heat and light energy. In a similar way, cells use oxygen to release the chemical energy stored in sugar molecules. This process is called respiration. Through respiration, cells obtain the energy needed for all life processes. Air-breathing animals aren't the only organisms that need oxygen. Plants, some bacteria, algae, fish, and other organisms need oxygen for respiration.

Figure 2 Water is an important abiotic factor in deserts and rain forests.

Life in deserts is limited to species that can survive for long periods without water.

Thousands of species can live in lush rain forests where rain falls almost every day.

Water

Water is essential to life on Earth. It is a major ingredient of the fluid inside the cells of all organisms. In fact, most organisms are 50 percent to 95 percent water. Respiration, digestion, photosynthesis, and many other important life processes can take place only in the presence of water. As **Figure 2** shows, environments that have plenty of water usually support a greater diversity of and a larger number of organisms than environments that have little water.

2 Teach

Use Science Words

Word Origin Write the words *biotic* and *abiotic* on the board. Ask students to find the meanings of *bio-*, *sphere*, and *a-*. *bio:* life; *sphere:* area; *a:* without, or not L2

Activity

Factors for Growth Have students make a list of the biotic and abiotic factors that would affect the population growth of frogs near a pond. L2 LS **Naturalist**

Quick Demo

Releasing Energy

Materials candle, bell jar, lighter

Estimated Time five minutes

Procedure Light a candle and explain that oxygen helps release heat and light energy from the wax. In a similar way, oxygen in living things helps release the energy from food. Put a bell jar over the candle and direct students to observe. As the flame goes out, explain that without oxygen, energy cannot be released. LS **Visual-Spatial**

Virtual Labs

Abiotic Factors *How do organisms react to changes in abiotic factors?*

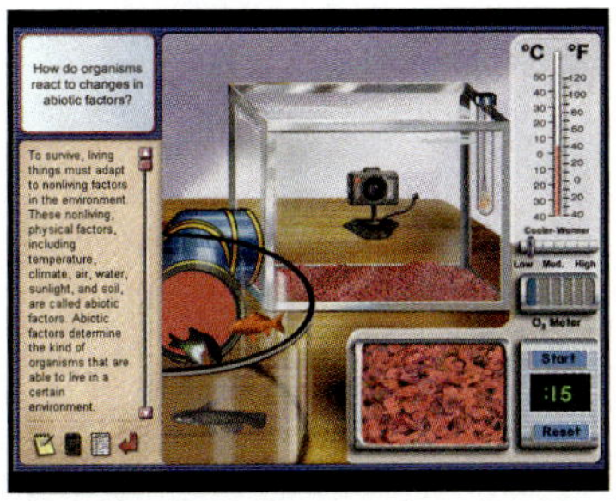

Differentiated Instruction

Challenge Refer to a world map of biomes. Have students compare areas of similar latitude that have different biomes. Varying amounts of precipitation is most likely the primary difference between these areas. L3 LS **Visual-Spatial**

Learning Disabled Ask students to comment on this question: We can't see air, so how do we know it exists? A variety of evidence can be discussed. We can feel the air when we move our hands very fast. We can see the effect of air when we blow it into a balloon. L1 LS **Naturalist**

Fun Fact

One gram of soil may contain hundreds of millions of organisms.

Mini LAB

Purpose Students investigate the components that make up soil. L2 LS **Kinesthetic**

Materials two cups soil, large jar with lid, water, alum or dishwashing liquid, spoon, metric ruler

Teaching Strategy Review the components of soil before students carry out the activity.

Analysis

1. Clay is likely suspended in the water.
2. Answers will vary. It is likely that silt will form the greatest component.

Assessment

Oral Why did you examine the sample at different times? Different-sized particles settle at different rates. Use **Performance Assessment in the Science Classroom,** p. 89. L2

Try at Home

Caption Answer

Figure 3 These fish often depend on biotic materials that fall from above.

Mini LAB

Determining Soil Makeup

Procedure

1. Collect 2 cups of **soil.** Remove large pieces of debris and break up clods.
2. Put the soil in a **quart jar** or **similar container that has a lid.**
3. Fill the container with **water** and add 1 teaspoon of **dishwashing liquid.**
4. Put the lid on tightly and shake the container.
5. After 1 min, measure and record the depth of sand that settled on the bottom.
6. After 2 h, measure and record the depth of silt that settles on top of the sand.
7. After 24 h, measure and record the depth of the layer between the silt and the floating organic matter.

Analysis

1. Clay particles are so small that they can remain suspended in water. Where is the clay in your sample?
2. Is sand, silt, or clay the greatest part of your soil sample?

Try at Home

Soil

Soil is a mixture of mineral and rock particles, the remains of dead organisms, water, and air. It is the topmost layer of Earth's crust, and it supports plant growth. Soil is formed, in part, of rock that has been broken down into tiny particles.

Soil is considered an abiotic factor because most of it is made up of nonliving rock and mineral particles. However, soil also contains living organisms and the decaying remains of dead organisms. Soil life includes bacteria, fungi, insects, and worms. The decaying matter found in soil is called humus. Soils contain different combinations of sand, clay, and humus. The type of soil present in a region has an important influence on the kinds of plant life that grow there.

Sunlight

All life requires energy, and sunlight is the energy source for almost all life on Earth. During photosynthesis, producers convert light energy into chemical energy that is stored in sugar molecules. Consumers are organisms that cannot make their own food. Energy is passed to consumers when they eat producers or other consumers. As shown in **Figure 3,** photosynthesis cannot take place if light is never available.

Shady forest

Bottom of deep ocean

Figure 3 Photosynthesis requires light. Little sunlight reaches the shady forest floor, so plant growth beneath trees is limited. Sunlight does not reach into deep lake or ocean waters. Photosynthesis can take place only in shallow water or near the water's surface.
Infer *how fish that live at the bottom of the deep ocean obtain energy.*

LAB DEMONSTRATION

Purpose to determine the amount of humus in soil

Materials Bunsen burner, crucible, topsoil rich in humus, goggles, scale

Preparation Obtain and dry topsoil before class. Explain to students that humus can burn, but the rest of soil cannot.

Procedure Weigh a crucible, fill it with soil, and then weigh the soil-filled crucible. Heat it on high for 10 min. Allow it to cool and reweigh it. Determine the mass of humus and its percentage of original soil mass.

Expected Outcome Topsoils are 2–15% humus.

Assessment

Give students the following measurements and have them calculate the mass and percentage of humus in soil.

initial mass = 10 grams; final mass = 9 grams

mass of humus = 1 gram; 10% of original soil L2

Figure 4 Temperature is an abiotic factor that can affect an organism's survival.

The penguin has a thick layer of fat to hold in heat and keep the bird from freezing. These emperor penguins huddle together for added warmth.

The Arabian camel stores fat only in its hump. This way, the camel loses heat from other parts of its body, which helps it stay cool in the hot desert.

Temperature

Sunlight supplies life on Earth with light energy for photosynthesis and heat energy for warmth. Most organisms can survive only if their body temperatures stay within the range of 0°C to 50°C. Water freezes at 0°C. The penguins in **Figure 4** are adapted for survival in the freezing Antarctic. Camels can survive the hot temperatures of the Arabian Desert because their bodies are adapted for staying cool. The temperature of a region depends in part on the amount of sunlight it receives. The amount of sunlight depends on the land's latitude and elevation.

Reading Check *What does sunlight provide for life on Earth?*

Latitude In this chapter's Launch Lab, you discovered that temperature is affected by latitude. You found that cities located at latitudes farther from the equator tend to have colder temperatures than cities at latitudes nearer to the equator. As **Figure 5** shows, polar regions receive less of the Sun's energy than equatorial regions. Near the equator, sunlight strikes Earth directly. Near the poles, sunlight strikes Earth at an angle, which spreads the energy over a larger area.

Figure 5 Because Earth is curved, latitudes farther from the equator are colder than latitudes near the equator.

Visual Learning

Figure 4 Have students identify other adaptations camels have to their native desert environment. Possible answers: Their feet are widespread to enable them to walk easily on sand; to protect them from blowing sand, they have a double row of eyelashes, haired ear openings, and the ability to close their nostrils; they also have keen senses of sight and smell. L2

Reading Check

Answer Sunlight provides light and heat. Students may also say that, through the process of photosynthesis, sunlight provides energy for life processes.

Inquiry Lab

Water Cycle

Purpose to explore the water cycle

Possible Materials plastic bag, clear sealed container

Estimated Time four weeks

Teaching Strategies

- Students could gather topsoil and add water, or collect mud from a pond and seal it in a clear container.
- Students could predict and observe how water moves throughout the sealed environment condensing on the inner surface and evaporating.
- Students can predict and observe what influence the plants have on the cycle.
- Allow students to explore other questions that arise.

For additional inquiry activities, see *Science Inquiry Labs*.

Differentiated Instruction

Challenge Have students research the high and low temperatures at the north pole and the equator. Have students compare the findings with the area where they live and graph the results. L3

Curriculum Connection

Health Sunlight prevents rickets because it enables the skin to produce vitamin D. Ingesting vitamin D in food will also prevent rickets. Have students research which foods are good sources of vitamin D. Foods rich in vitamin D include milk, eggs, fortified breakfast cereals, sardines, salmon, beef, and margarine. L2

Applying Math

National Math Standards

Correlation to Mathematics Objectives

1, 2, 4, 5, 6, 9, 10

Answer to Practice Problem

Students should predict temperatures of around 23°C at 550 m and 27°C at 375 m.

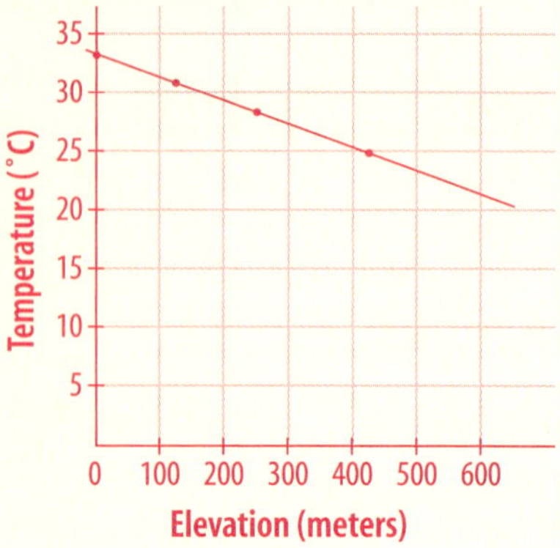

Teacher FYI

Altitude Sickness Altitude sickness is a condition resulting from the lack of oxygen at elevations above 2,400 meters (8,000 ft). Mountain climbers might be affected if climbing without supplemental oxygen. Symptoms may include shortness of breath, headaches, decreased coordination, impaired sight and hearing, nosebleeds, decreased mental clarity, dizziness, and hallucinations. If untreated, the condition can be fatal. In most cases, returning to a lower altitude and receiving oxygen quickly reverses the symptoms.

Figure 6 The stunted growth of these trees is a result of abiotic factors.

Elevation If you have climbed or driven up a mountain, you probably noticed that the temperature got cooler as you went higher. A region's elevation, or distance above sea level, affects its temperature. Earth's atmosphere acts as insulation that traps the Sun's heat. At higher elevations, the atmosphere is thinner than it is at lower elevations. Air becomes warmer when sunlight heats molecules in the air. Because there are fewer molecules at higher elevations, air temperatures there tend to be cooler.

At higher elevations, trees are shorter and the ground is rocky, as shown in **Figure 6.** Above the timberline—the elevation beyond which trees do not grow—plant life is limited to low-growing plants. The tops of some mountains are so cold that no plants can survive. Some mountain peaks are covered with snow year-round.

Applying Math — Solve for an Unknown

TEMPERATURE CHANGES You climb a mountain and record the temperature every 1,000 m of elevation. The temperature is 30°C at 304.8 m, 25°C at 609.6 m, 20°C at 914.4 m, 15°C at 1,219.2 m, and 5°C at 1,828.8 m. Make a graph of the data. Use your graph to predict the temperature at an altitude of 2,133.6 m.

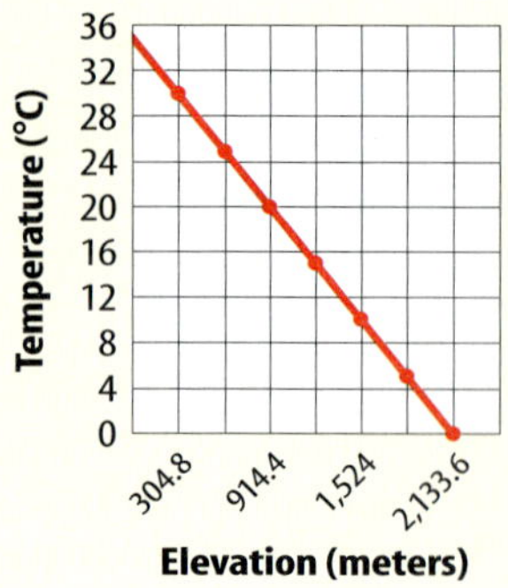

Solution

1. *This is what you know:* The data can be written as ordered pairs (elevation, temperature). The ordered pairs for these data are (304.8, 30), (609.6, 25), (914.4, 20), (1,219.2, 15), (1,828.8, 5).
2. *This is what you want to find:* Predict the temperature at an elevation of 2,133.6 m.
3. *This is what you need to do:* Graph the data by plotting elevation on the x-axis and temperature on the y-axis.
4. *Predict the temperature at 2,133.6 m:* Extend the graph line to predict the temperature at 2,133.6 m.

Practice Problems

1. Temperatures on another mountain are 33°C at sea level, 31°C at 125 m, 29°C at 250 m, and 26°C at 425 m. Graph the data and predict the temperature at 550 m.
2. Predict what the temperature would be at 375 m.

Science Online — For more practice, visit booke.msscience.com/math_practice

Cultural Diversity

Kenyan Runners Many great long-distance runners are from the Kalenjin tribe in a high-altitude area of Kenya. Why might great runners come from this area? Possible answers: Because of lower oxygen levels at high altitudes, these Kenyans may have adaptations for increased cardiovascular efficiency; their high-altitude training may help them produce more red blood cells; there may also be social reasons. L2

Climate

In Fairbanks, Alaska, winter temperatures may be as low as −52°C, and more than a meter of snow might fall in one month. In Key West, Florida, snow never falls and winter temperatures rarely dip below 5°C. These two cities have different climates. **Climate** refers to an area's average weather conditions over time, including temperature, rainfall or other precipitation, and wind.

For the majority of living things, temperature and precipitation are the two most important components of climate. The average temperature and rainfall in an area influence the type of life found there. Suppose a region has an average temperature of 25°C and receives an average of less than 25 cm of rain every year. It is likely to be the home of cactus plants and other desert life. A region with similar temperatures that receives more than 300 cm of rain every year is probably a tropical rain forest.

Wind Heat energy from the Sun not only determines temperature, but also is responsible for the wind. The air is made up of molecules of gas. As the temperature increases, the molecules spread farther apart. As a result, warm air is lighter than cold air. Colder air sinks below warmer air and pushes it upward, as shown in **Figure 7.** These motions create air currents that are called wind.

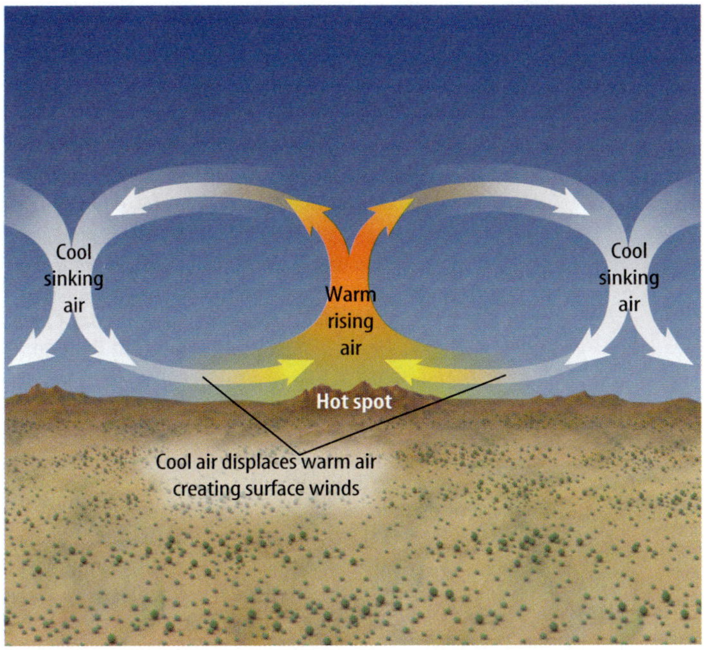

Figure 7 Winds are created when sunlight heats some portions of Earth's surface more than others. In areas that receive more heat, the air becomes warmer. Cold air sinks beneath the warm air, forcing the warm air upward.

INTEGRATE Career

Farmer Changes in weather have a strong influence in crop production. Farmers sometimes adapt by changing planting and harvesting dates, selecting a different crop, or changing water use. In your Science Journal, describe another profession affected by climate.

Science Online

Topic: Weather Data
Visit booke.msscience.com for Web links to information about recent weather data for your area.

Activity In your Science Journal, describe how these weather conditions affect plants or animals that live in your area.

Discussion

Weather and Climate Weather refers to the day-to-day atmospheric conditions in an area. Climate takes a long-term view and describes the general weather conditions associated with an area. Discuss with students the weather and climate differences for their environment.

INTEGRATE Career

Farmer Answers will vary but should describe how climate affects the profession. An example is a fisherman who can find it difficult to navigate in adverse weather. Fish movements and populations also can be influenced by weather.

Research Have students investigate historical methods of predicting weather conditions. For example, how did farmers know when to plant crops?

Quick Demo

Rain Shadow Effect

Materials sponge

Estimated Time one minute

Procedure Show the rain shadow effect by moving a damp sponge over an imaginary mountain. As air rises, explain that it cools and water vapor condenses into rain. Squeeze the sponge to show rain. As air descends on the other side of the mountain, it warms. Any remaining water in the air changes back to water vapor. L2 ELL
LS **Visual-Spatial**

Active Reading

Quickwrites This strategy, sometimes called freewrites, lets students use spontaneous writing to discover what they already know. Have students write a list of ideas about a topic, then share these ideas with the class. Next, have students write their ideas without worrying about punctuation, spelling, and grammar. Have students use a Quickwrite to share ideas about abiotic factors. L2

Differentiated Instruction

Learning Disabled Show students a globe and explain that since areas near the equator receive more direct sunshine, they are much warmer than areas toward the poles. As hot air is pushed aloft near the equator, cooler air rushes in to take its place. This causes wind, which is responsible for precipitation patterns. L1 LS **Visual-Spatial**

Teacher FYI

Cascade Mountains The Cascade Mountains, part of the Pacific mountain system, extend from California to Canada. The tallest peak in this range is Mt. Ranier, which reaches 4,392 m (14,410 ft). Most of the peaks are extinct volcanoes, although some have erupted in the recent past. Mount St. Helens is one of these, with eruptions in 1980 and 1981.

3 Assess

DAILY INTERVENTION

Check for Understanding

Challenge Have students investigate how professional greenhouses increase the levels of carbon dioxide and the temperature to achieve an optimal plant growing environment. L3

Reteach

Abiotic Factors Have students explain how water, air, and temperature can affect life. L1 LS **Naturalist**

Assessment

Portfolio Have students cut out pictures from magazines that illustrate abiotic factors in an ecosystem and use the pictures to make posters. Direct students to label each abiotic factor. Use **Performance Assessment in the Science Classroom,** p. 145. L2 P

Figure 8 In Washington State, the western side of the Cascade Mountains receives an average of 101 cm of rain each year. The eastern side of the Cascades is in a rain shadow that receives only about 25 cm of rain per year.

INTEGRATE Earth Science **The Rain Shadow Effect** The presence of mountains can affect rainfall patterns. As **Figure 8** shows, wind blowing toward one side of a mountain is forced upward by the mountain's shape. As the air nears the top of the mountain, it cools. When air cools, the moisture it contains falls as rain or snow. By the time the cool air crosses over the top of the mountain, it has lost most of its moisture. The other side of the mountain range receives much less precipitation. It is not uncommon to find lush forests on one side of a mountain range and desert on the other side.

section 1 review

Summary

Environmental Factors

- Organisms depend on one another as well as sunlight, air, water, and soil.

Air, Water, and Soil

- Some of the gases in air provide substances to support life.
- Water is a major component of the cells in all organisms.
- Soil supports plant growth.

Sunlight, Temperature, and Climate

- Light energy supports almost all life on Earth.
- Most organisms require temperature between 0°C and 50°C to survive.
- For most organisms, temperature and precipitation are the two most important components of climate.

Self Check

1. **Compare and contrast** biotic factors and abiotic factors in ecosystems.
2. **Explain** why soil is considered an abiotic factor and a biotic factor.
3. **Think Critically** On day 1, you hike in shade under tall trees. On day 2, the trees are shorter and farther apart. On day 3, you see small plants but no trees. On day 4, you see snow. What abiotic factors might contribute to these changes?

Applying Math

4. **Use an Electronic Spreadsheet** Obtain two months of temperature and precipitation data for two cities in your state. Enter the data in a spreadsheet and calculate average daily temperature and precipitation. Use your calculations to compare the two climates.

Science Online booke.msscience.com/self_check_quiz

section 1 review

1. Biotic factors are, or once were, living. Abiotic factors are nonliving.
2. While it is mostly nonliving material, there are many things living in soil, as well as decaying organic matter called humus.
3. Changes in elevation, which influence temperature, are probably responsible. This could occur on a mountain, as you reach and surpass the timberline.
4. Spreadsheets should result in average temperature and rainfall for two different cities.

LAB

Humus Farm

Besides abiotic factors, such as rock particles and minerals, soil also contains biotic factors, including bacteria, molds, fungi, worms, insects, and decayed organisms. Crumbly, dark brown soil contains a high percentage of humus that is formed primarily from the decayed remains of plants, animals, and animal droppings. In this lab, you will cultivate your own humus.

Real-World Question

How does humus form?

Goals

- **Observe** the formation of humus.
- **Observe** biotic factors in the soil.
- **Infer** how humus forms naturally.

Materials

widemouthed jar	water
soil	marker
grass clippings or green leaves	metric ruler
	graduated cylinder

Safety Precautions

Wash your hands thoroughly after handling soil, grass clippings, or leaves.

Humus Formation	
Date	Observations
	Answers will vary.

Procedure

1. Copy the data table below into your Science Journal.
2. Place 4 cm of soil in the jar. Pour 30 mL of water into the jar to moisten the soil.
3. Place 2 cm of grass clippings or green leaves on top of the soil in the jar.
4. Use a marker to mark the height of the grass clippings or green leaves in the jar.
5. Put the jar in a sunny place. Every other day, add 30 mL of water to it. In your Science Journal, write a prediction of what you think will happen in your jar.
6. **Observe** your jar every other day for four weeks. Record your observations in your data table.

Conclude and Apply

1. **Describe** what happened during your investigation.
2. **Infer** how molds and bacteria help the process of humus formation.
3. **Infer** how humus forms on forest floors or in grasslands.

Communicating Your Data

Compare your humus farm with those of your classmates. With several classmates, write a recipe for creating the richest humus. Ask your teacher to post your recipe in the classroom. **For more help, refer to the Science Skill Handbook.**

Communicating Your Data

Students may want to consult cookbooks to see how the ingredients for recipes are written and arranged.

Assessment

Process Ask students to discuss how forest ecosystems would be affected if leaf litter did not become humus. Use **Performance Assessment in the Science Classroom,** p. 93. L2

BENCH TESTED

Real-World Question

Purpose Students observe humus formation over several weeks. L2 LS **Visual-Spatial**

Process Skills observe, describe, infer

Time Required 25 minutes for setup; 5-minute observation periods every other day for four weeks

Alternate Materials If green leaves are used instead of grass clippings, the leaves must be chopped in order for the activity to work.

Safety Precautions Instruct students to wash their hands when finished.

Procedure

Teaching Strategy Encourage students to bring in leaf litter to be used in the activity.

Troubleshooting To be certain the humus farms get enough water, assign a person from each group to water the jars every other day. Keep grass clippings moist but not waterlogged. Do not seal jars.

Conclude and Apply

1. Students first observe the growth of mold. As the mold and decomposing bacteria break down the clippings or litter, large particles of organic matter will form. These particles are the beginning of humus.
2. Molds and bacteria decompose materials into organic particles.
3. Organic materials fall to the ground and are decomposed into organic particles.

section 2 Cycles in Nature

1 Motivate

Bellringer

INTERACTIVE CHALKBOARD PowerPoint® Presentations

Section Focus Transparencies also are available on the Interactive Chalkboard CD-ROM. L2 ELL

SECTION 2 Section Focus Transparency **A Vital Job**

All living things generate waste products. Luckily, some living things decompose and recycle that waste. Fungi, a group of plants that includes mushrooms, are great decomposers. The decomposing actions of fungi release nutrients back into the environment.

1. What are these mushrooms living on?
2. If there were no decomposers, what might happen?
3. Describe the role of fungi in recycling nutrients.

The Nonliving Environment

Tie to Prior Knowledge

Water Cycle Most students will have some familiarity with the water cycle. Ask students to explain how water cycles through the environment. Ask them to think of other materials that also cycle. L2

as you read

***What* You'll Learn**

- **Explain** the importance of Earth's water cycle.
- **Diagram** the carbon cycle.
- **Recognize** the role of nitrogen in life on Earth.

***Why* It's Important**

The recycling of matter on Earth demonstrates natural processes.

Review Vocabulary
biosphere: the part of the world in which life can exist

New Vocabulary
- evaporation
- condensation
- water cycle
- nitrogen fixation
- nitrogen cycle
- carbon cycle

The Cycles of Matter

Imagine an aquarium containing water, fish, snails, plants, algae, and bacteria. The tank is sealed so that only light can enter. Food, water, and air cannot be added. Will the organisms in this environment survive? Through photosynthesis, plants and algae produce their own food. They also supply oxygen to the tank. Fish and snails take in oxygen and eat plants and algae. Wastes from fish and snails fertilize plants and algae. Organisms that die are decomposed by the bacteria. The organisms in this closed environment can survive because the materials are recycled. A constant supply of light energy is the only requirement. Earth's biosphere also contains a fixed amount of water, carbon, nitrogen, oxygen, and other materials required for life. These materials cycle through the environment and are reused by different organisms.

The Water Cycle

If you leave a glass of water on a sunny windowsill, the water will evaporate. **Evaporation** takes place when liquid water changes into water vapor, which is a gas, and enters the atmosphere, shown in **Figure 9.** Water evaporates from the surfaces of lakes, streams, puddles, and oceans. Water vapor enters the atmosphere from plant leaves in a process known as transpiration (trans puh RAY shun). Animals release water vapor into the air when they exhale. Water also returns to the environment from animal wastes.

Figure 9 Water vapor is a gas that is present in the atmosphere.

Section 2 Resource Manager

Chapter *FAST FILE* Resources
- Transparency Activity, p. 45
- Directed Reading for Content Mastery, p. 21
- MiniLAB, p. 4
- Lab Activity, pp. 13–16
- Enrichment, p. 31
- Reinforcement, p. 28

Reading and Writing Skill Activities, p. 29
Physical Science Critical Thinking/Problem Solving, pp. 12, 17
Life Science Critical Thinking/Problem Solving, pp. 12, 22

Condensation Water vapor that has been released into the atmosphere eventually comes into contact with colder air. The temperature of the water vapor drops. Over time, the water vapor cools enough to change back into liquid water. The process of changing from a gas to a liquid is called **condensation.** Water vapor condenses on particles of dust in the air, forming tiny droplets. At first, the droplets clump together to form clouds. When they become large and heavy enough, they fall to the ground as rain or other precipitation. As the diagram in **Figure 10** shows, the **water cycle** is a model that describes how water moves from the surface of Earth to the atmosphere and back to the surface again.

Figure 10 The water cycle involves evaporation, condensation, and precipitation. Water molecules can follow several pathways through the water cycle.
Identify *as many water cycle pathways as you can from this diagram.*

Water Use Data about the amount of water people take from reservoirs, rivers, and lakes for use in households, businesses, agriculture, and power production is shown in **Table 1.** These actions can reduce the amount of water that evaporates into the atmosphere. They also can influence how much water returns to the atmosphere by limiting the amount of water available to plants and animals.

Table 1 U.S. Estimated Water Use in 1995

Water Use	Millions of Gallons per Day	Percent of Total
Homes and Businesses	41,600	12.2
Industry and Mining	28,000	8.2
Farms and Ranches	139,200	40.9
Electricity Production	131,800	38.7

2 Teach

Identifying Misconceptions

Movement of Water Vapor Students may think that the same water falls in a given area again and again. Emphasize that water vapor moves great distances in the air all over Earth; that water in lakes, rivers, and streams is always moving; that currents in the ocean are constantly moving; and that groundwater may move great distances.

Caption Answer

Figure 10 Pathways include precipitation, evaporation, transpiration, seeping into groundwater, and runoff from the ground into other bodies of water.

Visual Learning

Table 1 Which type of water use consumes the most water? farms and ranches Suggest a reason why. Possible answers: irrigation of crops or grasses, watering of livestock

Quick Demo

Condensation

Materials glass cup, ice

Estimated Time five minutes

Procedure Place some water and ice in a glass cup and place it on a table. Wait briefly and examine the outside of the glass. Explain that when warm air that contains water vapor contacts a cool surface, the water vapor condenses into liquid water.

Teacher FYI

Balancing Systems Although it is possible to achieve a closed aquarium system, it is not easy. The system must be perfectly balanced between producer and consumer, and this balance must remain constant over time.

Science Journal

Water Stories Have students imagine they are molecules of water. They should each make up a story about how they travel and what they encounter as they go through one complete cycle of the water cycle. L2 LS **Linguistic**

Use an Analogy

Sugars What would happen to a student who ate only candy? They would ingest a lot of sugar but not enough protein. Sugar provides energy but no protein to build or repair tissues. In a similar way, if a plant did not have enough nitrogen it could make sugars, but its growth and functioning would be impaired because it could not make enough proteins. L2

LS **Logical-Mathematical**

Reading Check

Answer the changing of atmospheric nitrogen into nitrogen compounds that plants can use

Discussion

Nitrogen Deficiency What would a plant with a nitrogen deficiency look like? It may have yellow leaves and/or stunted growth.

Caption Answer

Figure 11 Decaying organs; accept all correct responses.

The Nitrogen Cycle

The element nitrogen is important to all living things. Nitrogen is a necessary ingredient of proteins. Proteins are required for the life processes that take place in the cells of all organisms. Nitrogen is also an essential part of the DNA of all organisms. Although nitrogen is the most plentiful gas in the atmosphere, most organisms cannot use nitrogen directly from the air. Plants need nitrogen that has been combined with other elements to form nitrogen compounds. Through a process called **nitrogen fixation,** some types of soil bacteria can form the nitrogen compounds that plants need. Plants absorb these nitrogen compounds through their roots. Animals obtain the nitrogen they need by eating plants or other animals. When dead organisms decay, the nitrogen in their bodies returns to the soil or to the atmosphere. This transfer of nitrogen from the atmosphere to the soil, to living organisms, and back to the atmosphere is called the **nitrogen cycle,** shown in **Figure 11.**

Reading Check *What is nitrogen fixation?*

Figure 11 During the nitrogen cycle, nitrogen gas from the atmosphere is converted to a soil compound that plants can use.
State *one source of recycled nitrogen.*

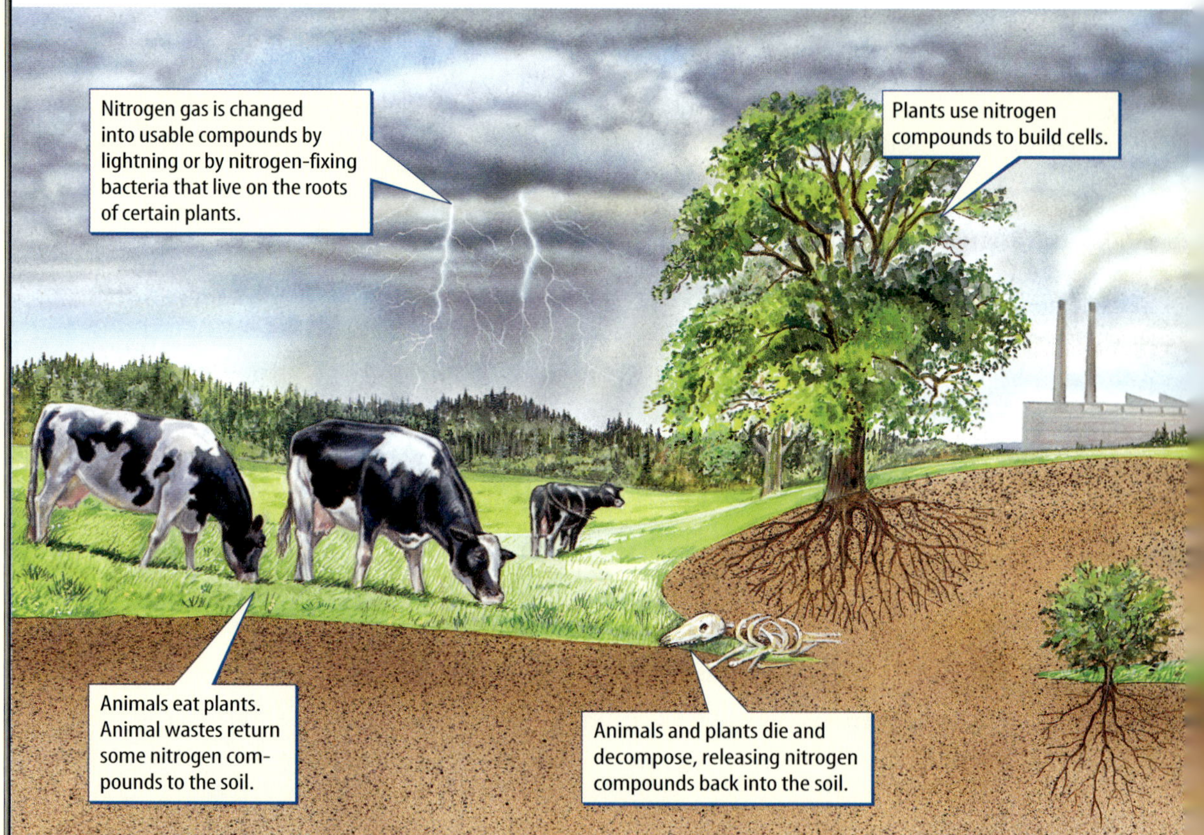

Differentiated Instruction

Challenge Have students research how nitrogen moves through an aquarium ecosystem and present a model of what they find.

Fish waste, uneaten food, plant matter → Breaks down to ammonia → Bacteria convert ammonia to nitrites → Other bacteria convert nitrites to nitrates → Nitrates used by plants or removed when water is changed L3

Learning Disabled Help students realize that as plants grow, they use minerals from the soil. As plants die or drop leaves, these minerals are restored. In farming, however, the nutrients are taken away from the soil. If this is ongoing, the soil will become less fertile. In organic farming, humus and animal wastes are added to the soil to restore fertility. Other farmers also use chemical fertilizers. L1 LS **Naturalist**

46 ◆ E CHAPTER 2 The Nonliving Environment

Figure 12 The swollen nodules on the roots of soybean plants contain colonies of nitrogen-fixing bacteria that help restore nitrogen to the soil. The bacteria depend on the plant for food, while the plant depends on the bacteria to form the nitrogen compounds the plant needs.

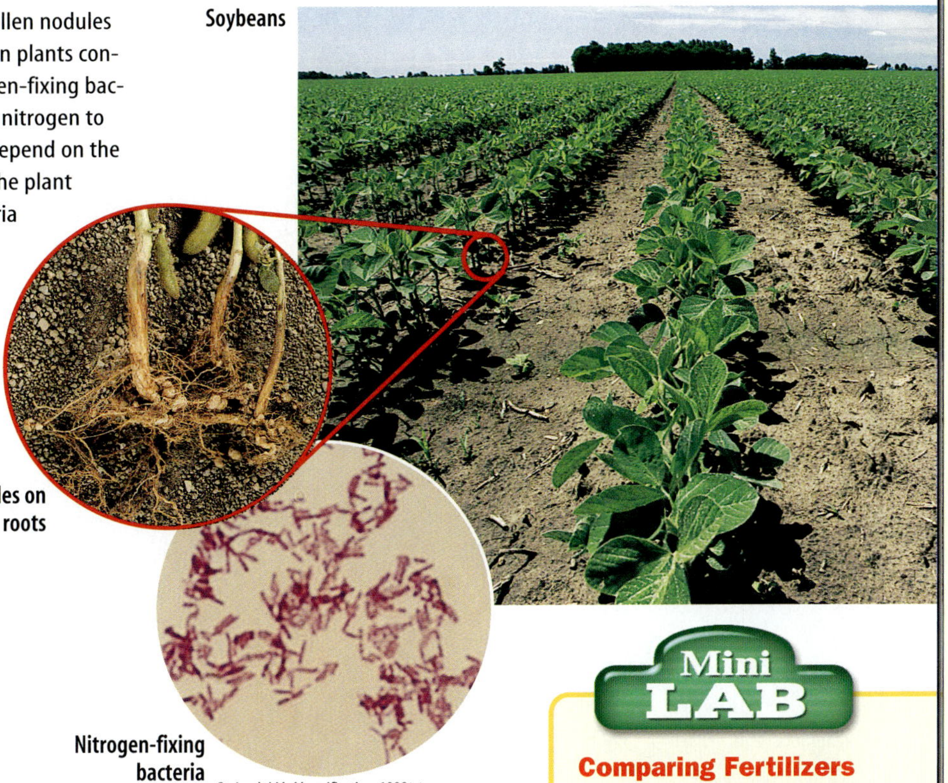

Soil Nitrogen Human activities can affect the part of the nitrogen cycle that takes place in the soil. If a farmer grows a crop, such as corn or wheat, most of the plant material is taken away when the crop is harvested. The plants are not left in the field to decay and return their nitrogen compounds to the soil. If these nitrogen compounds are not replaced, the soil could become infertile. You might have noticed that adding fertilizer to soil can make plants grow greener, bushier, or taller. Most fertilizers contain the kinds of nitrogen compounds that plants need for growth. Fertilizers can be used to replace soil nitrogen in crop fields, lawns, and gardens. Compost and animal manure also contain nitrogen compounds that plants can use. They also can be added to soil to improve fertility.

Another method farmers use to replace soil nitrogen is to grow nitrogen-fixing crops. Most nitrogen-fixing bacteria live on or in the roots of certain plants. Some plants, such as peas, clover, and beans, including the soybeans shown in **Figure 12,** have roots with swollen nodules that contain nitrogen-fixing bacteria. These bacteria supply nitrogen compounds to the soybean plants and add nitrogen compounds to the soil.

Mini LAB

Comparing Fertilizers

Procedure

1. Examine the three numbers (e.g., 5-10-5) on the **labels of three brands of houseplant fertilizer.** The numbers indicate the percentages of nitrogen, phosphorus, and potassium, respectively, that the product contains.
2. Compare the prices of the three brands of fertilizer.
3. Compare the amount of each brand needed to fertilize a typical houseplant.

Analysis

1. **Identify** the brand with the highest percentage of nitrogen.
2. **Calculate** which brand is the most expensive source of nitrogen. The least expensive.

Mini LAB

Purpose Students compare the nutrient amounts in different fertilizer brands. L2

LS **Kinesthetic**

Materials labels and pricing information for three different brands of fertilizer

Teaching Strategy Have students ask their parents if they can bring from home empty fertilizer bags that have price tags on them.

Analysis

1. Answers will vary; the nitrogen percentage is listed first on the label.
2. Answers will vary depending on brands examined.

Assessment

Performance Give students the following information. Brand Q is 5-2-6 and Brand R is 10-4-1. Which would be the better fertilizer for soil that has adequate amounts of nitrogen and phosphorus but that is low in potassium? Brand Q Use **Performance Assessment in the Science Classroom,** p. 99. L2

Visual Learning

Figure 11 Ask students to use the diagram as a basis for suggesting to farmers one thing they can do to increase the nitrogen in their soil. Add decaying organic matter (compost).

Cultural Diversity

Beans Nitrogen-fixing legumes are so important in traditional agriculture that the food they produce has become part of many cultures. Ask students to name dishes they eat at home that are made from beans. L2

NATIONAL GEOGRAPHIC

Visualizing the Carbon Cycle

Have students examine the pictures and read the captions. Then ask the following questions.

What activities release carbon dioxide into the atmosphere? The burning of fossil fuels, the decomposition of carbon-containing molecules by decomposers, and the breakdown of sugar molecules in plants and by other organisms releases carbon dioxide into the atmosphere.

Why is this process called a cycle? Possible answer: Plants use carbon dioxide from the atmosphere to make sugars. The plants then are consumed by other organisms and the carbon dioxide is released back into the atmosphere as a waste product.

Activity

Carbon Cycle Have students make their own diagram of the carbon cycle, without looking at this page, and label each part. L2 LS **Visual-Spatial** P

NATIONAL GEOGRAPHIC **VISUALIZING THE CARBON CYCLE**

Figure 13

Carbon—in the form of different kinds of carbon-containing molecules—moves through an endless cycle. The diagram below shows several stages of the carbon cycle. It begins when plants and algae remove carbon from the environment during photosynthesis. This carbon returns to the atmosphere via several carbon-cycle pathways.

A Air contains carbon in the form of carbon dioxide gas. Plants and algae use carbon dioxide to make sugars, which are energy-rich, carbon-containing compounds.

B Organisms break down sugar molecules made by plants and algae to obtain energy for life and growth. Carbon dioxide is released as a waste.

C Burning fossil fuels and wood releases carbon dioxide into the atmosphere.

D When organisms die, their carbon-containing molecules become part of the soil. The molecules are broken down by fungi, bacteria, and other decomposers. During this decay process, carbon dioxide is released into the air.

E Under certain conditions, the remains of some dead organisms may gradually be changed into fossil fuels such as coal, gas, and oil. These carbon compounds are energy rich.

Differentiated Instruction

English-Language Learners Have students work in groups to prepare a concept map that describes the carbon cycle. Have them include terms such as *consumers*, *photosynthesis*, *respiration*, and *producers*. L2 ELL COOP LEARN LS **Visual-Spatial**

The Carbon Cycle

Carbon atoms are found in the molecules that make up living organisms. Carbon is an important part of soil humus, which is formed when dead organisms decay, and it is found in the atmosphere as carbon dioxide gas (CO_2). The **carbon cycle** describes how carbon molecules move between the living and nonliving world, as shown in **Figure 13.**

The carbon cycle begins when producers remove CO_2 from the air during photosynthesis. They use CO_2, water, and sunlight to produce energy-rich sugar molecules. Energy is released from these molecules during respiration—the chemical process that provides energy for cells. Respiration uses oxygen and releases CO_2. Photosynthesis uses CO_2 and releases oxygen. These two processes help recycle carbon on Earth.

Reading Check *How does carbon dioxide enter the atmosphere?*

Human activities also release CO_2 into the atmosphere. Fossil fuels such as gasoline, coal, and heating oil are the remains of organisms that lived millions of years ago. These fuels are made of energy-rich, carbon-based molecules. When people burn these fuels, CO_2 is released into the atmosphere as a waste product. People also use wood for construction and for fuel. Trees that are harvested for these purposes no longer remove CO_2 from the atmosphere during photosynthesis. The amount of CO_2 in the atmosphere is increasing. Extra CO_2 could trap more heat from the Sun and cause average temperatures on Earth to rise.

Science Online

Topic: Life Processes
Visit booke.msscience.com for Web links to information about chemical equations that describe photosynthesis and respiration.

Activity Use these equations to explain how respiration is the reverse of photosynthesis.

section 2 review

Summary

The Cycles of Matter

- Earth's biosphere contains a fixed amount of water, carbon, nitrogen, oxygen, and other materials that cycle through the environment.

The Water Cycle

- Water cycles through the environment using several pathways.

The Nitrogen Cycle

- Some types of bacteria can form nitrogen compounds that plants and animals can use.

The Carbon Cycle

- Producers remove CO_2 from the air during photosynthesis and produce O_2.
- Consumers remove O_2 and produce CO_2.

Self Check

1. **Describe** the water cycle.
2. **Infer** how burning fossil fuels might affect the makeup of gases in the atmosphere.
3. **Explain** why plants, animals, and other organisms need nitrogen.
4. **Think Critically** Most chemical fertilizers contain nitrogen, phosphorous, and potassium. If they do not contain carbon, how do plants obtain carbon?

Applying Skills

5. **Identify and Manipulate Variables and Controls** Describe an experiment that would determine whether extra carbon dioxide enhances the growth of tomato plants.

Science Online booke.msscience.com/self_check_quiz

Reading Check

Answer as a by-product of cellular respiration

3 Assess

DAILY INTERVENTION

Check for Understanding

Challenge Suppose you covered all the bodies of water on Earth with oil. How would this affect the water cycle? Possible answers: water could not evaporate; the water cycle would be disrupted; a drought may result. L3

Reteach

Cycles of Matter Ask students to explain the different paths water vapor, carbon, and nitrogen in the air would follow before they could enter a plant. L2 LS **Naturalist**

Assessment

Portfolio Have students draw a diagram of the nitrogen cycle. Use **Performance Assessment in the Science Classroom,** p. 127. L2

section 2 review

1. Cycles should include evaporation, condensation, and precipitation.
2. It could result in an increase in the concentration of carbon dioxide.
3. to make amino acids for proteins and nucleic acids
4. Plants can get carbon from the atmosphere.
5. Answers will vary, but should include one group with normal CO_2 and one with extra CO_2. An operational definition of the growth of the tomato plants should be given, along with plans for taking measurements.

1 Motivate

INTERACTIVE CHALKBOARD

PowerPoint® Presentations

Bellringer

Section Focus Transparencies also are available on the Interactive Chalkboard CD-ROM. L2 ELL

Section 3 Section Focus Transparency **What's for dinner?**

Living things need energy to survive. Plants can get the energy they need directly from the Sun. Other organisms feed on plants or animals in order to get energy. This family is having a nutritious dinner. How many different kinds of food can you see in this picture?

1. How do living things use energy?
2. How do apple trees get energy? How do fish get energy?
3. Some organisms are said to be at the top of the food chain. What does this expression mean? Does a food chain ever end? Explain.

The Nonliving Environment

Tie to Prior Knowledge

Depleted Energy Students may have seen athletes "hit the wall." Ask why this happens, and if they have ever "run out of energy." Explain that energy is necessary to fuel life. L2

section 3

Energy Flow

as you read

What You'll Learn

- **Explain** how organisms produce energy-rich compounds.
- **Describe** how energy flows through ecosystems.
- **Recognize** how much energy is available at different levels in a food chain.

Why It's Important

All living things, including people, need a constant supply of energy.

Review Vocabulary

energy: the capacity for doing work

New Vocabulary

- chemosynthesis
- food web
- energy pyramid

Converting Energy

All living things are made of matter, and all living things need energy. Matter and energy move through the natural world in different ways. Matter can be recycled over and over again. The recycling of matter requires energy. Energy is not recycled, but it is converted from one form to another. The conversion of energy is important to all life on Earth.

Photosynthesis During photosynthesis, producers convert light energy into the chemical energy in sugar molecules. Some of these sugar molecules are broken down as energy. Others are used to build complex carbohydrate molecules that become part of the producer's body. Fats and proteins also contain stored energy.

Chemosynthesis Not all producers rely on light for energy. During the 1970s, scientists exploring the ocean floor were amazed to find communities teeming with life. These communities were at a depth of almost 3.2 km and living in total darkness. They were found near powerful hydrothermal vents like the one shown in **Figure 14.**

Figure 14 Chemicals in the water that flows from hydrothermal vents provide bacteria with a source of energy. The bacterial producers use this energy to make nutrients through the process of chemosynthesis. Consumers, such as tubeworms, feed on the bacteria.

50 ◆ E CHAPTER 2

Section 3 Resource Manager

Chapter FAST FILE Resources
- Transparency Activity, pp. 46, 47–48
- Directed Reading for Content Mastery, pp. 21, 22
- Enrichment, p. 32
- Reinforcement, p. 29
- Lab Worksheet, pp. 7–8

Mathematics Skill Activities, p. 9

Lab Management and Safety, p. 38

Hydrothermal Vents A hydrothermal vent is a deep crack in the ocean floor through which the heat of molten magma can escape. The water from hydrothermal vents is extremely hot from contact with molten rock that lies deep in Earth's crust.

Because no sunlight reaches these deep ocean regions, plants or algae cannot grow there. How do the organisms living in this community obtain energy? Scientists learned that the hot water contains nutrients such as sulfur molecules that bacteria use to produce their own food. The production of energy-rich nutrient molecules from chemicals is called **chemosynthesis** (kee moh SIHN thuh sus). Consumers living in the hydrothermal vent communities rely on chemosynthetic bacteria for nutrients and energy. Chemosynthesis and photosynthesis allow producers to make their own energy-rich molecules.

Reading Check *What is chemosynthesis?*

Energy Transfer

Energy can be converted from one form to another. It also can be transferred from one organism to another. Consumers cannot make their own food. Instead, they obtain energy by eating producers or other consumers. The energy stored in the molecules of one organism is transferred to another organism. That organism can oxidize food to release energy that it can use for maintenance and growth or is transformed into heat. At the same time, the matter that makes up those molecules is transferred from one organism to another.

Food Chains A food chain is a way of showing how matter and energy pass from one organism to another. Producers—plants, algae, and other organisms that are capable of photosynthesis or chemosynthesis—are always the first step in a food chain. Animals that consume producers such as herbivores are the second step. Carnivores and omnivores—animals that eat other consumers—are the third and higher steps of food chains. One example of a food chain is shown in **Figure 15.**

INTEGRATE Earth Science

Hydrothermal Vents The first hydrothermal vent community discovered was found along the Galápagos rift zone. A rift zone forms where two plates of Earth's crust are spreading apart. In your Science Journal, describe the energy source that heats the water in the hydrothermal vents of the Galápagos rift zone.

Figure 15 In this food chain, grasses are producers, marmots are herbivores that eat the grasses, and grizzly bears are consumers that eat marmots. The arrows show the direction in which matter and energy flow.
Infer *what might happen if grizzly bears disappeared from this ecosystem.*

2 Teach

INTEGRATE Earth Science

Hydrothermal Vents The source of heat in vents is geothermal energy. The hot magna heats the water to 400°C but it does not boil because of the pressure at these depths. The water quickly cools to about 2°C.

Reading Check

Answer Chemosynthesis is the production of energy-rich food molecules from chemical energy.

Discussion

Flow of Matter and Energy What do the arrows in food chains represent? the flow of matter and energy from one organism to another Ask students to generate other food chains. Answers will vary. L2

Activity

Deep-Sea Food Chain Have students make a food chain for a deep-sea community. Food chains should be similar to that of a simple aquatic system, except that chemosynthetic bacteria will replace aquatic plants as producers.
L2 LS **Visual-Spatial** P

Caption Answer

Figure 15 The population of marmots will increase to a level that can be supported by the food supply.

Differentiated Instruction

Challenge Have students research how matter and energy flows through a food web. Have students design a poster of a food web with arrows that show the flow of energy and arrows that show the flow of matter. L3

Learning Disabled Have students who understand the concepts of producer and consumer work with students who are having difficulty. Have partners group the terms *producer, photosynthesis, plants, consumer,* and *animal.* L1 COOP LEARN LS **Interpersonal**

SECTION 3 Energy Flow E ◆ 51

Quick Demo

Food Energy

Materials marshmallow, lighter, fork

Estimated Time one minute

Procedure Place a marshmallow on the fork and light it. Explain that energy in the food is being transformed into heat and light energy. When someone consumes a marshmallow or other food items, some of the energy is used for growth and some is released as heat.

Visual Learning

Figure 16 Have students draw three different food chains based on the food web. Then have students assess the impact on other organisms if one organism from the food web were removed. The impact varies. If species X feeds only on species Y, when species Y dies out, so will species X. However, if species X feeds on many species, the loss of one will not cause species X to die out. L2 P

Make a Model

Insect Food Find a grasshopper or cricket and weigh it. Then have students collect ten times the weight of the insect in grass. For example, if the insect weighs 1 g, students should gather 10 g of grass. Place the insect on top of the pile of grass to make a model of a biomass pyramid. L1 LS **Logical-Mathematical**

Use an Analogy

Energy Loss Help students understand why so much energy is lost from one level to another. If they raised a cat from a kitten, ask them to imagine how much food the cat has eaten during its life. It is far more than the weight of the cat. Most of the food was used for energy. If a coyote ate their cat, it would not get all the food the cat ate, just the caloric value of the cat itself. L1

Food Webs A forest community includes many feeding relationships. These relationships can be too complex to show with a food chain. For example, grizzly bears eat many different organisms, including berries, insects, chipmunks, and fish. Berries are eaten by bears, birds, insects, and other animals. A bear carcass might be eaten by wolves, birds, or insects. A **food web** is a model that shows all the possible feeding relationships among the organisms in a community. A food web is made up of many different food chains, as shown in **Figure 16.**

Energy Pyramids

Food chains usually have at least three links, but rarely more than five. This limit exists because the amount of available energy is reduced as you move from one level to the next in a food chain. Imagine a grass plant that absorbs energy from the Sun. The plant uses some of this energy to grow and produce seeds. Some of the energy is stored in the seeds.

Figure 16 Compared to a food chain, a food web provides a more complete model of the feeding relationships in a community.

52 ◆ E CHAPTER 2 The Nonliving Environment

Teacher FYI

Conservation of Energy The law of conservation of energy states that energy is not created or destroyed, but only converted into other forms of energy. In most ecosystems the energy starts as light energy, which is converted through photosynthesis into chemical energy.

Differentiated Instruction

Visually Impaired Obtain or make a pyramid-shaped object. Allow students to handle the model to identify its shape. Explain that the model depicts the relative amount of energy available as one moves up feeding levels in a food web. L2 LS **Kinesthetic**

Available Energy When a mouse eats grass seeds, energy stored in the seeds is transferred to the mouse. However, most of the energy the plant absorbed from the Sun was used for the plant's growth. The mouse uses energy from the seed for its own life processes, including respiration, digestion, and growth. Some of this energy was given off as heat. A hawk that eats the mouse obtains even less energy. The amount of available energy is reduced from one feeding level of a food chain to another.

An **energy pyramid,** like the one in **Figure 17,** shows the amount of energy available at each feeding level in an ecosystem. The bottom of the pyramid, which represents all of the producers, is the first feeding level. It is the largest level because it contains the most energy and the largest number of organisms. As you move up the pyramid, the transfer of energy is less efficient and each level becomes smaller. Only about ten percent of the energy available at each feeding level of an energy pyramid is transferred to the next higher level.

Figure 17 An energy pyramid shows that each feeding level has less energy than the one below it.
Describe *what would happen if the hawks and snakes outnumbered the rabbits and mice in this ecosystem.*

Reading Check ***Why does the first feeding level of an energy pyramid contain the most energy?***

section 3 review

Summary

Converting Energy

- Most producers convert light energy into chemical energy.
- Some producers can produce their own food using energy in chemicals such as sulfur.

Energy Transfer

- Producers convert energy into forms that other organisms can use.
- Food chains show how matter and energy pass from one organism to another.

Energy Pyramids

- Energy pyramids show the amount of energy available at each feeding level.
- The amount of available energy decreases from the base to the top of the energy pyramid.

Self Check

1. **Compare and contrast** a food web and an energy pyramid.
2. **Explain** why there is a limit to the number of links in a food chain.
3. **Think Critically** Use your knowledge of food chains and the energy pyramid to explain why the number of mice in a grassland ecosystem is greater than the number of hawks.

Applying Math

4. **Solve One-Step Equations** A forest has 24,055,000 kilocalories (kcals) of producers, 2,515,000 kcals of herbivores, and 235,000 kcals of carnivores. How much energy is lost between producers and herbivores? Between herbivores and carnivores?

Science Online booke.msscience.com/self_check_quiz

Caption Answer

Figure 17 Less energy would be available to hawks and snakes. Eventually they would begin to die off until a balance was achieved unless they moved out of the ecosystem.

Reading Check

Answer because little energy has been lost in maintaining body systems

3 Assess

DAILY INTERVENTION

Check for Understanding

Challenge Ask students to research the following organisms and design a food web that shows their relationships: *corn, rabbit, hawk, grass, wheat, rat, fox, human,* and *cow.* L3

Reteach

Food Webs and Humans Write the word *humans* on the board. Have students write the names of all the animals we eat below this word. Below this write all the plants that we eat or that are eaten by the animals we eat. Connect lines to make food chains. Then draw more lines to show a food web. L1 LS **Visual-Spatial**

Assessment

Performance Use the overhead projector to project a scene from nature. Help students identify the organisms in the picture. Then have them design food chains based on the picture and their knowledge of the organisms. Use **Performance Assessment in the Science Classroom,** p. 127. L2

section 3 review

1. A food web shows the path of energy and matter in an ecosystem. An energy pyramid shows the amount of energy available at each level.
2. because a significant amount of energy is lost at each level
3. Hawks eat mice. For the mice to support the hawk population, there must be far more energy at the mouse level than at the hawk level. Since mice are smaller than hawks, this means that there must be many more mice than hawks.
4. producers and herbivores: 21,540,000 kcal; herbivores and carnivores: 2,280,000 kcal

SECTION 3 Energy Flow E ◆ 53

BENCH TESTED

Real-World Question

Purpose Students perform a variation of Johannes Baptista Van Helmont's famous experiment to determine if plants get their mass from the soil. L2 LS **Logical-Mathematical**

Process Skills observe, measure, predict, interpret data, use numbers

Time Required 45 minutes to set up; three weeks for the plants to grow; 1 hour for the final measurements, calculations, and cleanup

Procedure

Safety Precautions Caution students to wash their hands after handling plants and soil.

Teaching Strategy This experiment helps uncover and correct the misconception that the material for a mass increase in a plant comes from the soil. Prior to getting results, ask students if most of the mass gained by the radish plants will come from the soil. If they don't think it will come from the soil, ask where they think it will come from.

LAB

Where does the mass of a plant come from?

Goals

- **Measure** the mass of soil before and after radish plants have been grown in it.
- **Measure** the mass of radish plants grown in the soil.
- **Analyze** the data to determine whether the mass gained by the plants equals the mass lost by the soil.

Materials

8-oz plastic or paper cup
potting soil to fill cup
scale or balance
radish seeds (4)
water
paper towels

Safety Precautions

Real-World Question

An enormous oak tree starts out as a tiny acorn. The acorn sprouts in dark, moist soil. Roots grow down through the soil. Its stem and leaves grow up toward the light and air. Year after year, the tree grows taller, its trunk grows thicker, and its roots grow deeper. It becomes a towering oak that produces thousands of acorns of its own. An oak tree has much more mass than an acorn. Where does this mass come from? The soil? The air? In this activity, you'll find out by conducting an experiment with radish plants. Does all of the matter in a radish plant come from the soil?

54

Differentiated Instruction

Learning Disabled Help students understand the purpose of the investigation. Hand students a piece of two-by-four lumber so they can feel how heavy it is. Ask them where the weight of the tree came from. The weight came from the food the tree produced and utilized to build its body. L1

Alternative Inquiry Lab

Plant Growth What factors influence the mass of a plant? Partner with a local farmer or your soil conservation district to plan an investigation of the compounds used to increase plant mass and farm profits. Students should research all of the issues surrounding these compounds, conduct the research, and present their findings. PBL

Using Scientific Methods

Procedure

1. Copy the data table into your Science Journal.
2. Fill the cup with dry soil.
3. Find the mass of the cup of soil and record this value in your data table.
4. Moisten the soil in the cup. Plant four radish seeds 2 cm deep in the soil. Space the seeds an equal distance apart. Wash your hands.
5. Add water to keep the soil barely moist as the seeds sprout and grow.
6. When the plants have developed four to six true leaves, usually after two to three weeks, carefully remove the plants from the soil. Gently brush the soil off the roots. Make sure all the soil remains in the cup.
7. Spread the plants out on a paper towel. Place the plants and the cup of soil in a warm area to dry out.
8. When the plants are dry, measure their mass and record this value in your data table. Write this number with a plus sign in the Gain or Loss column.
9. When the soil is dry, find the mass of the cup of soil. Record this value in your data table. Subtract the End mass from the Start mass and record this number with a minus sign in the Gain or Loss column.

Mass of Soil and Radish Plants

	Start	End	Gain (+) or Loss (−)
Mass of dry soil and cup	75.8 g	75.7 g	0.1 g −
Mass of dried radish plants	0 g	2.1 g	2.1 g +

Analyze Your Data

1. **Calculate** how much mass was gained or lost by the soil. By the radish plants.
2. Did the mass of the plants come completely from the soil? How do you know?

Conclude and Apply

1. In the early 1600s, a Belgian scientist named J. B. van Helmont conducted this experiment with a willow tree. What is the advantage of using radishes instead of a tree?
2. **Predict** where all of the mass gained by the plants came from.

Communicating Your Data

Compare your conclusions with those of other students in your class. **For more help, refer to the Science Skill Handbook.**

Analyze Your Data

Expected Outcome The radish plants' mass gain will be more than the soil's mass loss.

Answers to Questions

1. Answers will vary, but the soil will have a very small mass loss and the plants will have a much larger mass gain.
2. No, because the mass loss of the soil is smaller than the mass gain of the plants.

Error Analysis Some errors may occur because the plants and soil are not completely dry. However, this usually does not affect the conclusion.

Conclude and Apply

1. Radishes grow faster and use less space. You can get a quantitative amount of mass gained and lost.
2. Answers will vary, but should include carbon dioxide from the air and water from the soil.

Assessment

Content Pretend that you grew a willow tree for five years like Van Helmont. Write down how much mass the tree might have gained and how much mass the soil might have lost. Use **Performance Assessment in the Science Classroom,** p. 89. L2

Communicating Your Data

Have students post their data tables on a bulletin board for other students to see.

SCIENCE Stats

Content Background

The term *climate* refers to the average weather of a region over a period of time. Earth has many climates. The main influences on the climate of a region include:

- wind, which distributes moisture and heat
- distance from the equator, which influences solar radiation
- large bodies of water, which decreases temperature variation
- altitude, which affects temperature
- presence of mountains, which influence precipitation

Discussion

Temperature How would the annual temperature variation on an ocean island compare to the mainland at the same latitude? Students should infer that the large body of water would decrease annual temperature variation on the island.

Activity

Daily Record Have students work as a class to record and graph daily high and low temperatures each day for a month. L2

Applying Math

241.3 cm per month

Graph It

Have students graph the average monthly rainfall in your area. Compare the precipitation received in your location to that received in a tropical rain forest. Ask students to predict how the local environment would change if the amount of precipitation increased. L2

SCIENCE Stats

Extreme Climates

Did you know...

... The greatest snowfall in one year occurred at Mount Baker in Washington State. Approximately 2,896 cm of snow fell during the 1998–99, 12-month snowfall season. That's enough snow to bury an eight-story building.

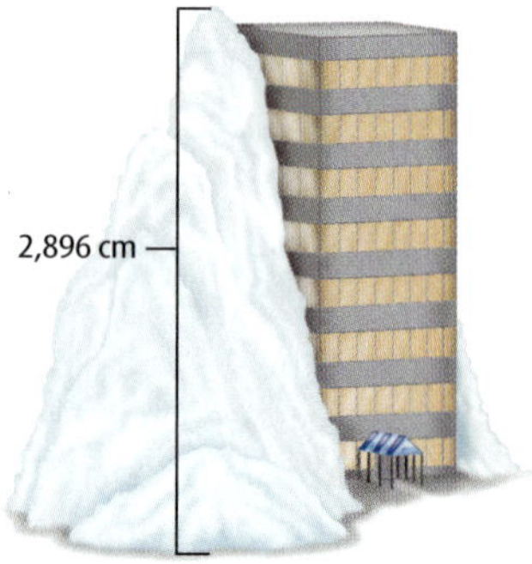

Applying Math What was the average monthly snowfall at Mount Baker during the 1998–99 snowfall season?

... The hottest climate in the United States is found in Death Valley, California. In July 1913, Death Valley reached approximately 57°C. As a comparison, a comfortable room temperature is about 20°C.

... The record low temperature of a frigid −89°C was set in Antarctica in 1983. As a comparison, the temperature of a home freezer is about −15°C.

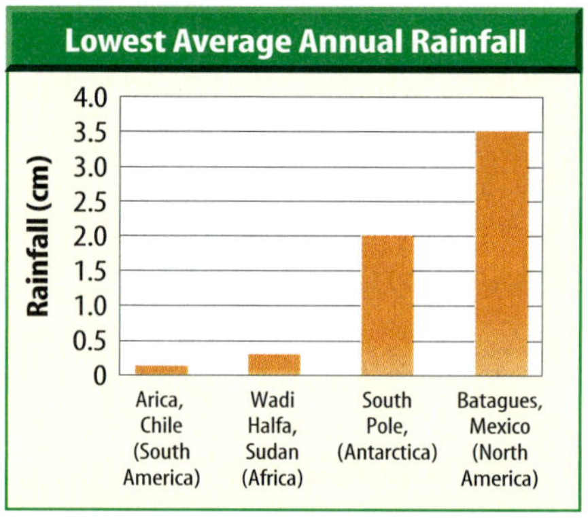

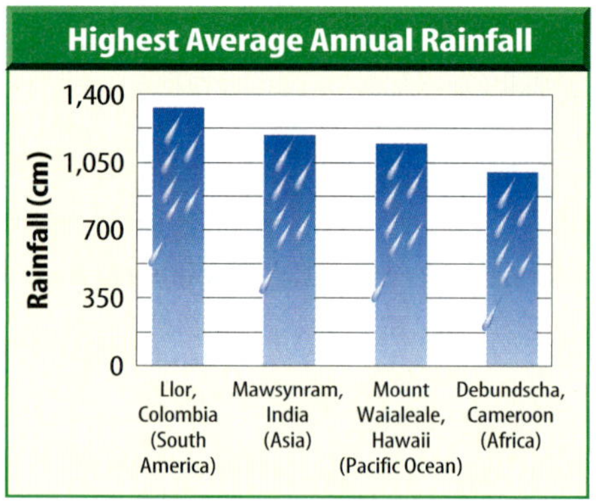

Graph It

Visit booke.msscience.com/science_stats to find the average monthly rainfall in a tropical rain forest. Make a line graph to show how the amount of precipitation changes during the 12 months of the year.

56 ◆ E CHAPTER 2 The Nonliving Environment

Visual Learning

Lowest Average Annual Rainfall, Highest Average Annual Rainfall Have each student choose one location represented on each graph and use the annual rainfall number to calculate an average monthly rainfall number for each location. Then, have students cut a strip of paper the same height as the average amount of monthly rainfall in each of the locations they have chosen. Use the paper strips for comparison. L2 LS **Visual-Spatial**

Reviewing Main Ideas

Section 1 Abiotic Factors

1. Abiotic factors include air, water, soil, sunlight, temperature, and climate.
2. The availability of water and light influences where life exists on Earth.
3. Soil and climate have an important influence on the types of organisms that can survive in different environments.
4. High latitudes and elevations generally have lower average temperatures.

Section 2 Cycles in Nature

1. Matter is limited on Earth and is recycled through the environment.
2. The water cycle involves evaporation, condensation, and precipitation.
3. The carbon cycle involves photosynthesis and respiration.
4. Nitrogen in the form of soil compounds enters plants, which then are consumed by other organisms.

Section 3 Energy Flow

1. Producers make energy-rich molecules through photosynthesis or chemosynthesis.
2. When organisms feed on other organisms, they obtain matter and energy.
3. Matter can be recycled, but energy cannot.
4. Food webs are models of the complex feeding relationships in communities.
5. Available energy decreases as you go to higher feeding levels in an energy pyramid.

Visualizing Main Ideas

This diagram represents photosynthesis in a leaf. Match each letter with one of the following terms: light, carbon dioxide, *or* oxygen.

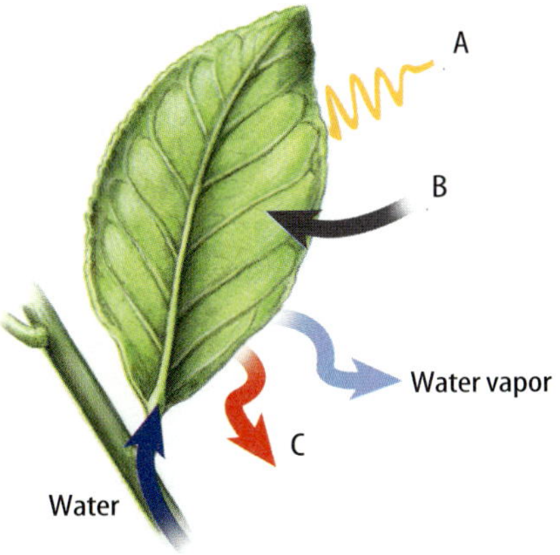

Science Online book.msscience.com/interactive_tutor

chapter 2 Study Guide

Reviewing Main Ideas

Summary statements can be used by students to review the major concepts of the chapter.

Visualizing Main Ideas

A—light; B—carbon dioxide; C—oxygen

Science Online

Visit book.msscience.com
/self_check_quiz
/interactive_tutor
/vocabulary_puzzlemaker
/chapter_review
/standardized_test

Assessment Transparency

For additional assessment questions, use the *Assessment Transparency* located in the transparency book.

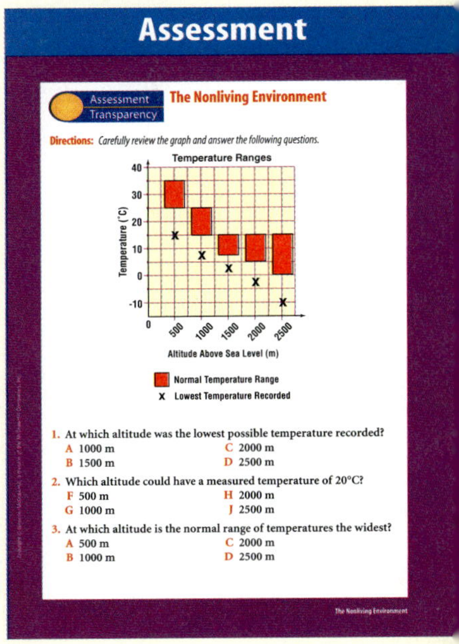

Assessment

Assessment Transparency **The Nonliving Environment**

Directions: *Carefully review the graph and answer the following questions.*

1. At which altitude was the lowest possible temperature recorded?
   A 1000 m C 2000 m
   B 1500 m D 2500 m
2. Which altitude could have a measured temperature of 20°C?
   F 500 m H 2000 m
   G 1000 m J 2500 m
3. At which altitude is the normal range of temperatures the widest?
   A 500 m C 2000 m
   B 1000 m D 2500 m

The Nonliving Environment

Using Vocabulary

1. evaporation
2. nitrogen fixation
3. nitrogen cycle
4. chemosynthesis
5. carbon cycle
6. condensation
7. water cycle
8. carbon cycle

Checking Concepts

9. B
10. B
11. D
12. A
13. A
14. B
15. A
16. C
17. B

chapter 2 Review

Using Vocabulary

abiotic p. 36	energy pyramid p. 53
atmosphere p. 37	evaporation p. 44
biotic p. 36	food web p. 52
carbon cycle p. 49	nitrogen cycle p. 46
chemosynthesis p. 51	nitrogen fixation p. 46
climate p. 41	soil p. 38
condensation p. 45	water cycle p. 45

Which vocabulary word best corresponds to each of the following events?

1. A liquid changes to a gas.
2. Some types of bacteria form nitrogen compounds in the soil.
3. Decaying plants add nitrogen to the soil.
4. Chemical energy is used to make energy-rich molecules.
5. Decaying plants add carbon to the soil.
6. A gas changes to a liquid.
7. Water flows downhill into a stream. The stream flows into a lake, and water evaporates from the lake.
8. Burning coal and exhaust from automobiles release carbon into the air.

Checking Concepts

Choose the word or phrase that best answers the question.

9. Which of the following is an abiotic factor?
   A) penguins
   B) rain
   C) soil bacteria
   D) redwood trees

Use the equation below to answer question 10.

$$CO_2 + H_2O \xrightarrow[\text{energy}]{\text{light}} \text{sugar} + O_2$$

10. Which of the following processes is shown in the equation above?
    A) condensation
    B) photosynthesis
    C) burning
    D) respiration

11. Which of the following applies to latitudes farther from the equator?
    A) higher elevations
    B) higher temperatures
    C) higher precipitation levels
    D) lower temperatures

12. Water vapor forming droplets that form clouds directly involves which process?
    A) condensation
    B) respiration
    C) evaporation
    D) transpiration

13. Which one of the following components of air is least necessary for life on Earth?
    A) argon
    B) nitrogen
    C) carbon dioxide
    D) oxygen

14. Which group makes up the largest level of an energy pyramid?
    A) herbivores
    B) producers
    C) decomposers
    D) carnivores

15. Earth receives a constant supply of which of the following items?
    A) light energy
    B) carbon
    C) nitrogen
    D) water

16. Which of these is an energy source for chemosynthesis?
    A) sunlight
    B) moonlight
    C) sulfur molecules
    D) carnivores

Use the illustration below to answer question 17.

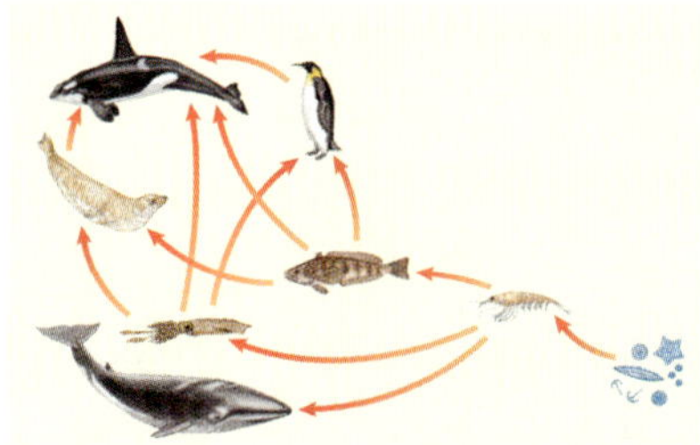

17. What is the illustration above an example of?
    A) food chain
    B) food web
    C) energy pyramid
    D) carbon cycle

Science Online booke.msscience.com/vocabulary_puzzlemaker

Use the ExamView® Pro Testmaker CD-ROM to:

- create multiple versions of tests
- create modified tests with one mouse click for inclusion students
- edit existing questions and add your own questions
- build tests aligned with state standards using built-in State Curriculum Tags
- change English tests to Spanish with one mouse click and vice versa

Thinking Critically

18. Draw a Conclusion A country has many starving people. Should they grow vegetables and corn to eat, or should they grow corn to feed cattle so they can eat beef? Explain.

19. Explain why a food web is a better model of energy flow than a food chain.

20. Infer Do bacteria need nitrogen? Why or why not?

21. Describe why it is often easier to walk through an old, mature forest of tall trees than through a young forest of small trees.

22. Explain why giant sequoia trees grow on the west side of California's Inyo Mountains and Death Valley, a desert, is on the east side of the mountains.

23. Concept Map Copy and complete this food web using the following information: *caterpillars and rabbits eat grasses, raccoons eat rabbits and mice, mice eat grass seeds,* and *birds eat caterpillars.*

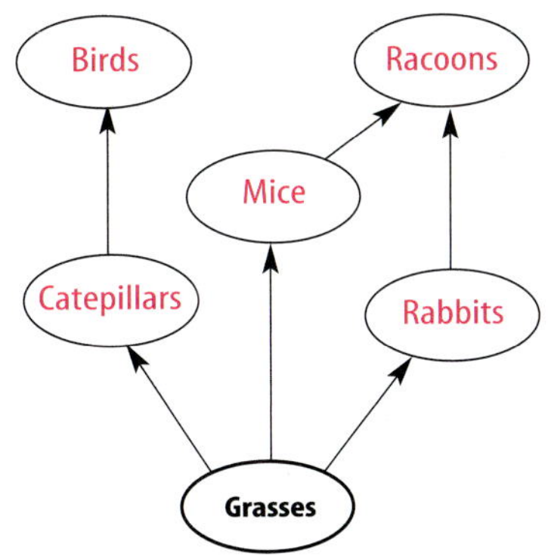

24. Form a Hypothesis For each hectare of land, ecologists found 10,000 kcals of producers, 10,000 kcals of herbivores, and 2,000 kcals of carnivores. Suggest a reason why producer and herbivore levels are equal.

25. Recognize Cause and Effect A lake in Kenya has been taken over by a floating weed. How could you determine if nitrogen fertilizer runoff from farms is causing the problem?

Performance Activities

26. Poster Use magazine photographs to make a visual representation of the water cycle.

Applying Math

27. Energy Budget Raymond Lindeman, from the University of Minnesota, was the first person to calculate the total energy budget of an entire community at Cedar Bog Lake in MN. He found the total amount of energy produced by producers was 1,114 kilocalories per meter squared per year. About 20% of the 1,114 kilocalories were used up during respiration. How many kilocalories were used during respiration?

28. Kilocalorie Use Of the 600 kilocalories of producers available to a caterpillar, the caterpillar consumes about 150 kilocalories. About 25% of the 150 kilocalories is used to maintain its life processes and is lost as heat, while 16% cannot be digested. How many kilocalories are lost as heat? What percentage of the 600 kilocalories is available to the next feeding level?

Use the table below to answer question 29.

Mighty Migrators

Species	Distance (km)
Desert locust	4,800
Caribou	800
Green turtle	1,900
Arctic tern	35,000
Gray whale	19,000

29. Make and Use Graphs Climate can cause populations to move from place to place. Make a bar graph of migration distances shown above.

Assessment Resources

Reproducible Masters

Chapter *Fast File* Resources
Chapter Review, pp. 39–40
Chapter Tests, pp. 41–44
Assessment Transparency Activity, p. 51

Glencoe Science Web site
Chapter Review Test
Standardized Test Practice

Glencoe Technology
Assessment Transparency
Exam*View*® Pro Testmaker
MindJogger Videoquiz
Interactive Chalkboard

chapter 2 Review

Thinking Critically

18. The country should grow corn and other vegetables for the people to eat. If they grow corn to feed cattle, most of the energy of the corn will be lost.

19. Webs show complex interactions rather than the simple interactions of a chain.

20. Yes; nitrogen is an important component of DNA and proteins.

21. In the mature forest, the tall trees block most of the sunlight from reaching the floor. There is therefore little plant growth, making it easy to walk.

22. As the air blows over the mountains from the ocean on the west side, it cools and loses moisture. Thus there is a forest on one side, but a desert on the other.

23. See student page.

24. Reasons may include a poor growth year for plants or a very good year for herbivore reproduction.

25. Possible answer: Ask farmers to restrict their use of nitrogen fertilizer and see if this clears up the lake.

Performance Activities

26. Posters will vary, but should show how the cycle continues. Use **Performance Assessment in the Science Classroom,** p. 145.

Applying Math

National Math Standards

1, 5, 6, 9

27. 222.8 kilocalories

28. 37.5 kilocalories were lost as heat, and 14.75% of the 600 kilocalories are available to the next feeding level.

29. From highest bar to lowest: Arctic tern, gray whale, desert locust, green turtle, caribou

FAST FILE

Answer Sheet A practice answer sheet can be found at booke.msscience.com/answer_sheet.

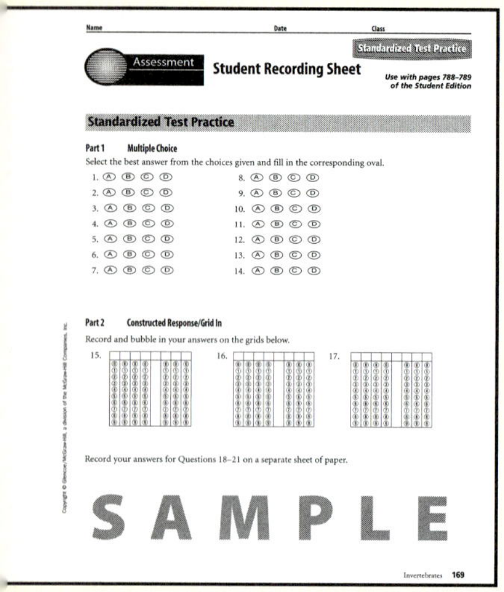
Name Date Class

Assessment **Student Recording Sheet** Standardized Test Practice

Use with pages 788–789 of the Student Edition

Standardized Test Practice

Part 1 Multiple Choice

Select the best answer from the choices given and fill in the corresponding oval.

1. Ⓐ Ⓑ Ⓒ Ⓓ 8. Ⓐ Ⓑ Ⓒ Ⓓ
2. Ⓐ Ⓑ Ⓒ Ⓓ 9. Ⓐ Ⓑ Ⓒ Ⓓ
3. Ⓐ Ⓑ Ⓒ Ⓓ 10. Ⓐ Ⓑ Ⓒ Ⓓ
4. Ⓐ Ⓑ Ⓒ Ⓓ 11. Ⓐ Ⓑ Ⓒ Ⓓ
5. Ⓐ Ⓑ Ⓒ Ⓓ 12. Ⓐ Ⓑ Ⓒ Ⓓ
6. Ⓐ Ⓑ Ⓒ Ⓓ 13. Ⓐ Ⓑ Ⓒ Ⓓ
7. Ⓐ Ⓑ Ⓒ Ⓓ 14. Ⓐ Ⓑ Ⓒ Ⓓ

Part 2 Constructed Response/Grid In

Record and bubble in your answers on the grids below.

15. 16. 17.

Record your answers for Questions 18–21 on a separate sheet of paper.

SAMPLE

Invertebrates 169

Part 1 Multiple Choice

1. B
2. B
3. B
4. C
5. A
6. D
7. C
8. D
9. C
10. B

Part 1 Multiple Choice

Record your answers on the answer sheet provided by your teacher or on a sheet of paper.

1. The abiotic factor that provides energy for nearly all life on Earth is

A. air. **C.** water.
B. sunlight. **D.** soil.

2. Which of the following is characteristic of places at high elevations?

A. fertile soil
B. fewer molecules in the air
C. tall trees
D. warm temperatures

Use the diagram below to answer questions 3 and 4.

3. The air at point C is

A. dry and warm.
B. dry and cool.
C. moist and warm.
D. moist and cool.

4. The air at point A is

A. dry and warm.
B. dry and cool.
C. moist and warm.
D. moist and cool.

5. What process do plants use to return water vapor to the atmosphere?

A. transpiration **C.** respiration
B. evaporation **D.** condensation

6. Clouds form as a result of what process?

A. evaporation **C.** respiration
B. transpiration **D.** condensation

Use the illustration of the nitrogen cycle below to answer questions 7 and 8.

7. Which of the following items shown in the diagram contribute to the nitrogen cycle by releasing AND absorbing nitrogen?

A. the decaying organism only
B. the trees only
C. the trees and the grazing cows
D. the lightning and the decaying organism

8. Which of the following items shown in the diagram contribute to the nitrogen cycle by ONLY releasing nitrogen?

A. the decaying organism only
B. the trees only
C. the trees and the grazing cows
D. the lightning and the decaying organism

9. Where is most of the energy found in an energy pyramid?

A. at the top level
B. in the middle levels
C. at the bottom level
D. all levels are the same

10. What organisms remove carbon dioxide gas from the air during photosynthesis?

A. consumers **C.** herbivores
B. producers **D.** omnivores

Part 2 Short Response/Grid In

11. Possible answers: water—essential part of all cells, required for many life processes; soil—provides nutrients for plants to grow, includes important bacteria, fungi, insets, and worms

12. farms and ranches

13. about 51%

14. on or in the roots of certain plants such as peas, clover, beans and soybeans

15. Possible answers: burning fossil fuels, during the decaying process of organisms, during respiration

16. They use sulfur molecules and other nutrients as an energy source to make their own food.

17. grasses → rabbit → weasel → coyote

Part 2 | Short Response/Grid In

Record your answers on the answer sheet provided by your teacher or on a sheet of paper.

11. Give two examples of abiotic factors and describe how each one is important to biotic factors.

Use the table below to answer questions 12 and 13.

U.S. Estimated Water Use in 1995		
Water Use	**Millions of Gallons per Day**	**Percent of Total**
Homes and Businesses	41,600	12.2
Industry and Mining	28,000	8.2
Farms and Ranches	139,200	40.9
Electricity Production	131,800	38.7

12. According to the table above, what accounted for the highest water use in the U.S. in 1995?

13. What percentage of the total amount of water use results from electricity production and homes and business combined?

14. Where are nitrogen-fixing bacteria found?

15. Describe two ways that carbon is released into the atmosphere.

16. How are organisms near hydrothermal vents deep in the ocean able to survive?

17. Use a diagram to represent the transfer of energy among these organisms: a weasel, a rabbit, grasses, and a coyote.

Test-Taking Tip

Answer All Questions Never leave any answer blank.

Part 3 | Open Ended

Record your answers on a sheet of paper.

18. Explain how a decrease in the amount of sunlight would affect producers that use photosynthesis, and producers that use chemosynthesis.

19. Describe how wind and wind currents are produced.

20. Use the water cycle to explain why beads of water form on the outside of a glass of iced water on a hot day.

21. Draw a flowchart that shows how soy beans, deer, and nitrogen-fixing bacteria help cycle nitrogen from the atmosphere, to the soil, to living organisms, and back to the atmosphere.

Use the diagram below to answer questions 22 and 23.

22. What term is used for the diagram above? Explain how the diagram represents energy transfer.

23. Explain how the grass and bear populations would be affected if the marmot population suddenly declined.

24. Compare and contrast an energy pyramid and a food web.

25. What happens to the energy in organisms at the top of an energy pyramid when they die?

22. food chain The grass is eaten by the marmot and the marmot is eaten by the bear. Energy from the grass is transferred to the marmot, and then to the bear.

23. With fewer marmots, the grass population would increase. The bear population would find a different source of food or its population would decline.

24. An energy pyramid shows the amount of energy available at each level of an ecosystem. A food web shows the feeding relationships among organisms in a community.

25. When an organism dies, the energy in it is used by decomposers.

Rubrics

For more help evaluating open-ended assessment questions, see the rubric on p. 10T.

Part 3 | Open Ended

18. Producers that use photosynthesis would not be able to produce as much food so they would decline. Producers that use chemosynthesis would not be affected since they do not require sunlight to make their food.

19. Wind occurs when sunlight unevenly heats Earth. As air is warmed, it rises and the colder air sinks beneath it. The movement of cold and warm air creates wind currents.

20. When the cold liquid in the glass comes into contact with the warmer air outside of it, the water vapor that it evaporating condenses on the outside of the glass and forms beads of water.

21. Nitrogen-fixing bacteria turn nitrogen from the air into useful compounds → soy beans use these nitrogen compounds to grow → deer consume nitrogen when they eat the soy beans → deer die and their decaying bodies return the nitrogen compounds to the soil

chapter 3 Organizer

<table>
<tr><th rowspan="2">Section/Objectives</th><th colspan="2">Standards</th><th rowspan="2">Labs/Features</th></tr>
<tr><th>National</th><th>State/Local</th></tr>
<tr><td>Chapter Opener</td><td>See pp. 16T–17T for a Key to Standards.</td><td></td><td>Launch Lab: What environment do houseplants need?, p. 63
Foldables, p. 63</td></tr>
<tr><td>Section 1 How Ecosystems Change
2 sessions
1 block
1. Explain how ecosystems change over time.
2. Describe how new communities begin in areas without life.
3. Compare pioneer communities and climax communities.</td><td>National Content Standards: UCP.1–UCP.5, A.1, A.2, C.1, C.3, C.4</td><td></td><td>Science Online, p. 65
Visualizing Secondary Succession, p. 66</td></tr>
<tr><td>Section 2 Biomes
2 sessions
1 block
4. Explain how climate influences land environments.
5. Identify seven biomes of Earth.
6. Describe the adaptations of organisms found in each biome.</td><td>National Content Standards: UCP.1, UCP.2, UCP.3, UCP.5, A.1, A.2, C.1, C.3, C.4</td><td></td><td>MiniLAB: Modeling Rain Forest Leaves, p. 72
Integrate Earth Science, p. 74
Lab: Studying a Land Ecosystem, p. 76</td></tr>
<tr><td>Section 3 Aquatic Ecosystems
3 sessions
1.5 blocks
7. Compare flowing freshwater and standing freshwater ecosystems.
8. Identify and describe important saltwater ecosystems.
9. Identify problems that affect aquatic ecosystems.</td><td>National Content Standards: UCP.1, UCP.2, UCP.3, UCP.5, A.1, A.2, C.1, C.3–C.5, F.2</td><td></td><td>MiniLAB: Modeling Freshwater Environments, p. 78
Integrate Career, p. 79
Applying Math: Convert Units, p. 80
Science Online, p. 81
Lab: Exploring Wetlands, p. 84
Science and Society: Creating Wetlands to Purify Wastewater, p. 86</td></tr>
</table>

Glencoe Exclusive! TeacherWorks™ All-In-One Planner and Resource Center

Lab Materials	Reproducible Resources	Section Assessment	Technology
Launch Lab: houseplant	**Chapter *FAST FILE* Resources** Foldables Worksheet, p. 15 Directed Reading Overview, p. 17 Note-taking Worksheets, pp. 31–33	Glencoe's Assessment Advantage	TeacherWorks includes: • Interactive Teacher Edition • Lesson Planner with calendar • Access to all program blacklines • Correlations to standards • Web links
Need materials? Contact Science Kit at 1-800-828-7777 or www.sciencekit.com on the Internet.	**Chapter *FAST FILE* Resources** Transparency Activity, p. 42 Enrichment, p. 28 Reinforcement, p. 25 Directed Reading, p. 18 Lab Activity, pp. 9–10	**Portfolio** Visual Learning, p. 66 **Performance** Applying Math, p. 67 **Content** Section Review, p. 67	Section Focus Transparency Virtual Labs CD-ROM Guided Reading Audio Program Interactive Chalkboard CD-ROM
MiniLAB: poster board, metric ruler, scissors, sink, water, spray bottle **Lab:** graph paper, thermometer, magnifying lens, notebook, binoculars, pencil, field guides, compass, tape measure	**Chapter *FAST FILE* Resources** MiniLAB, p. 3 Transparency Activity, p. 43 Enrichment, p. 29 Reinforcement, p. 26 Directed Reading, p. 19 Lab Worksheet, pp. 5–6 **Reading and Writing Skill Activities,** p. 1	**Portfolio** Assessment, p. 75 **Performance** MiniLAB, p. 72 Applying Skills, p. 75 **Content** Section Review, p. 75	Section Focus Transparency Virtual Labs CD-ROM Guided Reading Audio Program Interactive Chalkboard CD-ROM Video Lab
MiniLAB: pond sediment or debris; pond plants, pond water, and organisms; clear-plastic container with lid; net; magnifying lens **Lab:** Internet sites or reference sources on wetlands	**Chapter *FAST FILE* Resources** MiniLAB, p. 4 Transparency Activity, p. 44 Enrichment, p. 30 Reinforcement, p. 27 Directed Reading, pp. 19, 20 Lab Worksheet, pp. 7–8 Lab Activity, pp. 11–14 Transparency Activity, pp. 45–46 **Lab Management and Safety,** p. 37	**Portfolio** Science Journal, p. 80 **Performance** MiniLAB, p. 78 Applying Math, p. 80 Applying Skills, p. 83 **Content** Section Review, p. 83	Section Focus Transparency Teaching Transparency Virtual Labs CD-ROM Guided Reading Audio Program Interactive Chalkboard CD-ROM

End of Chapter Assessment

Glencoe's Assessment Advantage

Blackline Masters	Technology	Professional Series
Chapter *FAST FILE* Resources Chapter Review, pp. 35–36 Chapter Tests, pp. 37–40 **Standardized Test Practice,** pp. 15–18	MindJogger Videoquiz Virtual Labs CD-ROM Exam*View*® Pro Testmaker TeacherWorks CD-ROM Interactive Chalkboard CD-ROM	**Performance Assessment in the Science Classroom (PASC)**

Transparencies

Section Focus

Section 1 Section Focus Transparency **New Land, New Life**

Both photographs below are of Surtsey, an island near Iceland. Surtsey was created in 1963 by a volcanic eruption. Surtsey provides a wonderful opportunity for scientists to study how life starts in a new area.

1. Describe what you see in the first picture. Would you expect plants or animals to be able to live there?
2. What evidence of life do you see in the second picture?
3. How has Surtsey changed? How might it continue to change?

L2

Section 2 Section Focus Transparency **Sentinel**

This unusual tree is a type of quiver tree. This rare species is found in parts of South Africa and Namibia. This particular type of quiver tree is critically endangered; estimates show fewer than 200 full-grown trees left.

1. Describe the characteristics of the quiver tree from the photo.
2. What kind of environment does this tree lives in?
3. Hypothesize what factors might threaten the quiver tree?

L2

Section 3 Section Focus Transparency **Is there a Great Happy Swamp?**

Almost three-fourths of Earth's surface is covered by water. This water is in the Great Dismal Swamp located in Virginia and North Carolina.

1. Why were wetlands like the Great Dismal Swamp originally considered dismal and useless places? Why are they valued today?
2. Name some other water systems. How does human activity affect these systems?

L2

This is a representation of key blackline masters available in the Teacher Classroom Resources. See Resource Manager boxes within the chapter for additional information.

Assessment

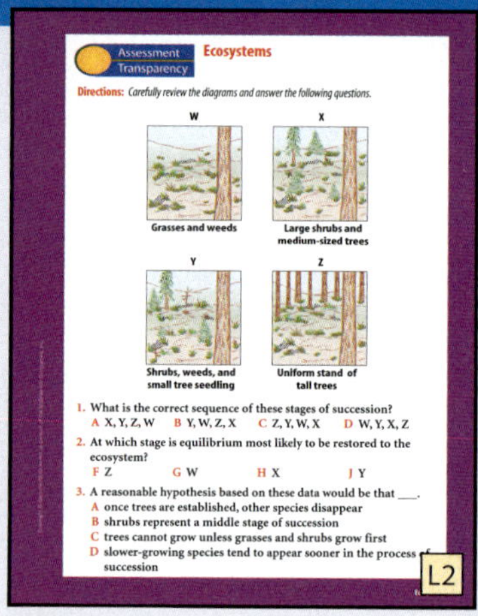
Assessment Transparency **Ecosystems**

Directions: Carefully review the diagrams and answer the following questions.

1. What is the correct sequence of these stages of succession?
   A X, Y, Z, W B Y, W, Z, X C Z, Y, W, X D W, Y, X, Z
2. At which stage is equilibrium most likely to be restored to the ecosystem?
   F Z G W H X J Y
3. A reasonable hypothesis based on these data would be that ___.
   A once trees are established, other species disappear
   B shrubs represent a middle stage of succession
   C trees cannot grow unless grasses and shrubs grow first
   D slower-growing species tend to appear sooner in the process of succession

L2

Teaching

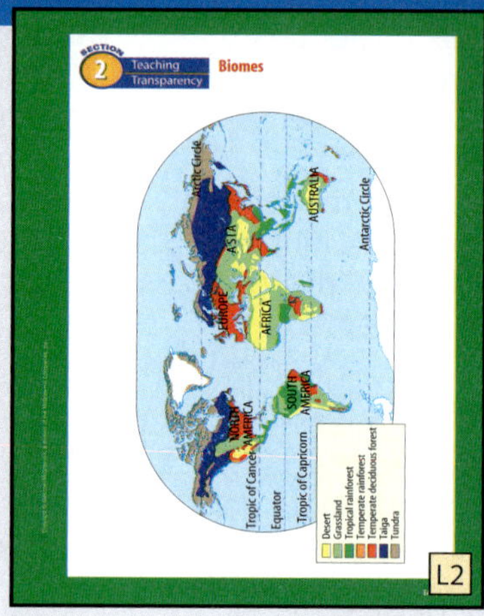
Section 2 Teaching Transparency **Biomes**

L2

Key to Teaching Strategies

The following designations will help you decide which activities are appropriate for your students.

- **L1** Level 1 activities should be appropriate for students with learning difficulties.
- **L2** Level 2 activities should be within the ability range of all students.
- **L3** Level 3 activities are designed for above-average students.
- **ELL** ELL activities should be within the ability range of English-Language Learners.
- **COOP LEARN** Cooperative Learning activities are designed for small group work.
- **LS** Multiple Learning Styles logos, as described on page 12T, are used throughout to indicate strategies that address different learning styles.
- **P** These strategies represent student products that can be placed into a best-work portfolio.
- **PBL** Problem-Based Learning activities apply real-world situations to learning.

Hands-on Activities

Student Text Lab Worksheet

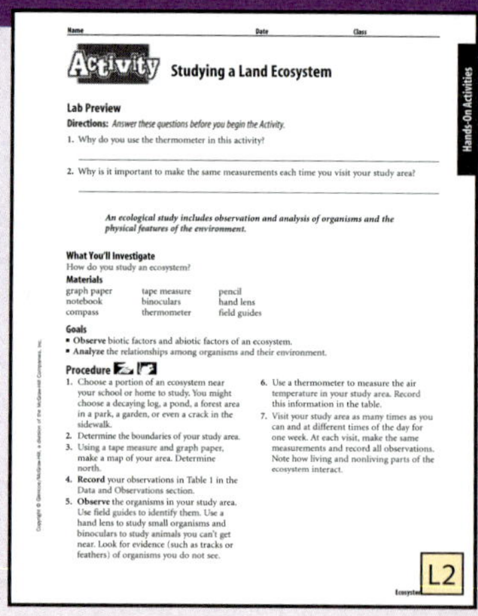
Name Date Class

Activity **Studying a Land Ecosystem**

Hands-On Activities

Lab Preview

Directions: Answer these questions before you begin the Activity.

1. Why do you use the thermometer in this activity?
2. Why is it important to make the same measurements each time you visit your study area?

An ecological study includes observation and analysis of organisms and the physical features of the environment.

What You'll Investigate

How do you study an ecosystem?

Materials

graph paper, notebook, compass, tape measure, binoculars, thermometer, pencil, hand lens, field guides

Goals

- Observe biotic factors and abiotic factors of an ecosystem.
- Analyze the relationships among organisms and their environment.

Procedure

1. Choose a portion of an ecosystem near your school or home to study. You might choose a decaying log, a pond, a forest area in a park, a garden, or even a crack in the sidewalk.
2. Determine the boundaries of your study area.
3. Using a tape measure and graph paper, make a map of your area. Determine north.
4. Record your observations in Table 1 in the Data and Observations section.
5. Observe the organisms in your study area. Use field guides to identify them. Use a hand lens to study small organisms and binoculars to study animals you can't get near. Look for evidence (such as tracks or feathers) of organisms you do not see.
6. Use a thermometer to measure the air temperature in your study area. Record this information in the table.
7. Visit your study area as many times as you can and at different times of the day for one week. At each visit, make the same measurements and record all observations. Note how living and nonliving parts of the ecosystem interact.

L2

Laboratory Activities

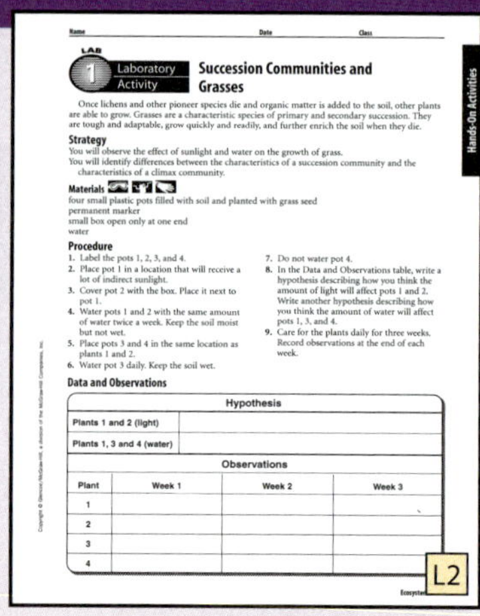
Name Date Class

Lab 1 Laboratory Activity **Succession Communities and Grasses**

Hands-On Activities

Once lichens and other pioneer species die and organic matter is added to the soil, other plants are able to grow. Grasses are a characteristic species of primary and secondary succession. They are tough and adaptable, grow quickly and readily, and further enrich the soil when they die.

Strategy

You will observe the effect of sunlight and water on the growth of grass.
You will identify differences between the characteristics of a succession community and the characteristics of a climax community.

Materials

four small plastic pots filled with soil and planted with grass seed
permanent marker
small box open only at one end
water

Procedure

1. Label the pots 1, 2, 3, and 4.
2. Place pot 1 in a location that will receive a lot of indirect sunlight.
3. Cover pot 2 with the box. Place it next to pot 1.
4. Water pots 1 and 2 with the same amount of water twice a week. Keep the soil moist but not wet.
5. Place pots 3 and 4 in the same location as plants 1 and 2.
6. Water pot 3 daily. Keep the soil wet.
7. Do not water pot 4.
8. In the Data and Observations table, write a hypothesis describing how you think the amount of light will affect pots 1 and 2. Write another hypothesis describing how you think the amount of water will affect pots 1, 3, and 4.
9. Care for the plants daily for three weeks. Record observations at the end of each week.

Data and Observations

Hypothesis			
Plants 1 and 2 (light)			
Plants 1, 3 and 4 (water)			
Observations			
Plant	Week 1	Week 2	Week 3
1			
2			
3			
4			

L2

Meeting Different Ability Levels

Content Outline

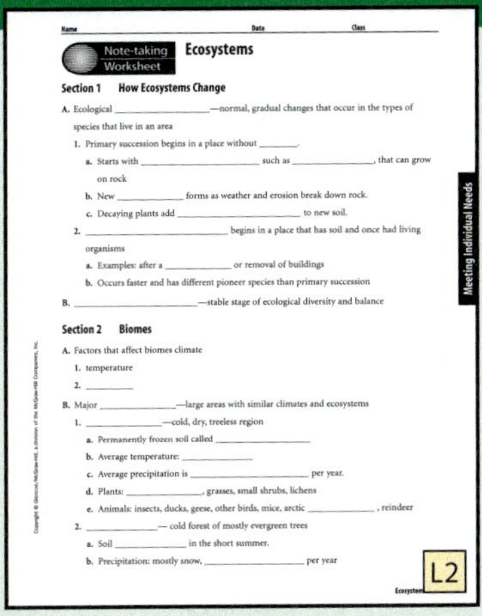

Reinforcement

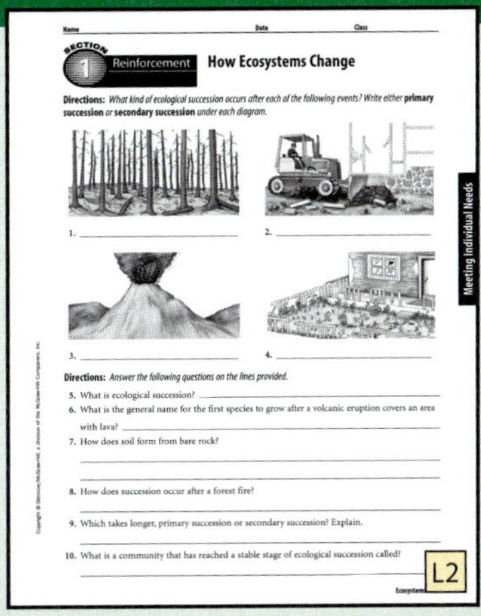

Enrichment

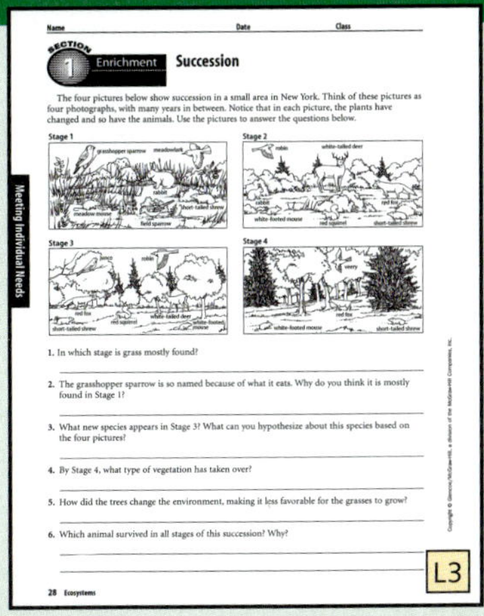

Directed Reading (English/Spanish)

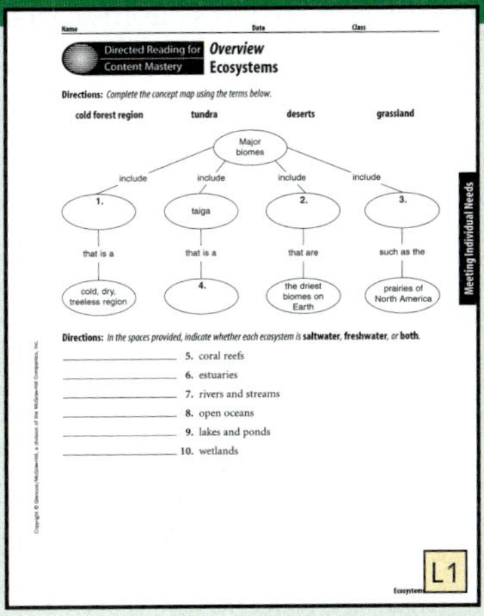

Study Guide

Reading Essentials

Assessment

Test Practice Workbook

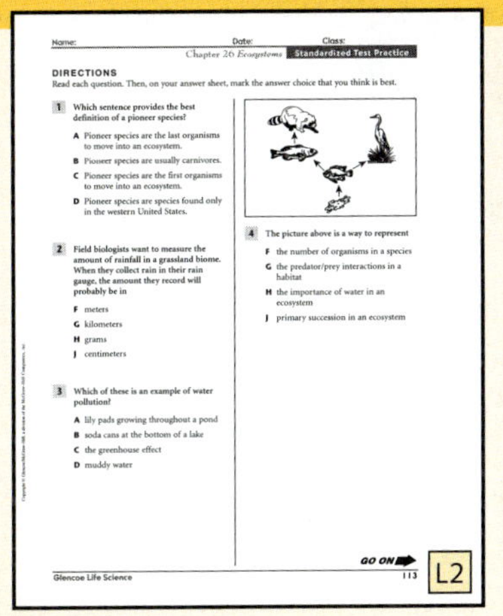

Chapter Review

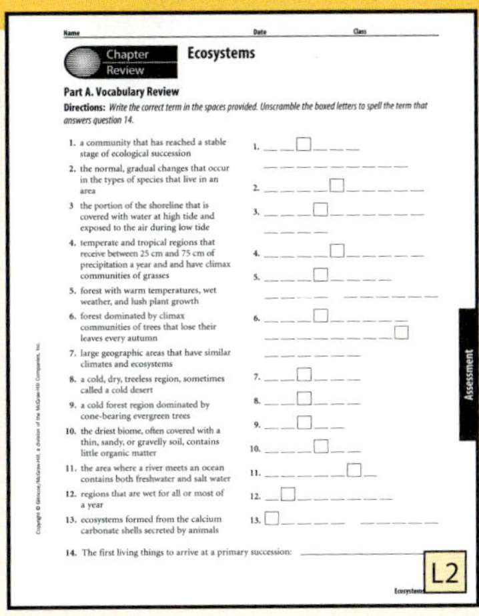

Chapter Tests

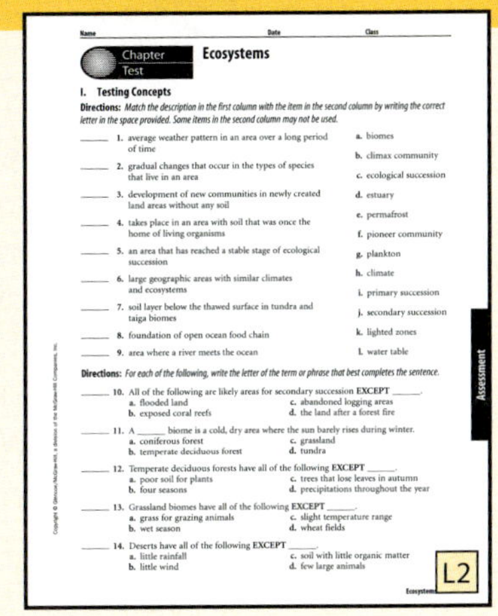

Science Content Background

section 1 How Ecosystems Change

Defining Ecosystems

A system is a group of parts that work together to form a functioning whole. Some common examples of systems include cars, watches, and digestive systems. An ecosystem is groups of organisms interacting with each other and their environment through a flow of energy and cycling of matter. It is the interdependence of all the parts that makes the whole a system.

section 2 Biomes

Tundra

The arctic tundra receives less energy from the Sun than does any other biome. The Sun's energy is dispersed over a greater area at higher latitudes. The Sun's rays lose much of their energy because they must travel diagonally through the atmosphere.

Desert

Desert covers less than 5% of North America. The North American deserts are the Mojave, Sonoran, Great Basin, and Chihuahuan. Temperatures in the desert undergo drastic changes over a 24-hour period. Daytime temperatures can be 30°C higher than nighttime readings. These fluctuations occur because of the lack of heat-retaining moisture in the desert air.

section 3 Aquatic Ecosystems

Nutrients in Aquatic Ecosystems

In aquatic ecosystems, nutrients tend to sink below the light zone where many organisms cannot use them. In the ocean, the process that moves the bottom water with its valuable nutrients up to the surface is called an upwelling. In an upwelling, winds and currents work together near the continental shelf. The winds carry water away from land. Bottom water with its nutrients is pulled up into the photic zone to replace water that has moved out to sea. Upwellings occur in only a few places. One such place is along the coast of Peru.

Kevin & Cat Sweeney/DRK Photo

chapter content resources

Internet Resources

For additional content background, visit **booke.msscience.com** to:

- access your book online
- find references to related articles in popular science magazines
- access Web links with related content background
- access current events with science journal topics

Print Resources

Blue Planet, by Andrew Byatt et. al., DK Publishing, 2001

Wetlands, by William Mitsch and James G. Gosselink, John Wiley & Sons, 2000

What is a Biome? (Science of Living Things), by Bobbie Kalman, Crabtree Publishing Co., 1999

Ecology of Desert Systems, by Walter Whitford, Academic Press, 2002

IDENTIFYING Misconceptions

Find Out What Students Think

Students may think that . . .

Systems only exist as parts of an organism or process.

Ecosystems consist of only a few interactions among living things in an area.

Students often do not understand the nature of systems in general, and they understand even less about the specific aspects of an ecosystem. Students may have learned that some things are called systems or ecosystems, but they may not be aware that ecosystems consist of parts, subsystems, and many interactions between biotic and abiotic factors.

Discussion

Show a photo of a bicycle, or bring in a bicycle for students to examine. Have students describe what they see. It is likely that students will respond that they see a bike. Point out that although there is one bicycle functioning as a whole unit, the bicycle consists of many components such as wheels, seat, chain, pedals, and so on. If any one of these components is removed, the bicycle will not function properly. The individual parts work together to do what a bicycle is supposed to do. In addition, in order for the bike to function, it needs an input of energy from a rider. Like a bicycle, an ecosystem is made up of many components, both abiotic (nonliving) and biotic (living), and it receives energy in the form of light and heat from the Sun.

Promote Understanding

Activity

Have students make a terrarium or display a completed terrarium.

- Discuss the biotic (living) and abiotic (nonliving) factors included in the terrarium, such as plants, insects, light, water, soil, and air.
- Have students diagram connections between the components. Then relate these connections to the flow of energy and cycling of materials. What do you think would happen if one of the components of the terrarium were removed? Explain that if a component such as light were removed, the plants would not be able to photosynthesize. Have students discuss the long-term effects of this event. The plants would die, the insects that eat the plants would not have food, and so on.
- Emphasize that the removal of light would cause the ecosystem to function improperly. Underscore this idea, using a diagram of the flow of energy and cycling of materials in the terrarium.

As an extension, have students choose another component of the terrarium and discuss how its removal would affect all other components of the terrarium.

Assess

After completing the chapter, see *Identifying Misconceptions* in the Study Guide at the end of the chapter.

chapter 3

Chapter Vocabulary

succession, p. 64
pioneer species, p. 64
climax community, p. 67
biome, p. 68
tundra, p. 69
taiga, p. 70
temperate deciduous forest, p. 71
temperate rain forest, p. 71
tropical rain forest, p. 72
desert, p. 74
grassland, p. 75
wetland, p. 79
coral reef, p. 81
intertidal zone, p. 82
estuary, p. 82

Science Journal Traits of trees would include thick bark and high branches. Shrub traits would include rhizomes and root systems that sprout after they've been burned back.

INTERACTIVE CHALKBOARD with Image Bank

PowerPoint® Presentations

This CD-ROM is an editable Microsoft® PowerPoint® presentation that includes:

- a pre-made presentation for every chapter
- interactive graphics
- animations
- audio clips
- image bank
- all new section and chapter questions
- Standardized Test Practice
- transparencies
- pre-lab questions for all labs
- Foldables directions
- links to booke.msscience.com

chapter 3

Ecosystems

chapter preview

sections

1 How Ecosystems Change
2 Biomes
Lab Studying a Land Ecosystem
3 Aquatic Ecosystems
Lab Exploring Wetlands
Virtual Lab What are the different types of land environments?

The Benefits of Wildfires

Ecosystems are places where organisms, including humans, interact with each other and with their physical environment. In some ecosystems, wildfires are an essential part of the physical environment. Organisms in these ecosystems are well adapted to the changes that fire brings, and can benefit from wildfires.

Science Journal What traits might plants on this burning Montana hillside have that enable them to survive?

Theme Connection

Stability and Change Ecosystems change in ways that can be predicted. In the process of succession, each community affects the environment, often producing conditions that result in the community's replacement.

About the Photo

Wildfire The 2000 fire season was long, difficult, and expensive. Over 120,000 fires burned more than 8 million acres in the United States, costing more than a billion dollars. The largest fire burned for over two months and blackened almost 300,000 acres in western Montana. Many ecosystems rely on fire in order to regenerate themselves, so fire is beneficial and necessary for many ecosystems.

Start-Up Activities

Launch LAB

What environment do houseplants need?

The plants growing in your classroom or home may not look like the same types of plants that you find growing outside. Many indoor plants don't grow well outside in most North American climates. Do the lab below to determine what type of environment most houseplants thrive in.

1. Examine a healthy houseplant in your classroom or home.
2. Describe the environmental conditions found in your classroom or home. For example, is the air humid or dry? Is the room warm or cool? Does the temperature stay about the same, or change during the day?
3. Using observations from step 1 and descriptions from step 2, hypothesize about the natural environment of the plants in your classroom or home.
4. **Think Critically** In your Science Journal, record the observations that led to your hypothesis. How would you design an experiment to test your hypothesis?

FOLDABLES Study Organizer

Primary and Secondary Succession Make the following Foldable to help you illustrate the main ideas about succession.

STEP 1 **Fold** a vertical sheet of paper in half from top to bottom.

STEP 2 **Fold** in half from side to side with the fold at the top.

STEP 3 **Unfold** the paper once. **Cut** only the fold of the top flap to make two tabs.

STEP 4 **Turn** the paper vertically and **label** on the front tabs as shown.

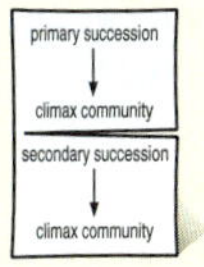

Illustrate and Label As you read the chapter, define terms and collect information under the appropriate tabs.

Science Online — Preview this chapter's content and activities at booke.msscience.com

E ◆ 63

Launch LAB

Purpose Students observe common houseplants and make hypotheses about the climate of the area they originally inhabited. L2 LS **Naturalist**

Preparation Establish successfully growing houseplants in your classroom several weeks before the start of this chapter.

Materials houseplants; optional thermometers and humidity meter

Teaching Strategies

- Help students describe the amount of light, heat, and moisture the plants receive.
- Have students compare their observations and hypotheses with other students. Discuss differences and similarities.

Think Critically

Houseplants generally grow best when they have moist soil, moderate temperatures, and medium to low light. These plants are typically found in the understories of tropical rain forests. Students might suggest placing different specimens of the same type of plant in bright and low light, or giving them varying amounts of water.

Assessment

Oral Have students explain how the local climate may affect houseplants if they were planted outside the school for one year. Use **Performance Assessment in the Science Classroom,** p. 89.

FOLDABLES Study Organizer **Dinah Zike Study Fold**

Student preparation materials for this Foldable are available in the **Chapter *Fast File* Resources.**

section 1 How Ecosystems Change

1 Motivate

Bellringer

INTERACTIVE CHALKBOARD PowerPoint® Presentations

Section Focus Transparencies also are available on the Interactive Chalkboard CD-ROM. L2 ELL

Section 1 Section Focus Transparency **New Land, New Life**

Both photographs below are of Surtsey, an island near Iceland. Surtsey was created in 1963 by a volcanic eruption. Surtsey provides a wonderful opportunity for scientists to study how life starts in a new area.

1. Describe what you see in the first picture. Would you expect plants or animals to be able to live there?
2. What evidence of life do you see in the second picture?
3. How has Surtsey changed? How might it continue to change?

Ecosystems

Tie to Prior Knowledge

Local Succession Relate the idea of succession to examples found in the local environment. For example, there may be abandoned lots or farm fields showing succession. Many communities are allowing roadside grass areas to revert back to their natural states.

as you read

What **You'll Learn**

- **Explain** how ecosystems change over time.
- **Describe** how new communities begin in areas without life.
- **Compare** pioneer species and climax communities.

Why **It's Important**

Understanding ecosystems and your role in them can help you manage your impact on them and predict the changes that may happen in the future.

Review Vocabulary

ecosystem: community of living organisms interacting with each other and their physical environment

New Vocabulary

- succession
- pioneer species
- climax community

Ecological Succession

What would happen if the lawn at your home were never cut? The grass would get longer, as in **Figure 1,** and soon it would look like a meadow. Later, larger plants would grow from seeds brought to the area by animals or wind. Then, trees might sprout. In fact, in 20 years or less you wouldn't be able to tell that the land was once a mowed lawn. An ecologist can tell you what type of ecosystem your lawn would become. If it would become a forest, they can tell you how long it would take and predict the type of trees that would grow there. **Succession** refers to the normal, gradual changes that occur in the types of species that live in an area. Succession occurs differently in different places around the world.

Primary Succession As lava flows from the mouth of a volcano, it is so hot that it destroys everything in its path. When it cools, lava forms new land composed of rock. It is hard to imagine that this land eventually could become a forest or grassland someday.

The process of succession that begins in a place previously without plants is called primary succession. It starts with the arrival of living things such as lichens (LI kunz). These living things, called **pioneer species,** are the first to inhabit an area. They survive drought, extreme heat and cold, and other harsh conditions and often start the soil-building process.

Figure 1 Open areas that are not maintained will become overgrown with grasses and shrubs as succession proceeds.

Section 1 Resource Manager

Chapter *FAST FILE* Resources

Transparency Activity, p. 42
Directed Reading for Content Mastery, pp. 17, 18
Note-taking Worksheets, pp. 31–33
Enrichment, p. 28
Reinforcement, p. 25
Lab Activity, pp. 9–10

64 ◆ E CHAPTER 3 Ecosystems

Figure 2 Lichens, like these in Colorado, are fragile and take many years to grow. They often cling to bare rock where many other organisms can't survive. **Describe** *how lichens form soil.*

New Soil During primary succession, shown in **Figure 2,** soil begins to form as lichens and the forces of weather and erosion help break down rocks into smaller pieces. When lichens die, they decay, adding small amounts of organic matter to the rock. Plants such as mosses and ferns can grow in this new soil. Eventually, these plants die, adding more organic material. The soil layer thickens, and grasses, wildflowers, and other plants begin to take over. When these plants die, they add more nutrients to the soil. This buildup is enough to support the growth of shrubs and trees. All the while, insects, small birds, and mammals have begun to move in. What was once bare rock now supports a variety of life.

Secondary Succession What happens when a fire, such as the one in **Figure 3,** disturbs a forest or when a building is torn down in a city? After a forest fire, not much seems to be left except dead trees and ash-covered soil. After the rubble of a building is removed, all that remains is bare soil. However, these places do not remain lifeless for long. The soil already contains the seeds of weeds, grasses, and trees. More seeds are carried to the area by wind and birds. Other wildlife may move in. Succession that begins in a place that already has soil and was once the home of living organisms is called secondary succession. Because soil already is present, secondary succession occurs faster and has different pioneer species than primary succession does.

Reading Check *Which type of succession usually starts without soil?*

Science Online

Topic: Eutrophication

Visit booke.msscience.com for Web links to information about eutrophication (yoo truh fih KAY shun)—secondary succession in an aquatic ecosystem.

Activity Using the information that you find, illustrate or describe in your Science Journal this process for a small freshwater lake.

2 Teach

Caption Answer

Figure 2 Lichens help to break down rocks into smaller pieces. When lichens die, they decay, adding small amounts of organic material to the soil.

Discussion

Wildfires Throughout much of the 20th century, fire suppression was thought to be good stewardship of the land. However, by the 1940s, many ecologists realized that fire was an important element of change for ecosystems. By the 1970s, Yellowstone and other national parks had instituted a "natural fire management plan" which ensured that naturally ignited wildfires would burn freely as an ecosystem process. The televised images of Yellowstone burning in 1988 brought emotional responses from the public. Many offered to donate seedlings to plant, including species that are more "fire-resistant". Explain why well-intentioned offers would damage Yellowstone Park's ecosystem in the long run. Possible answers: Introducing new tree species would change the system completely. Many organisms are dependent on those tree species that are not fire-resistant. Fire returns nutrients to the soil and encourages succession.

Reading Check

Answer primary succession

Quick Demo

Lichens

Materials rocks with a variety of lichens growing on them

Estimated Time 10 minutes

Procedure Show rocks and point out the different lichens. You may choose to pass the rocks around, but stress that lichens are fragile and remind students not to rub the lichens off. These are the organisms that help to form soil.

Teacher FYI

Secondary Pioneers The small size of lichens and mosses makes them successful in the harsh environments typical of primary succession. Fast reproduction is a characteristic of secondary pioneers. These plants, which often pop up in gardens, are usually called weeds.

Curriculum Connection

History As the eastern United States was settled by Europeans, large tracts of forest were converted into farmland. With increasing industrialization, many of the farms were abandoned and the land reverted back into forests. Ask students to determine the past extent of the eastern forests and draw a map of this region.

NATIONAL GEOGRAPHIC

Visualizing Secondary Succession

Have students examine the pictures and read the captions. Then ask the following questions.

What characteristics would make a seed likely to survive a fire? A seed that is surrounded by a thick seed coat is more likely to survive a fire than one with a thin seed coat.

As succession proceeds, what changes will occur in the population of plants found on the forest floor? At first, plants that need lots of sunlight can live on the forest floor, because there are no tall trees to block the sunlight. As trees grow and mature there will be less sunlight penetrating to the forest floor, so species that require less light will be found.

Activity

Song Lyrics Have the students work in teams to write song lyrics that describe secondary succession. Students can write lyrics that fit any melody they know and enjoy. The steps of secondary succession should be clearly and accurately described in the lyrics.

NATIONAL GEOGRAPHIC **VISUALIZING SECONDARY SUCCESSION**

Figure 3

In the summer of 1988, wind-driven flames like those shown in the background photo swept through Yellowstone National Park, scorching nearly a million acres. The Yellowstone fire was one of the largest forest fires in United States history. The images on this page show secondary succession—the process of ecological regeneration—triggered by the fire.

▶ After the fire, burned timber and blackened soil seemed to be all that remained. However, the fire didn't destroy the seeds that were protected under the soil.

◀ Within weeks, grasses and other plants were beginning to grow in the burned areas. Ecological succession was underway.

▶ Many burned areas in the park opened new plots for stands of trees. This picture shows young lodgepole pines in August 1999. The forest habitat of America's oldest national park is being restored gradually through secondary succession.

Visual Learning

Figure 3 Have students make an outline of the stages in secondary succession that are shown. Then, have them go back and fill in any stages that aren't shown. P

Differentiated Instruction

Challenge In some areas, under certain circumstances, natural fires are not put out. In many areas, resource managers do controlled burns. Have students research to find out why this is done. Students can take on the role of education officers and inform the rest of the class about the importance of fire in certain ecosystems. L3

Figure 4 This beech-maple forest is an example of a climax community.

Climax Communities A community of plants that is relatively stable and undisturbed and has reached an end stage of succession is called a **climax community.** The beech-maple forest shown in **Figure 4** is an example of a community that has reached the end of succession. New trees grow when larger, older trees die. The individual trees change, but the species remain stable. There are fewer changes of species in a climax community over time, as long as the community isn't disturbed by wildfire, avalanche, or human activities.

Primary succession begins in areas with no previous vegetation. It can take hundreds or even thousands of years to develop into a climax community. Secondary succession is usually a shorter process, but it still can take a century or more.

section 1 review

Summary

Ecological Succession

- Succession is the natural, gradual changes over time of species in a community.
- Primary succession occurs in areas that previously were without soil or plants.
- Secondary succession occurs in areas where soil has been disturbed.
- Climax communities have reached an end stage of succession and are stable.
- Climax communities have less diversity than communities in mid-succession.

Self Check

1. **Compare** primary and secondary succession.
2. **Describe** adaptations of pioneer species.
3. **Infer** the kind of succession that will take place on an abandoned, unpaved country road.
4. **Think Critically** Show the sequence of events in primary succession. Include the term *climax community.*

Applying Math

5. **Solve One-Step Equations** A tombstone etched with 1802 as the date of death has a lichen on it that is 6 cm in diameter. If the lichen began growing in 1802, calculate its average yearly rate of growth.

Science Online booke.msscience.com/self_check_quiz

IDENTIFYING Misconceptions

Ecosystems Students may think that ecosystems consist of only a few interactions among living things in an area. Refer to page F at the beginning of this chapter for teaching strategies that address this misconception.

3 Assess

DAILY INTERVENTION

Check for Understanding

Visual-Spatial Have students diagram succession of a pond ecosystem to a forest in their area. L2

Reteach

Forest Succession Arrange the stages of succession for a forest community out of sequence in a diagram. Photocopy the diagram and have students use numbers to put the pictures in the proper sequence. L2

LS **Logical-Mathematical**

Assessment

Content Have students write a paragraph identifying how each of the following is undergoing succession: a cornfield after a flood, the area around a volcanic eruption, and a newly formed sandbar in a river. Use **Performance Assessment in the Science Classroom,** p. 159.

section 1 review

1. Primary succession occurs in areas without soil and usually previously without plants. Secondary succession occurs in areas with soil and plants previously.
2. Pioneer species are adapted to survive drought, extreme heat and cold, and other harsh conditions.
3. secondary succession
4. More sunlight and water reach the forest floor, enabling more species of plants to survive. As trees grow into a climax forest, the environmental conditions are more limiting and fewer species are adapted to live in those conditions.
5. Growth per year = 6 cm / (current year − 1802). For example, if the year is 2002 then 6 cm / (2002 − 1802) = 0.03 cm per yr.

1 Motivate

Bellringer

INTERACTIVE CHALKBOARD
PowerPoint® Presentations

Section Focus Transparencies also are available on the Interactive Chalkboard CD-ROM. L2 ELL

Section 2 Section Focus Transparency — Sentinel

This unusual tree is a type of quiver tree. This rare species is found in parts of South Africa and Namibia. This particular type of quiver tree is critically endangered; estimates show fewer than 200 full-grown trees left.

1. Describe the characteristics of the quiver tree from the photo.
2. What kind of environment does this tree lives in?
3. Hypothesize what factors might threaten the quiver tree?

Ecosystems

Tie to Prior Knowledge

Biome Organisms Choose a biome that students are likely to have some knowledge about, but don't live in, such as a desert or a tropical rain forest. Have students make a list of organisms that live in that biome and the biome in which they live. Then have students hypothesize why different plants and animals live in these different biomes.

section 2 Biomes

as you read

What You'll Learn

- **Explain** how climate influences land environments.
- **Identify** seven biomes of Earth.
- **Describe** the adaptations of organisms found in each biome.

Why It's Important

Resources that you need to survive are found in a variety of biomes.

Review Vocabulary
climate: the average weather conditions of an area over many years

New Vocabulary

- biome
- tundra
- taiga
- temperate deciduous forest
- temperate rain forest
- tropical rain forest
- desert
- grassland

Factors That Affect Biomes

Does a desert in Arizona have anything in common with a desert in Africa? Both have heat, little rain, poor soil, water-conserving plants with thorns, and lizards. Even widely separated regions of the world can have similar biomes because they have similar climates. Climate is the average weather pattern in an area over a long period of time. The two most important climatic factors that affect life in an area are temperature and precipitation.

Major Biomes

Large geographic areas that have similar climates and ecosystems are called **biomes** (BI ohmz). Seven common types of land biomes are mapped in **Figure 5.** Areas with similar climates produce similar climax communities. Tropical rain forests are climax communities found near the equator, where temperatures are warm and rainfall is plentiful. Coniferous forests grow where winter temperatures are cold and rainfall is moderate.

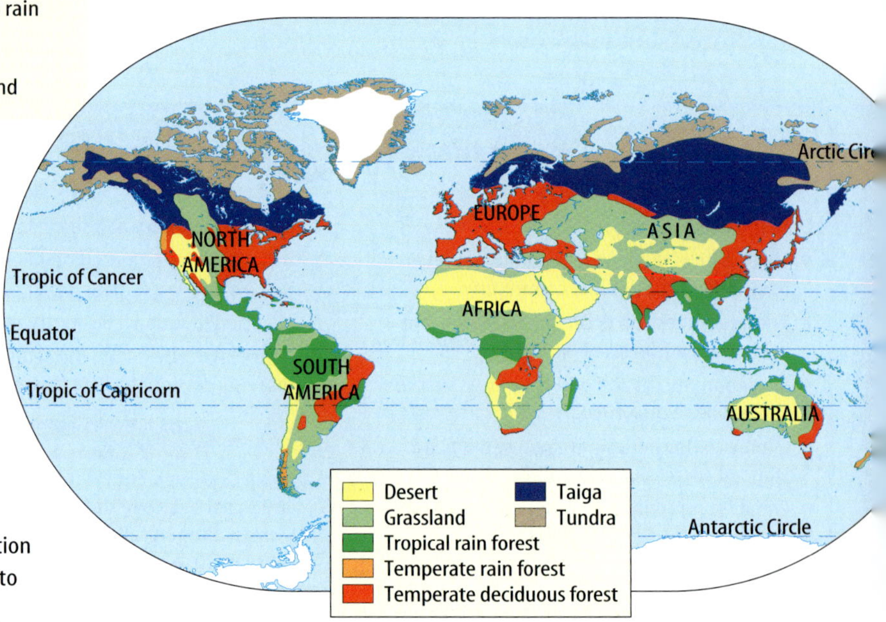

Figure 5 The land portion of Earth can be divided into seven biomes.

Section 2 Resource Manager

Chapter *FAST FILE* Resources
Transparency Activity, p. 43
Enrichment, p. 29
Directed Reading for Content Mastery, p. 19
MiniLAB, p. 3
Reinforcement, p. 26
Lab Worksheet, pp. 5–6

Reading and Writing Skill Activities, p. 1
Earth Science Critical Thinking/Problem Solving, p. 6
Life Science Critical Thinking/Problem Solving, p. 5

68 ◆ E CHAPTER 3 Ecosystems

Tundra At latitudes just south of the north pole or at high elevations, a biome can be found that receives little precipitation but is covered with ice most of the year. The **tundra** is a cold, dry, treeless region, sometimes called a cold desert. Precipitation averages less than 25 cm per year. Winters in the Arctic can be six to nine months long. For some of these months, the Sun never appears above the horizon and it is dark 24 hours a day. The average daily temperature is about –12°C. For a few days during the short, cold summer, the Sun is always visible. Only the top portion of soil thaws in the summer. Below the thawed surface is a layer of permanently frozen soil called permafrost, shown in **Figure 6.** Alpine tundra, found above the treeline on high mountains, have similar climates. Tundra soil has few nutrients because the cold temperatures slow the process of decomposition.

Figure 6 This permafrost in Alaska is covered by soil that freezes in the winter and thaws in the summer.
Infer *what types of problems this might cause for people living in this area.*

Tundra Life Tundra plants are adapted to drought and cold. They include mosses, grasses, and small shrubs, as seen in **Figure 7.** Many lichens grow on the tundra. During the summer, mosquitoes, blackflies, and other biting insects fill the air. Migratory birds such as ducks, geese, shorebirds, and songbirds nest on the Arctic tundra during the summer. Other inhabitants include hawks, snowy owls, and willow grouse. Mice, voles, lemmings, arctic hares, caribou, reindeer, and musk oxen also are found there.

People are concerned about overgrazing by animals on the tundra. Fences, roads, and pipelines have disrupted the migratory routes of some animals and forced them to stay in a limited area. Because the growing season is so short, plants and other vegetation can take decades to recover from damage.

Figure 7 Lichens, mosses, grasses, and small shrubs thrive on the tundra. Ptarmigan also live on the tundra. In winter, their feathers turn white. Extra feathers on their feet keep them warm and prevent them from sinking into the snow.

Tundra

Ptarmigan

2 Teach

Quick Demo

Arctic Tundra Light

Materials globe, flashlight

Estimated Time 10 minutes

Procedure Use the flashlight to model the rays of the Sun shining on Earth. The globe should be at a 23.5% angle (to model Earth's tilt). Students should be able to observe why some winter days in the Arctic tundra have 24 hours of darkness and some summer days have 24 hours of light.

Caption Answer

Figure 6 Permafrost makes it impossible to farm the land.

Discussion

Environment Choice Would you prefer to live in a house with ten acres of land in the tundra or in a small apartment in New York City? Have students defend their responses.

Fun Fact

The nose botfly has unique adaptations for surviving the cold winter. It lays its eggs in the nose of a caribou. The maggots emerge and travel to the caribou's throat where they are kept warm. In the spring, caribou cough the maggots up. The maggots pupate and then become flies.

Teacher FYI

Lichens Limited light and heat, and a short growing season are factors that limit producer growth in the tundra. Lichen species here grow extremely slowly—only 1/6 cm per year. It can take 40 years for lichens to recover from damage caused by a passing herd of caribou.

Cultural Diversity

Inuit Life Have students research the Inuit, who live in the Arctic tundra, and report on how they have adapted to the harsh climate. L2
LS **Linguistic**

SECTION 2 Biomes E ◆ 69

Use Science Words

Word Origin "Taiga" comes from the Russian word *taiga*, which means "swampy forest." Ask students why this is a good description of the taiga biome. The taiga is a cold forest region that is warmer and wetter than the tundra.

Caption Answer

Figure 8 Wide feet enable the lynx to run across the top of snow.

IDENTIFYING Misconceptions

Climate and Biomes Students may think that the location of a particular biome on Earth is permanent. They may not realize that changes in climate have affected the locations of biomes and probably will continue to do so in the future.

Quick Demo

Biome Precipitation

Materials strips of paper cut to appropriate length to represent the amount of precipitation in centimeters for each biome, tape

Estimated Time 5 minutes

Procedure Tape the strips of paper to the board as you discuss the biomes. Start with a strip of paper that represents the amount of rainfall in your area.

Figure 8 The taiga is dominated by cone-bearing trees. The lynx, a mammal adapted to life in the taiga, has broad, heavily furred feet that act like snowshoes to prevent it from sinking in the snow.
Infer *why "snowshoe feet" are important for a lynx.*

Taiga South of the tundra—between latitudes 50°N and 60°N and stretching across North America, northern Europe, and Asia—is the world's largest biome. The **taiga** (TI guh), shown in **Figure 8,** is a cold, forest region dominated by cone-bearing evergreen trees. Although the winter is long and cold, the taiga is warmer and wetter than the tundra. Precipitation is mostly snow and averages 35 cm to 100 cm each year.

Most soils of the taiga thaw completely during the summer, making it possible for trees to grow. However, permafrost is present in the extreme northern regions of the taiga. The forests of the taiga might be so dense that little sunlight penetrates the trees to reach the forest floor. However, some lichens and mosses do grow on the forest floor. Moose, lynx, shrews, bears, and foxes are some of the animals that live in the taiga.

Temperate Deciduous Forest Temperate regions usually have four distinct seasons each year. Annual precipitation ranges from about 75 cm to 150 cm and is distributed throughout the year. Temperatures range from below freezing during the winter to 30°C or more during the warmest days of summer.

Figure 9 White-tailed deer are one of many species that you can find in a deciduous forest. In autumn, the leaves on deciduous trees change color and fall to the ground.

Science Journal

Metric Measurements Students should become familiar with metric measurements, such as those used in this section for temperature. Have them convert the following Celsius temperatures to Fahrenheit:

10°C (50°F); 0°C (32°F); 20°C (68°F); 30°C (86°F)

Differentiated Instruction

Learning Disabled A concept map of "The Forests of the World" may help students visualize the relationship between taiga, temperate deciduous forest, temperate rain forest, and tropical rain forest. L1 LS **Visual-Spatial**

Temperate Forest Life Many evergreen trees grow in the temperate regions of the world. However, most of the temperate forests in Europe and North America are dominated by climax communities of deciduous trees, which lose their leaves every autumn. These forests, like the one in **Figure 9,** are called **temperate deciduous forests.** In the United States, most of them are located east of the Mississippi River.

When European settlers first came to America, they cut trees to create farmland and to supply wood. As forests were cut, organisms lost their habitats. When agriculture shifted from the eastern to the midwestern and western states, secondary succession began, and trees eventually returned to some areas. Now, nearly as many trees grow in the New England states as did before the American Revolutionary War. Many trees are located in smaller patches. Yet, the recovery of large forests such as those in the Adirondack Mountains in New York State shows the result of secondary succession.

Figure 10 In the Olympic rain forest in Washington State, mosses and lichens blanket the ground and hang from the trees. Wet areas are perfect habitats for amphibians like the Pacific giant salamander above.

Temperate Rain Forest New Zealand, southern Chile, and the Pacific Northwest of the United States are some of the places where **temperate rain forests,** shown in **Figure 10,** are found. The average temperature of a temperate rain forest ranges from 9°C to 12°C. Precipitation ranges from 200 cm to 400 cm per year.

Trees with needlelike leaves dominate these forests, including the Douglas fir, western red cedar, and spruce. Many grow to great heights. Animals of the temperate rain forest include the black bear, cougar, bobcat, northern spotted owl, and marbled murrelet. Many species of amphibians also inhabit the temperate rain forest, including salamanders.

The logging industry in the Northwest provides jobs for many people. However, it also removes large parts of the temperate rain forest and destroys the habitat of many organisms. Many logging companies now are required to replant trees to replace the ones they cut down. Also, some rain forest areas are protected as national parks and forests.

Make a Model

Land Use Have students:

(a) use long wooden dowels representing trees to show how the land in the Northeastern United States looked before European settlers,

(b) remove the dowels and put up short dowel fences and black paper roads to represent how the land looked when it was made into farms,

(c) return many of the long dowels but keep the fences and roads to show the return of trees. Make the point that although trees are present, many animals are restricted in their movement.

Virtual Labs

Environment *What are the different types of land environments?*

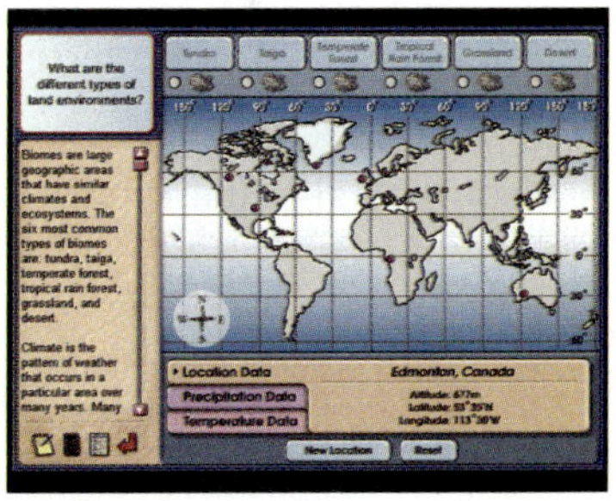

Differentiated Instruction

Challenge Six common land biomes are presented in this chapter. Students may be assigned other biomes to research, such as tidal wetlands, extreme deserts, bogs, chapparal, continental slope, temperate oceans, tropic savannahs, or coastal arctic tundra. Have them present their findings to the class as posters or other visuals.
L3 LS **Visual-Spatial**

Teacher FYI

Spotted Owls The spotted owl is an endangered species that lives in old-growth temperate forests. These forests are over 100 years old. Clear-cutting the forest and planting new trees destroys the necessary habitat of these birds.

SECTION 2 Biomes E ◆ 71

Mini LAB

Purpose Students observe rain forest leaves, analyze their structures, and relate leaf structure to leaf function. L2

LS **Kinesthetic**

Materials poster board, pencil, scissors, sink, water, spray bottle

Teaching Strategy Show students philodendron or ficus leaves so they can see real drip tips.

Analysis

1. the leaf with the drip tip
2. Removing water from the leaf's surface prohibits growth of mosses, fungi, or other moisture-loving organisms that could compete with the plant.

Assessment

Process Ask students how they could perform a similar experiment but obtain quantitative results. Answers should involve the measurement of the amount of water falling off the leaves using a graduated cylinder. Use **Performance Assessment in the Science Classroom,** p. 105.

Try at Home

Reading Check

Answer forest floor, understory, canopy, emergents

Figure 11 Tropical rain forests are lush environments that contain such a large variety of species that many have not been discovered.

Mini LAB

Modeling Rain Forest Leaves

Procedure

1. Draw an oval leaf about 10 cm long on a piece of **poster board.** Cut it out.
2. Draw a second leaf the same size but make one end pointed. This is called a drip tip. Cut this leaf out.
3. Hold your hands palm-side up over a **sink** and have someone lay a leaf on each one. Point the drip tip away from you. Tilt your hands down but do not allow the leaves to fall off.
4. Have someone gently spray water on the leaves and observe what happens.

Analysis

1. From which leaf does water drain faster?
2. Infer why it is an advantage for a leaf to get rid of water quickly in a rain forest.

Try at Home

Tropical Rain Forest Warm temperatures, wet weather, and lush plant growth are found in **tropical rain forests.** These forests are warm because they are near the equator. The average temperature, about 25°C, doesn't vary much between night and day. Most tropical rain forests receive at least 200 cm of rain annually. Some receive as much as 600 cm of rain each year.

Tropical rain forests, like the one in **Figure 11,** are home to an astonishing variety of organisms. They are one of the most biologically diverse places in the world. For example, one tree in a South American rain forest might contain more species of ants than exist in all of the British Isles.

Tropical Rain Forest Life Different animals and plants live in different parts of the rain forest. Scientists divide the rain forest into zones based on the types of plants and animals that live there, just as a library separates books about different topics onto separate shelves. The zones include: forest floor, understory, canopy, and emergents, as shown in **Figure 12.** These zones often blend together, but their existence provide different habitats for many diverse organisms to live in the tropical rain forest.

Reading Check *What are the four zones of a tropical rain forest?*

Although tropical rain forests support a huge variety of organisms, the soil of the rain forest contains few nutrients. Over the years, nutrients have been washed out of the soil by rain. On the forest floor, decomposers immediately break down organic matter, making nutrients available to the plants again.

LAB DEMONSTRATION

Purpose to show the stabilizing effect of buttress roots on rain forest trees

Materials 3 paper towel tubes, scissors, tape, trough with thin layer of sand

Preparation Cut out four buttresses (which resemble rocket fins) from a paper towel tube, and tape them to one tube.

Procedure Rain forest tree roots don't grow very deep. Push the paper towel tubes (one with butresses, one without) 1 cm into the sand. Blow on the tubes to see which falls first.

Expected Outcome The tube without buttresses will fall first.

Assessment

Why do buttresses make trees more stable? They give a tree a wider base. Why would a rain forest tree have buttresses but not a temperate forest tree? Rain forest tree roots do not penetrate the soil as deeply.

72 ◆ E CHAPTER 3 Ecosystems

Human Impact Farmers that live in tropical areas clear the land to farm and to sell the valuable wood. After a few years, the crops use up the nutrients in the soil, and the farmers must clear more land. As a result, tropical rain forest habitats are being destroyed. Through education, people are realizing the value and potential value of preserving the species of the rain forest. In some areas, logging is prohibited. In other areas, farmers are taught new methods of farming so they do not have to clear rain forest lands continually.

Figure 12 Tropical rain forests contain abundant and diverse organisms.

Use an Analogy

Rain Forest v. Library The layers of a rain forest are comparable to the shelves of a library. Just as the shelves provide more space for books, the layers of the rain forest provide more niches for supporting life. This is one of the reasons rain forests are biologically diverse.

Activity

Rain Forest Zones Have students make charts of the main features of the four major zones of the tropical rain forest: the emergents, canopy, understory, and forest floor. L2 LS **Naturalist**

Visual Learning

Figure 12 Ask students to explain how conditions above the canopy differ from conditions below the canopy. much more sunlight and wind above the canopy, darker and more humid below the canopy

INTEGRATE Earth Science

Desertification Areas with heavy agriculture in the Southwest are the most likely to be affected by desertification in the United States. Changes in climate and human activities also contribute to desertification, which is the loss of productivity of land in arid or semi-arid areas.

Research Have students research the expansion of the Sahara in Africa. Create a map that shows the change in its boundaries over the last 50 years.

Reading Check

Answer because of lack of water

IDENTIFYING Misconceptions

Cold Deserts A common idea about deserts is that they are hot, dry places. While this is true of the majority of deserts, there are some cold deserts such as Patagonia in Argentina. Cold deserts are usually caused by their remoteness from the coast, or by mountains that block moist coastal winds. The largest cold desert is in central Asia.

Fun Fact

At one time, the grassland biome covered nearly half of the land on Earth. Today, most of the grassland is used by humans for growing crops, raising animals, and housing.

Desert The driest biome on Earth is the **desert.** Deserts receive less than 25 cm of rain each year and support little plant life. Some desert areas receive no rain for years. When rain does come, it quickly drains away. Any water that remains on the ground evaporates rapidly.

Most deserts, like the one in **Figure 13,** are covered with a thin, sandy, or gravelly soil that contains little organic matter. Due to the lack of water, desert plants are spaced far apart and much of the ground is bare. Barren, windblown sand dunes are characteristics of the driest deserts.

Reading Check *Why is much of a desert bare ground?*

Desert Life Desert plants are adapted for survival in the extreme dryness and hot and cold temperatures of this biome. Most desert plants are able to store water. Cactus plants are probably the most familiar desert plants of the western hemisphere. Desert animals also have adaptations that help them survive the extreme conditions. Some, like the kangaroo rat, never need to drink water. They get all the moisture they need from the breakdown of food during digestion. Most animals are active only during the night, late afternoon, or early morning when temperatures are less extreme. Few large animals are found in the desert.

In order to provide water for desert cities, rivers and streams have been diverted. When this happens, wildlife tends to move closer to cities in their search for food and water. Education about desert environments has led to an awareness of the impact of human activities. As a result, large areas of desert have been set aside as national parks and wilderness areas to protect desert habitats.

INTEGRATE Earth Science

Desertification When vegetation is removed from soil in areas that receive little rain, the dry, unprotected surface can be blown away. If the soil remains bare, a desert might form. This process is called desertification. Look on a biome map and hypothesize about which areas of the United States are most likely to become deserts.

Figure 13 Desert plants, like these in the Sonoran Desert, are adapted for survival in the extreme conditions of the desert biome. The giant hairy scorpion found in some deserts has a venomous sting.

Cultural Diversity

Biomes and Dress Discuss how the dress of people in different areas is influenced by the biome in which they live. For example, Arabic people may wear long robes to shield themselves from the desert heat. Inuits may dress in animal skins to insulate themselves from the cold. Encourage interested students to present a show of "Biome Fashions Around the World." COOP LEARN

Grasslands Temperate and tropical regions that receive between 25 cm and 75 cm of precipitation each year and are dominated by climax communities of grasses are called **grasslands.** Most grasslands have a dry season, when little or no rain falls. This lack of moisture prevents the development of forests. Grasslands are found in many places around the world, and they have a variety of names. The prairies and plains of North America, the steppes of Asia, the savannas of Africa shown in **Figure 14,** and the pampas of South America are types of grasslands.

Figure 14 Animals such as zebras and wildebeests are adapted to life on the savannas in Africa.

Grasslands Life The most noticeable animals in grassland biomes are usually mammals that graze on the stems, leaves, and seeds of grass plants. Kangaroos graze in the grasslands of Australia. In Africa, communities of animals such as wildebeests, impalas, and zebras thrive in the savannas.

Grasslands are perfect for growing many crops such as wheat, rye, oats, barley, and corn. Grasslands also are used to raise cattle and sheep. However, overgrazing can result in the death of grasses and the loss of valuable topsoil from erosion. Most farmers and ranchers take precautions to prevent the loss of valuable habitats and soil.

section 2 review

Summary

Major Biomes

- Tundra, sometimes called a cold desert, can be divided into two types: arctic and alpine.
- Taiga is the world's largest biome. It is a cold forest region with long winters.
- Temperate regions have either a deciduous forest biome or a rain forest biome.
- Tropical rain forests are one of the most biologically diverse biomes.
- Humans have a huge impact on tropical rain forests.
- The driest biome is the desert. Desert organisms are adapted for extreme dryness and temperatures.
- Grasslands provide food for wildlife, livestock, and humans.

Self Check

1. **Determine** which two biomes are the driest.
2. **Compare and contrast** tundra organisms and desert organisms.
3. **Identify** the biggest climatic difference between a temperate rain forest and a tropical rain forest.
4. **Explain** why the soil of tropical rain forests make poor farmland.
5. **Think Critically** If you climb a mountain in Arizona, you might reach an area where the trees resemble the taiga trees in northern Canada. Why would a taiga forest exist in Arizona?

Applying Skills

6. **Record Observations** Animals have adaptations that help them survive in their environments. Make a list of animals that live in your area, and record the physical or behavioral adaptations that help them survive.

Science Online booke.msscience.com/self_check_quiz

3 Assess

DAILY INTERVENTION

Check for Understanding

Logical-Mathematical Have students draw a four-organism food chain for any five of the six biomes presented in this section. L3

Reteach

Biome Characteristics Make a table showing precipitation, temperature, common plants, and common animals for each biome. Have students identify the name of each biome described. L2 ELL

LS **Naturalist**

Assessment

Portfolio Have each student write a story with one of the biomes in this section as the setting. The plot should revolve around an organism that lives in the biome they chose and its interactions with other organisms and the environment. Students can take turns reading their stories to the class. Use **Performance Assessment in the Science Classroom,** p. 155. P

section 2 review

1. desert and tundra
2. Tundra organisms are adapted to survive cold and drought. Desert organisms are adapted to survive extreme dryness, and large temperature fluctuations.
3. Tropical rain forests are warmer (have a higher average temperature) than temperate rain forests, and tend to have more precipitation.
4. The rain forest has very nutrient-poor soil that is easily eroded.
5. Temperatures cool with increased elevation, so places with higher altitude in warm areas can resemble the taiga.
6. Answers will vary but should include adaptations to survive conditions including cold or heat, humidity or dryness, and land or aquatic environments.

BENCH TESTED

Real-World Question

Purpose Students conduct a study of an ecosystem and determine how the living things interact with each other and the environment. L2

LS **Intrapersonal**

Process Skills observe, classify, communicate, define operationally, recognize cause and effect, predict

Time Required 50 minutes for setup, four weeks to make observations

Procedure

Teaching Strategies

- Arrange for students to check out binoculars, field guides, and other useful materials from the science classroom and library.
- Approve the ecosystems chosen by the students before they begin their observations.
- Offer students a set time each day to discuss any problems they may have concerning their method of study.

Conclude and Apply

1. Populations would be affected by changes in temperature, water, or other abiotic factors. The entire ecosystem could be disrupted.
2. Other populations could be affected.
3. Possible hypotheses: competition for available resources could make one population extinct or forced away; students may hypothesize about results from favorable or unfavorable interactions among the populations in response to the introduction of a new population.

LAB

Studying a Land Ecosystem

An ecological study includes observation and analysis of organisms and the physical features of the environment.

Real-World Question

How do you study an ecosystem?

Goals

- **Observe** biotic factors and abiotic factors of an ecosystem.
- **Analyze** the relationships among organisms and their environments.

Materials

graph paper
binoculars
thermometer
pencil
magnifying lens
field guides
notebook
compass
tape measure

Safety Precautions

Procedure

1. Choose a portion of an ecosystem to study. You might choose a decaying log, a pond, a garden, or even a crack in the sidewalk.
2. Determine the boundaries of your study area.
3. Using a tape measure and graph paper, make a map of your area. Determine north.
4. **Record** your observations in a table similar to the one shown on this page.
5. **Observe** the organisms in your study area. Use field guides to identify them. Use a magnifying lens to study small organisms and binoculars to study animals you can't get near. Look for evidence (such as tracks or feathers) of organisms you do not see.
6. Measure and record the air temperature in your study area.
7. Visit your study area many times and at different times of day for one week. At each visit, make the same measurements and record all observations. Note how the living and nonliving parts of the ecosystem interact.

Environmental Observations		
Date		
Time of day		
Temperature	Answers will vary.	
Organisms observed		
Comments		

Conclude and Apply

1. **Predict** what might happen if one or more abiotic factors were changed suddenly.
2. **Infer** what might happen if one or more populations of plants or animals were removed from the area.
3. **Form a hypothesis** to explain how a new population of organisms might affect your ecosystem.

Communicating Your Data

Make a classroom display of all data recorded. **For more help, refer to the Science Skill Handbook.**

Assessment

Performance Have students alter the ecosystem in a way that is not harmful. Have them observe how foreign materials affect the behavior of living things in the ecosystem. Use **Performance Assessment in the Science Classroom,** p. 97.

Communicating Your Data

You may choose to display the best student posters along one wall. Allow volunteers to present their poster and data to the class in a short presentation.

76 ◆ E CHAPTER 3 Ecosystems

section 3

Aquatic Ecosystems

Freshwater Ecosystems

In a land environment, temperature and precipitation are the most important factors that determine which species can survive. In aquatic environments, water temperature, the amount of sunlight present, and the amounts of dissolved oxygen and salt in the water are important. Earth's freshwater ecosystems include flowing water such as rivers and streams and standing water such as lakes, ponds, and wetlands.

Rivers and Streams Flowing freshwater environments vary from small, gurgling brooks to large, slow-moving rivers. Currents can quickly wash loose particles downstream, leaving a rocky or gravelly bottom. As the water tumbles and splashes, as shown in **Figure 15,** air from the atmosphere mixes in. Naturally fast-flowing streams usually have clearer water and higher oxygen content than slow-flowing streams.

Most nutrients that support life in flowing-water ecosystems are washed into the water from land. In areas where the water movement slows, such as in the pools of streams or in large rivers, debris settles to the bottom. These environments tend to have higher nutrient levels and more plant growth. They contain organisms that are not as well adapted to swiftly flowing water, such as freshwater mussels, minnows, and leeches.

as you read

What You'll Learn

- **Compare** flowing freshwater and standing freshwater ecosystems.
- **Identify** and describe important saltwater ecosystems.
- **Identify** problems that affect aquatic ecosystems.

Why It's Important

All of the life processes in your body depend on water.

Review Vocabulary
aquatic: growing or living in water

New Vocabulary
- wetland
- intertidal zone
- coral reef
- estuary

Figure 15 Streams like this one are high in oxygen because of the swift, tumbling water.
Determine *where most nutrients in streams come from.*

1 Motivate

Bellringer

INTERACTIVE CHALKBOARD PowerPoint® Presentations

Section Focus Transparencies also are available on the Interactive Chalkboard CD-ROM. L2 ELL

Tie to Prior Knowledge

Water Environments Most students have some experience with water environments. Have them explain what body of water they observed and which organisms live there.

Caption Answer

Figure 15 They are washed into the stream from the land.

Section 3 Resource Manager

Chapter *FAST FILE* Resources
- Transparency Activity, pp. 44, 45–46
- Directed Reading for Content Mastery, pp. 19, 20
- Enrichment, p. 30
- MiniLAB, p. 4
- Reinforcement, p. 27
- Lab Worksheet, pp. 7–8
- Lab Activity, pp. 11–14

Mathematics Skill Activities, p. 9
Home and Community Involvement, p. 45
Cultural Diversity, p. 7
Earth Science Critical Thinking/Problem Solving, p. 1
Lab Management and Safety, p. 37

2 Teach

MiniLAB

Purpose Students observe a freshwater ecosystem. L2

LS **Naturalist**

Materials clear plastic container, material from the bottom of a pond, pond organisms, net, pond water, magnifying lens

Teaching Strategy Make sure students don't disturb the container excessively while observing, as sediment stirred up will make observations difficult.

Safety Precaution Make sure to get permission from the land owner or appropriate authorities before collecting samples from the pond.

Analysis

Students' paragraphs will vary depending on organisms obtained.

Assessment

Oral Have students use their observations from the MiniLAB to identify adaptations of the organisms. Use **Performance Assessment in the Science Classroom,** p. 89.

MiniLAB

Modeling Freshwater Environments

Procedure

1. Obtain a sample of **pond sediment or debris, plants, water, and organisms** from your teacher.
2. Cover the bottom of a **clear-plastic container** with about 2 cm of the debris.
3. Add one or two plants to the container.
4. Carefully pour pond water into the container until it is about two-thirds full.
5. Use a **net** to add several organisms to the water. Seal the container.
6. Using a **magnifying lens,** observe as many organisms as possible. Record your observations. Return your sample to its original habitat.

Analysis

Write a short paragraph describing the organisms in your sample. How did the organisms interact with each other?

Human Impact People use rivers and streams for many activities. Once regarded as a free place to dump sewage and other pollutants, many people now recognize the damage this causes. Treating sewage and restricting pollutants have led to an improvement in the water quality in some rivers.

Lakes and Ponds When a low place in the land fills with rainwater, snowmelt, or water from an overflowing stream, a lake or pond might form. Pond or lake water hardly moves. It contains more plants than flowing-water environments contain.

Lakes, such as the one shown in **Figure 16,** are larger and deeper than ponds. They have more open water because most plant growth is limited to shallow areas along the shoreline. In fact, organisms found in the warm, sunlit waters of the shorelines often are similar to those found in ponds. If you were to dive to the bottom, you would discover few, if any, plants or algae growing. Colder temperatures and lower light levels limit the types of organisms that can live in deep lake waters. Floating in the warm, sunlit waters near the surface of freshwater lakes and ponds are microscopic algae, plants, and other organisms known as plankton.

A pond is a small, shallow body of water. Because ponds are shallow, they are filled with animal and plant life. Sunlight usually penetrates to the bottom. The warm, sunlit water promotes the growth of plants and algae. In fact, many ponds are filled almost completely with plant material, so the only clear, open water is at the center. Because of the lush growth in pond environments, they tend to be high in nutrients.

Figure 16 Ponds contain more vegetation than lakes contain. The population of organisms in the shallow water of lakes is high. Fewer types of organisms live in the deeper water.

Differentiated Instruction

Physically Challenged If students have difficulty with manual dexterity in the MiniLAB, mount the magnifying glass on a ring stand. Have them look through the magnifying glass and slowly move the plastic container.

Visually Impaired Have a small shallow container represent a pond and a larger container represent a lake. Form a "V" with a plastic sheet. Incline the V at a steep angle, and pour water down it to represent a stream. Hold the V at a slighter angle, and pour a greater volume of water down it. This represents a river. Have students put their hands into the four conditions to feel the differences.

Water Pollution Human activities can harm freshwater environments. Fertilizer-filled runoff from farms and lawns, as well as sewage dumped into the water, can lead to excessive growth of algae and plants in lakes and ponds. The growth and decay of these organisms reduces the oxygen level in the water, which makes it difficult for some organisms to survive. To prevent problems, sewage is treated before it is released. People also are being educated about problems associated with polluting lakes and ponds. Fines and penalties are issued to people caught polluting waterways. These controls have led to the recovery of many freshwater ecosystems.

Wetlands As the name suggests, **wetlands,** shown in **Figure 17,** are regions that are wet for all or most of a year. They are found in regions that lie between landmasses and water. Other names for wetlands include swamps, bogs, and fens. Some people refer to wetlands as biological supermarkets. They are fertile ecosystems, but only plants that are adapted to waterlogged soil survive there. Wetland animals include beavers, muskrats, alligators, and the endangered bog turtle. Many migratory bird populations use wetlands as breeding grounds.

Reading Check *Where are wetlands found?*

Wetlands once were considered to be useless, disease-ridden places. Many were drained and destroyed to make roads, farmland, shopping centers, and housing developments. Only recently have people begun to understand the importance of wetlands. Products that come from wetlands, including fish, shellfish, cranberries, and plants, are valuable resources. Now many developers are restoring wetlands, and in most states access to land through wetlands is prohibited.

Figure 17 Life in the Florida Everglades was threatened due to pollution, drought, and draining of the water. Conservation efforts are being made in an attempt to preserve this ecosystem.

INTEGRATE Career

Environmental Author
Rachel Carson (1907–1964) was a scientist that turned her knowledge and love of the environment into articles and books. After 15 years as an editor for the U.S. Fish and Wildlife Service, she resigned and devoted her time to writing. She probably is known best for her book *Silent Spring,* in which she warned about the long-term effects of the misuse of pesticides. In your Science Journal, compile a list of other authors who write about environmental issues.

IDENTIFYING Misconceptions

Freshwater Salts Freshwater is not water without salts; it is water that has fewer salts than ocean or estuary water. Distilled water has no salts or other impurities. Obtain a liter of freshwater and put it into a glass beaker. Allow it to evaporate so students can see the ample amount of salts and other materials left behind.

INTEGRATE Career

Environmental Author Rachel Carson was the Editor-in-Chief of all publications for the US Fish and Wildlife Service. In addition to *Silent Spring,* she wrote *Under the Sea-Wind, The Sea Around Us,* and, *The Edge of the Sea.* She died in 1964 of breast cancer. Students' lists of authors will vary, but may include: Aldo Leopold, Edward Abbey, Farley Mowat, Vandana Shiva, Ann Haymond Zwinger, John Muir, and Barry Lopez.

Reading Check

Answer in regions that lie between solid land and water

Teacher FYI

Pollution Slow-moving rivers and standing lakes and ponds are easily polluted. There is less dissolved oxygen in the water, which is needed by organisms that break down the pollutants, and no currents wash pollutants away.

Curriculum Connection

History Have students research the Seminole people of Florida and how they survived attacks by the United States Army by hiding in the Everglades. L2

Quick Demo

Saltwater v. Freshwater

Materials two identical natural corks, two containers, freshwater, salt water

Estimated Time 5 minutes

Procedure Float one cork in freshwater and the other in salt water. Students should observe that the cork in salt water floats higher than the one in freshwater. Explain that this occurs because salt water is more dense than freshwater.

Make a Model

Estuaries Ask students to use clay, colored water, or other materials to show the mixing of freshwater and salt water when a river and ocean combine to form an estuary.

Activity

Dissolved Oxygen From a science supply company, purchase kits that measure the dissolved oxygen in water. Compare different bodies of water such as lakes, streams, and oceans. L3

LS **Naturalist**

Applying Math

National Math Standards

Correlation to Mathematics Objectives

1, 2, 4, 9

Answers to Practice Problems

1. (78°F − 32) / 1.8 = 25.6°C
2. (12°C × 1.8) + 32 = 53.6°F
   (23°C × 1.8) + 32 = 73.4°F
   73.4°F − 53.6°F = 19.8°F

Saltwater Ecosystems

About 95 percent of the water on the surface of Earth contains high concentrations of various salts. The amount of dissolved salts in water is called salinity. The average ocean salinity is about 35 g of salts per 1,000 g of water. Saltwater ecosystems include oceans, seas, a few inland lakes such as the Great Salt Lake in Utah, coastal inlets, and estuaries.

Applying Math — Convert Units

TEMPERATURE Organisms that live around hydrothermal vents in the ocean deal with temperatures that range from 1.7°C to 371°C. You have probably seen temperatures measured in degrees Celsius (°C) and degrees Fahrenheit (°F). Which one are you familiar with? If you know the temperature in one system, you can convert it to the other.

You have a Fahrenheit thermometer and measure the water temperature of a pond at 59°F. What is that temperature in degrees Celsius?

Solution

1. *This is what you know:* water temperature in degrees Fahrenheit = 59°F
2. *This is what you need to find out:* The water temperature in degrees Celsius.
3. *This is the procedure you need to use:*
   - Solve the equation for degrees Celsius:
     (°C × 1.8) + 32 = °F
     °C = (°F − 32)/1.8
   - Substitute the known value:
     °C = (59°F − 32)/1.8 = 15°C
   - Water temperature that is 59°F is 15°C.
4. *Check your answer:* Substitute the Celsius temperature back into the original equation. You should get 59.

Practice Problems

1. The thermometer outside your classroom reads 78°F. What is the temperature in degrees Celsius?
2. If lake water was 12°C in October and 23°C in May, what is the difference in degrees Fahrenheit?

Science Online — For more practice, visit booke.msscience.com/math_practice

Science Journal

Marine Biomes Have students compare estuaries and oceans. As part of this comparison, ask students to focus on living and nonliving factors associated with each ecosystem. Have them write a paragraph about each biome in their Science Journals. P

Fun Fact

The largest tides occur when Earth, the Sun, and the Moon are in line. This is when the combined gravities of the Sun and Moon have their greatest effect on Earth's waters.

Open Oceans Life abounds in the open ocean. Scientists divide the ocean into different life zones, based on the depth to which sunlight penetrates the water. The lighted zone of the ocean is the upper 200 m or so. It is the home of the plankton that make up the foundation of the food chain in the open ocean. Below about 200 m is the dark zone of the ocean. Animals living in this region feed on material that floats down from the lighted zone, or they feed on each other. A few organisms are able to produce their own food.

Coral Reefs One of the most diverse ecosystems in the world is the coral reef. **Coral reefs** are formed over long periods of time from the calcium carbonate shells secreted by animals called corals. When corals die, their shells remain. Over time, the shell deposits form reefs such as the Great Barrier Reef off the coast of Australia, shown in **Figure 18.**

Reefs do not adapt well to long-term stress. Runoff from fields, sewage, and increased sedimentation from cleared land harm reef ecosystems. Organizations like the Environmental Protection Agency have developed management plans to protect the diversity of coral reefs. These plans treat a coral reef as a system that includes all the areas that surround the reef. Keeping the areas around reefs healthy will result in a healthy environment for the coral reef ecosystem.

Science Online

Topic: Coral Reefs
Visit booke.msscience.com for Web links to information about coral reef ecosystems.

Activity Construct a diorama of a coral reef. Include as many different kinds of organisms as you can for a coral reef ecosystem.

Figure 18 The lighter areas around this island are part of the Great Barrier Reef. It comprises about 3,000 reefs and about 900 islands. Reefs contain colorful fish and a large variety of other organisms.

IDENTIFYING Misconceptions

Anemones and Corals Students may think that sea anemones and the coral in coral reefs are plants rather than animals. Help them realize that they are animals that feed on small organisms in the water.

Quick Demo

Calcium Carbonate Skeleton

Materials piece or pieces of coral such as those found in saltwater aquariums

Estimated Time 5 minutes

Procedure Hold up or pass around the piece or pieces of coral. Explain that this is the calcium carbonate skeleton left behind by the once-living coral.

Discussion

Osmosis Because freshwater fish have saltier blood than the surrounding water, water tends to move into the fish by osmosis. Freshwater fish are covered with scales and a mucous secretion that slows the movement of water into their tissues. The kidneys are adapted to filter out excess water, and they excrete large amounts of dilute urine. Saltwater fish have less salt in their blood than the surrounding water. What direction would water tend to diffuse? outward What adaptations would you expect saltwater fish to have to prevent water loss from their body? Marine fish drink sea water and excrete the excess salts.

Active Reading

Jigsaw In this collaborative learning technique, individuals become experts on a portion of a text and share their expertise with a small group, called their home group. Everyone shares responsibility for learning the assigned reading. Assign each person in each home group an expert number (1 through 5, for example). Have students gather into the expert groups that correspond to the number they were assigned. Have them read, discuss, and master chapter concepts, and determine how best to teach them to their home groups. Have students return to their home groups and share the content they learned in their expert groups. Have students use the Jigsaw strategy with the text on aquatic ecosystems.

Visual Learning

Figure 19 Explain that sea stars have special adaptions that enable them to inhabit the intertidal zone. One of these adaptations is the presence of numerous small tube feet under each arm, which enable the sea star to resist being swept away by the tide.

Caption Answer

Figure 20 crabs, clams, oysters

Reading Check

Answer freshwater carries tons of nutrients from inland soils to estuaries

Inquiry Lab

Biome Investigation

Purpose to observe what affects organisms within biomes, by studying a student-made ecosystem

Possible Materials large jars, soda bottles, or aquaria; variety of living organisms; water; soil

Estimated Time 1 or 2 class sessions for research and set-up

Teaching Strategies

- Have student groups choose a biome to study. If space is an issue, have the class vote on 1 or 2 biomes to study.
- Students can construct a mini-ecosystem that would be part of the chosen biome. They should research to ensure their environment is accurate.
- Encourage students to make comparisons to the real world, contrasting their ecosystem to the biome in which you live.
- Allow them to explore questions that arise as they observe the system.

For additional inquiry activities, see *Science Inquiry Labs.*

Figure 19 As the tide recedes, small pools of seawater are left behind. These pools contain a variety of organisms such as sea stars and periwinkles.

INTEGRATE Earth Science

Seashores All of Earth's landmasses are bordered by ocean water. The shallow waters along the world's coastlines contain a variety of saltwater ecosystems, all of which are influenced by the tides and by the action of waves. The gravitational pull of the Moon, and to a lesser extent, the Sun, on Earth causes the tides to rise and fall each day. The height of the tides varies according to the phases of the Moon, the season, and the slope of the shoreline. The **intertidal zone** is the portion of the shoreline that is covered with water at high tide and exposed to the air during low tide. Organisms that live in the intertidal zone, such as those in **Figure 19,** must be adapted to dramatic changes in temperature, moisture, and salinity and must be able to withstand the force of wave action.

Estuaries Almost every river on Earth eventually flows into an ocean. The area where a river meets an ocean and contains a mixture of freshwater and salt water is called an **estuary** (ES chuh wer ee). Other names for estuaries include bays, lagoons, harbors, inlets, and sounds. They are located near coastlines and border the land. Salinity in estuaries changes with the amount of freshwater brought in by rivers and streams, and with the amount of salt water pushed inland by the ocean tides.

Estuaries, shown in **Figure 20,** are extremely fertile, productive environments because freshwater streams bring in tons of nutrients washed from inland soils. Therefore, nutrient levels in estuaries are higher than in freshwater ecosystems or other saltwater ecosystems.

Curriculum Connection

Geography Have students locate Australia on a globe and then identify the location of the Great Barrier Reef. L1 LS **Visual-Spatial**

Differentiated Instruction

Challenge Ask students to make up a story about a molecule of water in a cloud falling to Earth as a raindrop. Direct students to have it pass through three aquatic ecosystems before it evaporates and begins the process again. Have students share their stories with the class. L3

Figure 20 The Chesapeake Bay is an estuary rich in resources. Fish and shrimp are harvested by commercial fishing boats.
Describe *what other resources can be found in estuaries.*

Estuary Life Organisms found in estuaries include many species of algae, salt-tolerant grasses, shrimp, crabs, clams, oysters, snails, worms, and fish. Estuaries also serve as important nurseries for many species of ocean fish. Estuaries provide much of the seafood consumed by humans.

Reading Check *Why are estuaries more fertile than other aquatic ecosystems?*

section 3 review

Summary

Freshwater Ecosystems

- Temperature, light, salt, and dissolved oxygen are important factors.
- Rivers, streams, lakes, ponds, and wetlands are freshwater ecosystems.
- Human activities, such as too much lawn fertilizer, can pollute aquatic ecosystems.

Saltwater Ecosystems

- About 95 percent of Earth's water contains dissolved salts.
- Saltwater ecosystems include open oceans, coral reefs, seashores, and estuaries.
- Organisms that live on seashores have adaptations that enable them to survive dramatic changes in temperature, moisture, and salinity.
- Estuaries serve as nursery areas for many species of ocean fish.

Self Check

1. **Identify** the similarities and differences between a lake and a stream.
2. **Compare and contrast** the dark zone of the ocean with the forest floor of a tropical rain forest. What living or nonliving factors affect these areas?
3. **Explain** why fewer plants are at the bottom of deep lakes.
4. **Infer** what adaptations are necessary for organisms that live in the intertidal zone.
5. **Think Critically** Would you expect a fast moving mountain stream or the Mississippi River to have more dissolved oxygen? Explain.

Applying Skills

6. **Communicate** Wetlands trap and slowly release rain, snow, and groundwater. Describe in your Science Journal what might happen to a town located on a floodplain if nearby wetlands are destroyed.

Science Online booke.msscience.com/self_check_quiz

section 3 review

1. Both contain freshwater. A lake has standing water, low oxygen levels, and rich plant growth in shallow water. A stream has swift-moving water, high oxygen levels, and a rocky bottom with little plant life.
2. Both contain scavengers and decomposers; both are affected by temperature and lack of light. The dark zone is a cold envrionment, and the rain forest is warm.
3. Cold temperatures and low light levels limit the types of organisms that can grow there.
4. Such organisms must be able to adapt to drastic changes in temperature, moisture, and salinity; they must be able to withstand wave motion.
5. fast moving stream; more oxygen is mixed into the water
6. The town would flood much more often.

3 Assess

DAILY INTERVENTION

Check for Understanding

Kinesthetic Provide each student with a note card on which the name or picture of a wetland element has been placed. Examples might include: algae, perch, dragonfly nymph, rock, osprey, frog, or snake. Have students stand in a circle. Have one student hold onto the end of a skein of yarn. That student selects another student in the circle, and while still holding onto the yarn, passes the skein to that student and states how the two elements may interact. For example, a student with a dragonfly nymph card would pass the yarn to the student with the rock card and say the dragonfly perches on the rock. Have students continue passing the yarn and making connections until a web forms. Emphasize that everyone in the circle is connected either directly or indirectly, just like the elements in an ecosystem.

Reteach

Ocean Ecosystems Write on the chalkboard or on an overhead transparency a description of the three major ocean ecosystems. Have students identify each one. L2 LS **Naturalist**

Assessment

Process Assess students' understanding by having them compare and contrast the salinity of rivers, estuaries, and the open ocean in their Science Journals. Rivers have lower salinity than oceans. The amount of salt in an estuary varies, but they contain more salt than rivers, and less than oceans. Use **PASC**, p. 175.

BENCH TESTED

Real-World Question

Purpose

Internet Students use Internet sites that can be accessed through booke.msscience.com/internet_lab. L2

Non-Internet Sources Collect pamphlets from a variety of wetlands conservancy groups.

Time Required three to five days

Make a Plan

Preparation

Internet To run through the steps that the students will follow, visit booke.msscience.com/internet_lab.

Non-Internet Organize printed materials by wetlands regions.

Follow Your Plan

Teaching Strategy Remind students to investigate how the amount of wetlands areas has changed over the last 25 years, and how that has made an impact on native plants and animals.

LAB — Use the Internet

Exploring Wetlands

Goals

- **Identify** wetland regions in the United States.
- **Describe** the significance of the wetland ecosystem.
- **Identify** plant and animal species native to a wetland region.
- **Identify** strategies for supporting the preservation of wetlands.

Data Source

Science Online

Visit booke.msscience.com/internet_lab for more information about wetland environments and for data collected by other students.

Real-World Question

Wetlands, such as the one shown below, are an important part of the environment. These fertile ecosystems support unique plants and animals that can survive only in wetland conditions. The more you understand the importance of wetlands, the more you can do to preserve and protect them. Why are wetlands an important part of the ecosystem?

Alternative Inquiry Lab

Real-World Connection Have groups take on different roles associated with a problem about wetlands. For example, one group can be engineers asked to provide an environmental impact study on a proposed construction site in the wetlands area. Another group can be legislators determining whether to grant the construction permits while another group acts as business executives assessing the costs and benefits of building in this location. Students should conduct this research and present their final recommendations. PBL

Using Scientific Methods

Make a Plan

1. **Determine** where some major wetlands are located in the United States.
2. **Identify** one wetland area to study in depth. Where is it located? Is it classified as a marsh, bog, or something else?
3. **Explain** the role this ecosystem plays in the overall ecology of the area.
4. **Research information** about the plants and animals that live in the wetland environment you are researching.
5. **Investigate** what laws protect the wetland you are studying.

Follow Your Plan

1. Make sure your teacher approves your plan before you start.
2. Perform the investigation.
3. Post your data at the link shown below.

Analyze Your Data

1. **Describe** the wetland area you have researched. What region of the United States is it located in? What other ecological factors are found in that region?
2. **Outline** the laws protecting the wetland you are investigating. How long have the laws been in place?
3. **List** the plants and animals native to the wetland area you are researching. Are those plants and animals found in other parts of the region or the United States? What adaptations do the plants and animals have that help them survive in a wetland environment?

Conclude and Apply

1. **Infer** Are all wetlands the same?
2. **Determine** what the ecological significance of the wetland area that you studied for that region of the country is.
3. **Draw Conclusions** Why should wetland environments be protected?
4. **Summarize** what people can do to support the continued preservation of wetland environments in the United States.

Communicating Your Data

Find this lab using the link below. **Post** your data in the table provided. **Review** other students' data to learn about other wetland environments in the United States.

Science Online
booke.msscience.com/internet_lab

Assessment

Portfolio Have students make a wetlands field guide for the area they investigated. Include maps and pictures. Have them describe how to support wetlands preservation. Use **Performance Assessment in the Science Classroom,** p. 129.

Communicating Your Data

Have students organize their field guides with section names. Suggest Native Animals and Maps to help them arrange their information.

Analyze Your Data

1. Answers will be based on the students' individual research.
2. Most laws protecting wetlands are federal and may include: The Clean Water Act (1972), Federal Highway Act of 1968, Estuary Protection Act (1972), National Wildlife Refuge Acts (numerous), and North American Wetlands Conservation Act (1980). In addition, more than half of the states have wetland laws.
3. Answers will vary based on the students' individual research. Some wetland plants have structures to keep their photosynthetic cells near the surface of the water and in the sunlight. Wetland trees, such as the cypress, have structures like tent stakes that provide support. Some wetland birds have more hemoglobin in their blood for extra oxygen, enabling them to stay underwater. Amphibian eggs develop quickly through the tadpole stage in case the wetland dries out.

Conclude and Apply

1. Wetlands regions are not all the same. Depending on where they are located, one area may have different species of plants than another.
2. Answers will vary. Remind students to think about the plants and animals that live in that wetlands region and if those species are found elsewhere.
3. Answers will vary. Students should investigate the amount of open space in that region.
4. Answers will vary. Have students find out about conservancy groups and their work to preserve wetlands regions.

TIME

Content Background

The idea of using wetlands to process municipal wastewater was first conceived in the 1970s.

Many wetlands are being restored or constructed. A constructed wetland mimics the action of a natural marsh. A substructure prevents polluted inflow from leaching into the soil before the wetland has processed it. Troughs are filled with sand and soil and then planted with appropriate wetland vegetation. Inflow volume, local climate, and degree of pre-processing of the wastewater determine the size and nature of a constructed wetland.

In some areas, the aim of wetlands water treatment is to enable the direct reuse of the outflow. Most constructed wetlands in the U.S. are in the South and West where the warm climate ensures year-round plant growth, but there are experimental wetland projects in the northern U.S. and southern Canada.

Discussion

Wastewater Treatment Have students research municipal wastewater treatment and individual septic systems. Under what circumstances might wetlands be a better option? Possible answer: In arid regions where groundwater is depleted by well drilling and irrigation and low-lying areas where high water tables compromise standard leach fields. Are there any instances of natural wetlands being used for wastewater treatment? Possible answer: Arcata, California

Investigate the Issue

Sludge Students can research methods of dealing with "sludge." Have them present the findings in a graph format and discuss the implications of their data.

TIME SCIENCE AND Society

SCIENCE ISSUES THAT AFFECT YOU!

Creating Wetlands to Purify Wastewater

Pebbles were added to the Corrales wetlands to help with drainage.

Water irises thrived in the wetlands, less than a year after planting.

Students enjoy pure water from the Corrales wetlands after it is filtered.

When you wash your hands or flush the toilet, do you think about where the wastewater goes? In most places, it eventually ends up being processed in a traditional sewage-treatment facility. But some places are experimenting with a new method that processes wastewater by creating wetlands. Wetlands are home to filtering plants, such as cattails, and sewage-eating bacteria.

In 1996, school officials at the Corrales Elementary School in Albuquerque, New Mexico, faced a big problem. The old wastewater-treatment system had failed. Replacing it was going to cost a lot of money. Instead of constructing a new sewage-treatment plant, school officials decided to create a natural wetlands system. The wetlands system could do the job less expensively, while protecting the environment.

Today, this wetlands efficiently converts polluted water into cleaner water that's good for the environment. U.S. government officials are monitoring this alternative sewage-treatment system to see if it is successful. So far, so good!

Wetlands filter water through the actions of the plants and microorganisms that live there. When plants absorb water into their roots, some also take up pollutants. The plants convert the pollutants to forms that are not dangerous. At the same time, bacteria and other microorganisms are filtering water as they feed. Water moves slowly through wetlands, so the organisms have plenty of time to do their work. Wetlands built by people to filter small amounts of pollutants are called "constructed wetlands". In many places, constructed wetlands are better at cleaning wastewater than sewers or septic systems.

Visit and Observe Visit a wetlands and create a field journal of your observations. Draw the plants and animals you see. Use a field guide to help identify the wildlife. If you don't live near a wetlands, use resources to research wetlands environments.

Science Online
For more information, visit booke.msscience.com/time

Visit and Observe Lead a discussion on how a constructed wetland benefits wildlife. In coastal areas, the discussion may include the importance of marshes to the ocean food chain. Near river systems, the effect on wildlife can be discussed in light of the flood control function of wetlands. Discuss how the decline in total wetland acreage can affect the food chain in general.

Resources for Teachers and Students

Constructed Wetlands in the Sustainable Landscape by Craig S. Campbell and Michael H. Ogden. New York, NY: John Wiley & Sons, 1999

U.S. Department of Agriculture National Agricultural Library Water Quality Information Center 10301 Baltimore Avenue, Beltsville, MD 20705

Reviewing Main Ideas

Section 1 How Ecosystems Change

1. Ecological succession is the gradual change from one plant community to another.
2. Primary succession begins in a place where no plants were before.
3. Secondary succession begins in a place that has soil and was once the home of living organisms.
4. A climax community has reached a stable stage of ecological succession.

Section 2 Biomes

1. Temperature and precipitation help determine the climate of a region.
2. Large geographic areas with similar climax communities are called biomes.
3. Earth's land biomes include tundra, taiga, temperate deciduous forest, temperate rain forest, tropical rain forest, grassland, and desert.

Section 3 Aquatic Ecosystems

1. Freshwater ecosystems include streams, rivers, lakes, ponds, and wetlands.
2. Wetlands are areas that are covered with water most of the year. They are found in regions that lie between land-masses and water.
3. Saltwater ecosystems include estuaries, sea-shores, coral reefs, a few inland lakes, and the deep ocean.
4. Estuaries are fertile transitional zones between freshwater and saltwater environments.

Visualizing Main Ideas

Copy and complete this concept map about land biomes.

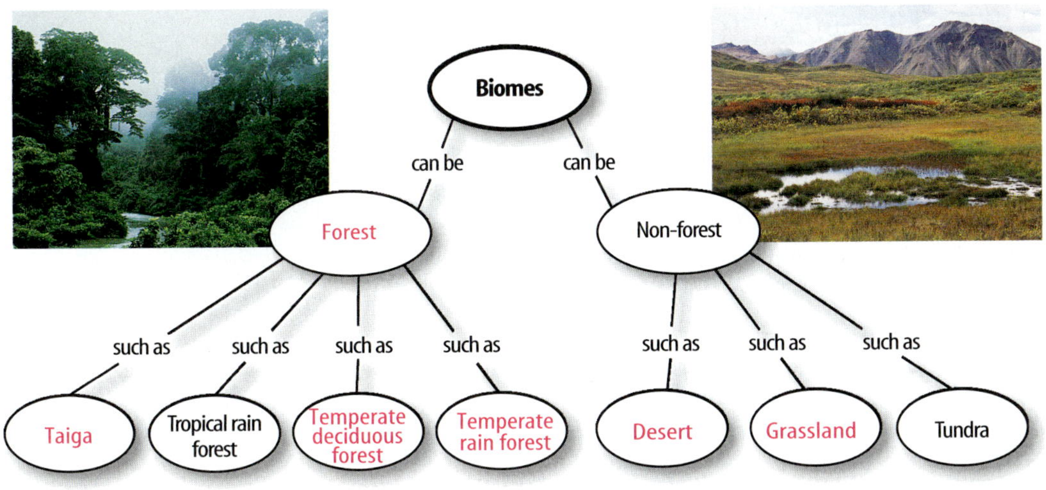

Identifying Misconceptions

Assess

Use the assessment as follow-up to page F at the beginning of this chapter.

Discussion Ask students to think of components that go into a particular ecosystem, such as a marine or forest ecosystem. Have them draw connections between the components. Have the students pick a few components to remove. Ask students to explain what happens to the flow of energy and/or cycling of materials if these components are removed. While removing some components may not cause the system to collapse, it may not be able to function optimally without them.

Reviewing Main Ideas

Summary statements can be used by students to review the major concepts of the chapter.

Visualizing Main Ideas

See student page.

Science Online

Visit booke.msscience.com
/self_check_quiz
/interactive_tutor
/vocabulary_puzzlemaker
/chapter_review
/standardized_test
/field_guide

Assessment Transparency

For additional assessment questions, use the *Assessment Transparency* located in the transparency book.

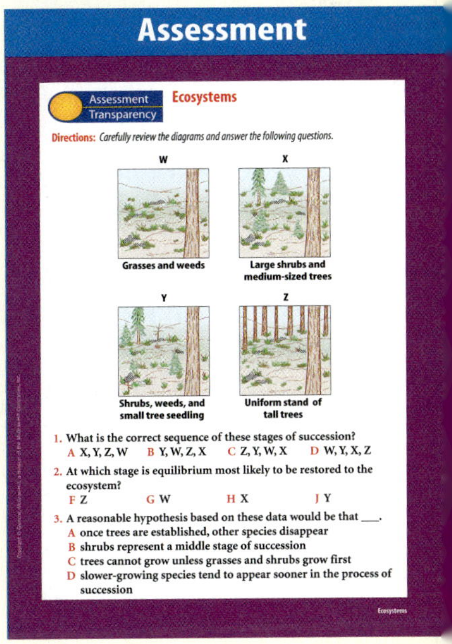

Assessment

Assessment Transparency **Ecosystems**

Directions: *Carefully review the diagrams and answer the following questions.*

W: Grasses and weeds

X: Large shrubs and medium-sized trees

Y: Shrubs, weeds, and small tree seedling

Z: Uniform stand of tall trees

1. What is the correct sequence of these stages of succession?
   A X, Y, Z, W B Y, W, Z, X C Z, Y, W, X D W, Y, X, Z
2. At which stage is equilibrium most likely to be restored to the ecosystem?
   F Z G W H X J Y
3. A reasonable hypothesis based on these data would be that ___.
   A once trees are established, other species disappear
   B shrubs represent a middle stage of succession
   C trees cannot grow unless grasses and shrubs grow first
   D slower-growing species tend to appear sooner in the process of succession

Ecosystems

Using Vocabulary

1. succession
2. climax community
3. temperate deciduous forest
4. temperate rain forest
5. tropical rain forests
6. estuary

Checking Concepts

7. B
8. A
9. B
10. C
11. B
12. A
13. B
14. D

chapter 3 Review

Using Vocabulary

biome p. 68	succession p. 64
climax community p. 67	taiga p. 70
coral reef p. 81	temperate deciduous forest p. 71
desert p. 74	temperate rain forest p. 71
estuary p. 82	tropical rain forest p. 72
grassland p. 75	tundra p. 69
intertidal zone p. 82	wetland p. 79
pioneer species p. 64	

Fill in the blanks with the correct vocabulary word or words.

1. ________ refers to the normal changes in the types of species that live in communities.
2. A(n) ________ is a group of organisms found in a stable stage of succession.
3. Deciduous trees are dominant in the ________.
4. The average temperature in ________ is between 9°C and 12°C.
5. ________ are the most biologically diverse biomes in the world.
6. A(n) ________ is an area where freshwater meets the ocean.

Checking Concepts

Choose the word or phrase that best answers the question.

7. What are tundra and desert examples of?
   A) ecosystems **C)** habitats
   B) biomes **D)** communities
8. What is a hot, dry biome called?
   A) desert **C)** coral reef
   B) tundra **D)** grassland
9. Where would organisms that are adapted to live in slightly salty water be found?
   A) lake **C)** open ocean
   B) estuary **D)** intertidal zone
10. Which biome contains mostly frozen soil called permafrost?
    A) taiga
    B) temperate rain forest
    C) tundra
    D) temperate deciduous forest
11. A new island is formed from a volcanic eruption. Which species probably would be the first to grow and survive?
    A) palm trees **C)** grasses
    B) lichens **D)** ferns
12. What would the changes in communities that take place on a recently formed volcanic island best be described as?
    A) primary succession
    B) secondary succession
    C) tertiary succession
    D) magma
13. What is the stable end stage of succession?
    A) pioneer species
    B) climax community
    C) limiting factor
    D) permafrost

Use the illustration below to answer question 14.

Observed Fire Danger Class—June, 2003

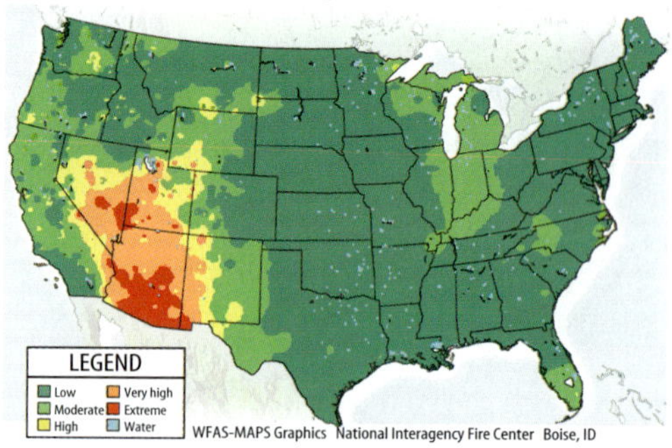

14. Which area of the U.S. had the highest observed fire danger on June 20, 2003?
    A) northeast **C)** northwest
    B) southeast **D)** southwest

Science Online booke.msscience.com/vocabulary_puzzlemaker

Use the Exam*View*® Pro Testmaker CD-ROM to:

- create multiple versions of tests
- create modified tests with one mouse click for inclusion students
- edit existing questions and add your own questions
- build tests aligned with state standards using built-in State Curriculum Tags
- change English tests to Spanish with one mouse click and vice versa

Thinking Critically

15. **Explain** In most cases, would a soil sample from a temperate deciduous forest be more or less nutrient-rich than a soil sample from a tropical rain forest?

16. **Explain** why some plant seeds need fire in order to germinate. How does this give these plants an advantage in secondary succession?

17. **Determine** A grassy meadow borders a beech-maple forest. Is one of these ecosystems undergoing succession? Why?

18. **Infer** why tundra plants are usually small.

19. **Make and Use a Table** Copy and complete the following table about aquatic ecosystems. Include these terms: *intertidal zone, lake, pond, coral reef, open ocean, river, estuary,* and *stream.*

Aquatic Ecosystems

Saltwater	Freshwater
intertidal zone	lake
coral reef	pond
open ocean	river
estuary	stream

20. **Recognize Cause and Effect** Wildfires like the one in Yellowstone National Park in 1988, cause many changes to the land. Determine the effect of a fire on an area that has reached its climax community.

Performance Activities

21. **Oral Presentation** Research a biome not in this chapter. Find out about its climate and location, and which organisms live there. Present this information to your class.

Applying Math

Use the table below to answer question 22.

Rainfall Amounts

Biome	Average Precipitation/Year (cm)
Taiga	50
Temperate rain forest	200
Tropical rain forest	400
Desert	25
Temperate deciduous forest	150
Tundra	25

22. **Biome Precipitation** How many times more precipitation does the tropical rain forest biome receive than the taiga or desert?

Use the graph below to answer question 23.

Average Temperature Range of Biomes

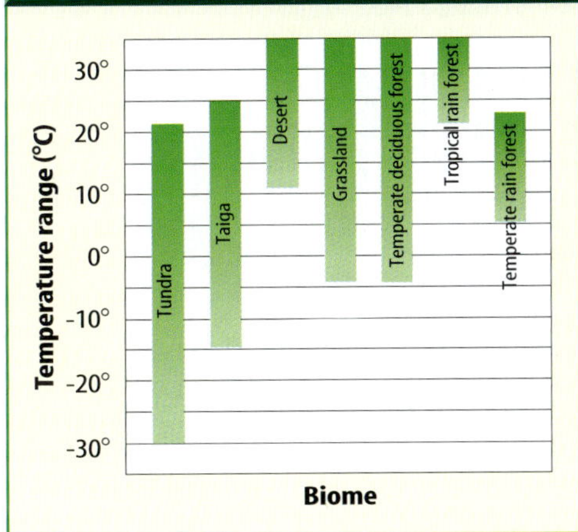

23. **Biome Temperatures** According to the graph, which biome has the greatest and which biome has the least variation in temperature throughout the year? Estimate the difference between the two.

Science Online booke.msscience.com/chapter_review

Assessment Resources

Reproducible Masters

Chapter *Fast File* Resources
- Chapter Review, pp. 35–36
- Chapter Tests, pp. 37–40
- Assessment Transparency Activity, p. 47

Glencoe Science Web site
- Chapter Review Test
- Standardized Test Practice

Glencoe Technology
- Assessment Transparency
- Exam*View*® Pro Testmaker
- MindJogger Videoquiz
- Interactive Chalkboard

chapter 3 Review

Thinking Critically

15. Temperate deciduous forest would have more nutrient-rich soil than the rain forest. Nutrients are cycled much faster in the rain forest.

16. Some seeds have thick seed coats that require scarring by fire, or are in cones that open and release the seeds only as a result of fire. When fire destroys other plants, these seeds sprout first and growth is rapid because there is less competition for resources.

17. The grassy meadow is undergoing succession towards becoming a forest.

18. Less than 25 cm of precipitation falls; sunlight is limited, and temperatures are too cold for plant growth most of the year.

19. See student page.

20. Fire will cause the climax community to undergo secondary succession and different plants can grow. Ecosystems not in climax communities are more diverse.

Performance Activities

21. Answers will vary.

Applying Math

National Math Standards

5

22. 8 times more than taiga, 16 times more than desert

23. tundra has the greatest variation, tropical rain forest has the least variation; about 34 cm/yr difference

FAST FILE

Answer Sheet A practice answer sheet can be found at booke.msscience.com/answer_sheet.

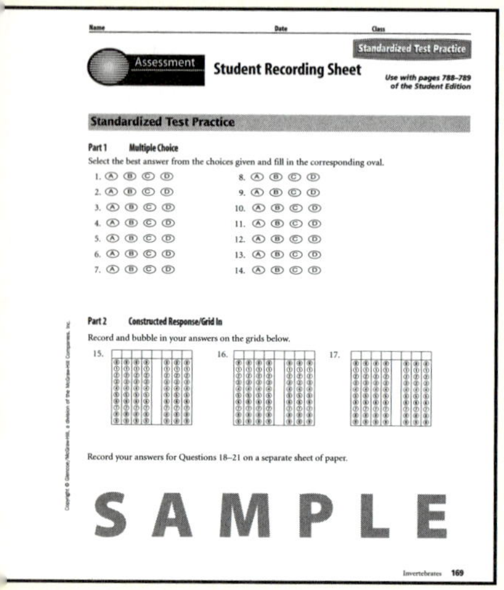

Part 1 Multiple Choice

1. B
2. B
3. A
4. D
5. C
6. C
7. B

chapter 3 Standardized Test Practice

Part 1 Multiple Choice

Record your answers on the answer sheet provided by your teacher or on a sheet of paper.

1. What two factors are most responsible for limiting life in a particular area?
   A. sunlight and temperature
   B. precipitation and temperature
   C. precipitation and sunlight
   D. soil conditions and precipitation

2. Which of the following forms during primary succession?
   A. trees
   B. soil
   C. wildlife
   D. grasses

Use the illustrations below to answer questions 3 and 4.

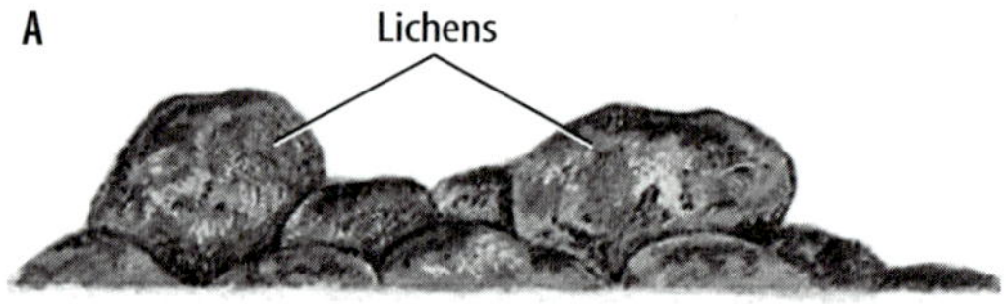

3. Which of the following statements best describes what is represented by A?
   A. Primary succession is occurring.
   B. Secondary succession is occurring.
   C. A forest fire has probably occurred.
   D. The climax stage has been reached.

4. Which of the following statements best describes what is represented by B?
   A. The climax stage has been reached.
   B. Pioneer species are forming soil.
   C. Bare rock covers most of the area.
   D. Secondary succession is occurring.

Use the map below to answer questions 5 and 6.

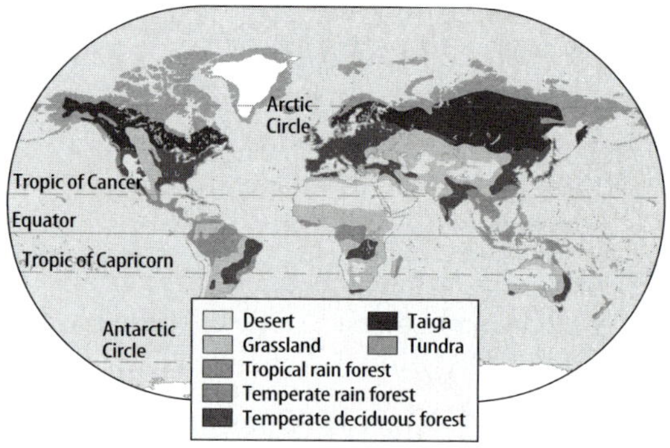

5. What biome is located in the latitudes just south of the north pole?
   A. taiga
   B. temperate deciduous rain forest
   C. tundra
   D. temperate rain forest

6. The tropical rainforest biome is found primarily near the
   A. Arctic Circle.
   B. Tropic of Cancer.
   C. equator.
   D. Tropic of Capricorn.

7. Which of the following is composed of a mix of salt water and freshwater?
   A. an intertidal zone
   B. an estuary
   C. a seashore
   D. a coral reef

Test-Taking Tip

Come Prepared Bring at least two sharpened No. 2 pencils and a good eraser to the test. Check to make sure that your eraser completely removes all pencil marks.

Part 2 Short Response/Grid In

8. cranberries, fish, shellfish
9. Primary succession takes longer because soil must be formed first before plants and other organisms can survive in the area.
10. taiga; The lynx's coat protects it from the cold temperatures, and its broad feet help it move through the snow.
11. desert; the scorpion does not require a lot of water and it is active at night, when temperatures are cooler.
12. most rainfall—tropical rain forest; least rainfall—desert and tundra
13. Grasslands do not receive enough precipitation for most forests to develop.
14. swamp, bog, marsh; fish, beavers, alligators, migratory birds
15. plankton; They are the foundation of the ocean food chain.

Part 2 Short Response/Grid In

Record your answers on the answer sheet provided by your teacher or on a sheet of paper.

8. Name two products that come from wetlands.

9. Which takes longer, primary succession or secondary succession? Why?

Use the photos below to answer questions 10 and 11.

10. In what biome would you most likely find A? How is this animal adapted to survive in its biome?

11. In what biome would you most likely find B? How is this animal adapted to survive in its biome?

12. Which biome receives the most rainfall per year? Which receives the least rainfall?

13. Why are forests unlikely to develop in grasslands?

14. What are two kinds of wetlands? What kinds of animals are found in wetlands?

15. What organisms inhabit the upper zone of the open ocean and why are they so important?

Part 3 Open Ended

Record your answers on a sheet of paper.

16. Explain how lichens contribute to the process of soil formation.

17. Compare and contrast a freshwater lake ecosystem with a freshwater pond ecosystem.

18. What special adaptations must all of the organisms that live in the intertidal zone have?

19. What are the differences between the temperate rain forest biome and the tropical rain forest biome?

Use the illustration below to answer questions 20 and 21.

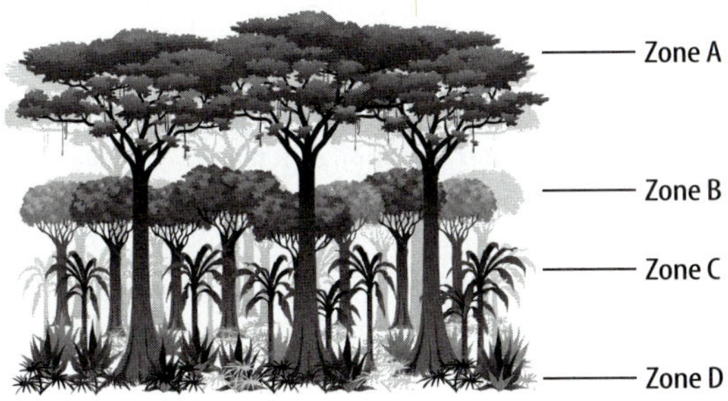

20. Identify and describe zone C in the diagram. What kinds of wildlife are found there?

21. Identify zone D and zone A. Describe the environment in each zone. Why might an organism that lives in zone A not be able to survive in zone D?

22. Discuss the effects of human impact on freshwater environments like lakes and ponds.

Part 3 Open Ended

16. They break down rocks. When the lichens die, they decay, adding organic matter to the rock. This is the beginning of new soil.

17. Both are aquatic ecosystems. Lakes are deeper and larger than ponds. Deep parts of the lake are colder than the shallow parts so different organisms live in each part of the lake. Ponds can be completely filled with plant material, with open water located at the center.

18. They must be able to withstand significant changes in temperature, moisture, and salinity, as well as the force of wave action.

19. The temperate rain forest receives less rainfall than the tropical rain forest and the temperature is not as high. There is greater biological diversity in the tropical rain forest.

20. Understory. It is dark and cool and provides a habitat for insects, reptiles, amphibians, and some mammals.

21. Zone A is the emergent zone. Organisms that live there require a lot of sunlight and space. Zone D, the forest floor, would be too dark for organisms that require a lot of sunlight to survive.

22. Runoff from fertilized lawns and farmland, along with sewage and other pollutants, eventually reaches lakes and ponds where algae and plants grow excessively. This reduces the oxygen level in the water, which makes it difficult for other organisms to live.

Rubrics

For more help evaluating open-ended assessment questions, see the rubric on p. 10T.

chapter 4 Organizer

Section/Objectives	Standards		Labs/Features
	National	**State/Local**	
Chapter Opener	See pp. 16T–17T for a Key to Standards.		**Launch Lab:** What happens when topsoil is left unprotected?, p. 93 **Foldables,** p. 93
Section 1 Resources 2 sessions 1 block 1. **Compare** renewable and nonrenewable resources. 2. **List** uses of fossil fuels. 3. **Identify** alternatives to fossil fuel use.	National Content Standards: UCP.2, UCP.3, A.1, A.2, F.2		**MiniLAB:** Observing Mineral Mining Effects, p. 96 **Integrate Social Studies,** p. 97 **Visualizing Solar Energy,** p. 101
Section 2 Pollution 2 sessions 1 block 4. **Describe** types of air pollution. 5. **Identify** causes of water pollution. 6. **Explain** methods that can be used to prevent erosion.	National Content Standards: UCP.2, UCP.3, A.1, A.2, F.2		**MiniLAB:** Measuring Acid Rain, p. 103 **Science Online,** p. 104 **Integrate Health,** p. 106 **Lab:** The Greenhouse Effect, p. 111
Section 3 The Three Rs of Conservation 3 sessions 1.5 blocks 7. **Recognize** ways you can reduce your use of natural resources. 8. **Explain** how you can reuse resources to promote conservation. 9. **Describe** how many materials can be recycled.	National Content Standards: UCP.2, UCP.3, A.1, A.2, F.2		**Science Online,** p. 114 **Applying Science:** What items are you recycling at home?, p. 114 **Lab:** Solar Cooking, p. 116 **Science and Language Arts:** Beauty Plagiarized, p. 118

Glencoe Exclusive! TeacherWorks™ All-In-One Planner and Resource Center

Lab Materials	Reproducible Resources	Section Assessment	Technology
Launch Lab: moist sand, potting soil, plastic basin or aluminum-foil baking pan, moss, water, spray bottle, beaker	**Chapter *FAST FILE* Resources** Foldables Worksheet, p. 17 Directed Reading Overview, p. 19 Note-taking Worksheets, pp. 33–35	Glencoe's Assessment Advantage	**TeacherWorks** includes: • Interactive Teacher Edition • Lesson Planner with calendar • Access to all program blacklines • Correlations to standards • Web links
MiniLAB: chocolate-chip cookie, paper plate, toothpick *Need materials?* Contact Science Kit at 1-800-828-7777 or www.sciencekit.com on the Internet.	**Chapter *FAST FILE* Resources** Transparency Activity, p. 44 MiniLAB, p. 3 Enrichment, p. 30 Reinforcement, p. 27 Directed Reading, p. 20 **Cultural Diversity,** p. 41 **Mathematics Skill Activities,** p. 9	**Portfolio** Activity, p. 101 **Performance** MiniLAB, p. 96 Applying Math, p. 100 **Content** Section Review, p. 100	Section Focus Transparency Virtual Labs CD-ROM Guided Reading Audio Program Interactive Chalkboard CD-ROM
MiniLAB: cup, rainwater, pH indicator paper, tap water, distilled water **Lab:** 1-L clear plastic soft-drink bottle with top cut off and label removed (2), 2 thermometers, potting soil, masking tape, plastic wrap, rubber band, lamp with 100-W lightbulb, watch or clock with second hand	**Chapter *FAST FILE* Resources** Transparency Activity, p. 45 MiniLAB, p. 4 Enrichment, p. 31 Reinforcement, p. 28 Directed Reading, p. 21 Lab Worksheet, pp. 5–6 Lab Activity, pp. 9–12 Transparency Activity, pp. 47–48	**Portfolio** Assessment, p. 110 **Performance** MiniLAB, p. 103 Applying Math, p. 106 **Content** Section Review, p. 106	Section Focus Transparency Teaching Transparency Virtual Labs CD-ROM Guided Reading Audio Program Interactive Chalkboard CD-ROM Video Lab
Lab: poster board, cardboard boxes, aluminum foil, string, wire coat hangers, clear plastic sheets, black cookware, thermometer, stopwatch, glue, tape, scissors	**Chapter *FAST FILE* Resources** Transparency Activity, p. 46 Enrichment, p. 32 Reinforcement, p. 29 Directed Reading, pp. 21, 22 Lab Worksheet, pp. 7–8 Lab Activity, pp. 13–16 **Lab Management and Safety,** p. 37	**Portfolio** Challenge, p. 114 **Performance** Applying Science, p. 114 Applying Skills, p. 115 **Content** Section Review, p. 115	Section Focus Transparency Virtual Labs CD-ROM Guided Reading Audio Program Interactive Chalkboard CD-ROM

End of Chapter Assessment

Glencoe's Assessment Advantage

Blackline Masters	Technology	Professional Series
Chapter *FAST FILE* Resources Chapter Review, pp. 37–38 Chapter Tests, pp. 39–42 **Standardized Test Practice,** pp. 19–22	MindJogger Videoquiz Virtual Labs CD-ROM Exam*View*® Pro Testmaker TeacherWorks CD-ROM Interactive Chalkboard CD-ROM	**Performance Assessment in the Science Classroom (PASC)**

chapter 4 Conserving Resources

Transparencies

Section Focus

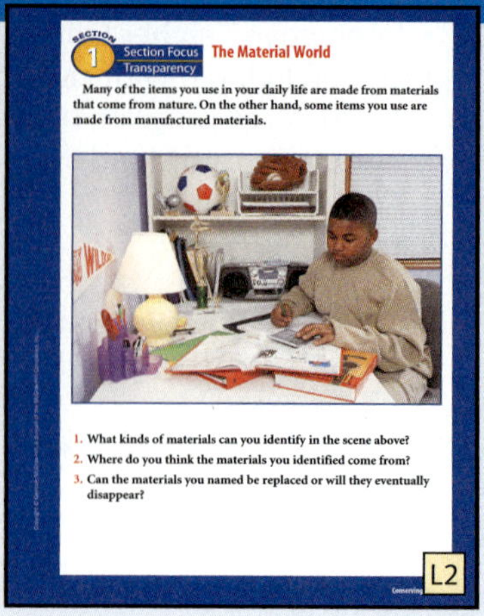
Section 1 Section Focus Transparency **The Material World**

Many of the items you use in your daily life are made from materials that come from nature. On the other hand, some items you use are made from manufactured materials.

1. What kinds of materials can you identify in the scene above?
2. Where do you think the materials you identified come from?
3. Can the materials you named be replaced or will they eventually disappear?

L2

Section 2 Section Focus Transparency **Ocean-Sized Ice Cubes**

One of the concerns with rising worldwide temperatures is the melting of ice at Earth's poles. Ice floating in water, as found in the Arctic Ocean, won't raise sea levels if it melts. But ice-sheets on land, as found in Greenland and Antarctica, do contain enough water to raise sea levels if they melt.

1. What is happening in the picture above?
2. What might cause ice sheets to melt?
3. Do you think human activity could affect the amount of ice that melts in Antarctica? Why or why not?

L2

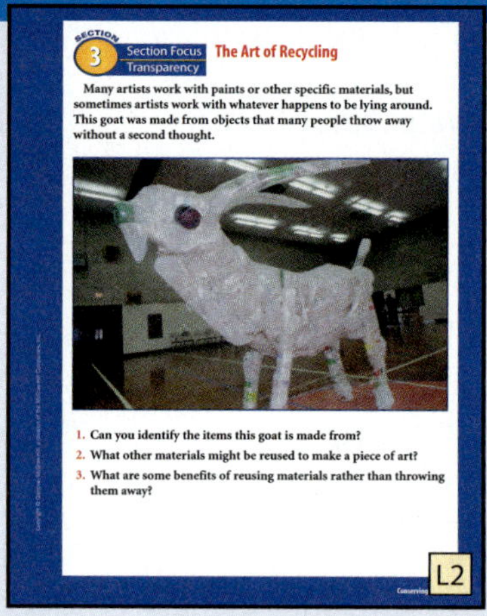
Section 3 Section Focus Transparency **The Art of Recycling**

Many artists work with paints or other specific materials, but sometimes artists work with whatever happens to be lying around. This goat was made from objects that many people throw away without a second thought.

1. Can you identify the items this goat is made from?
2. What other materials might be reused to make a piece of art?
3. What are some benefits of reusing materials rather than throwing them away?

L2

This is a representation of key blackline masters available in the Teacher Classroom Resources. See Resource Manager boxes within the chapter for additional information.

Assessment

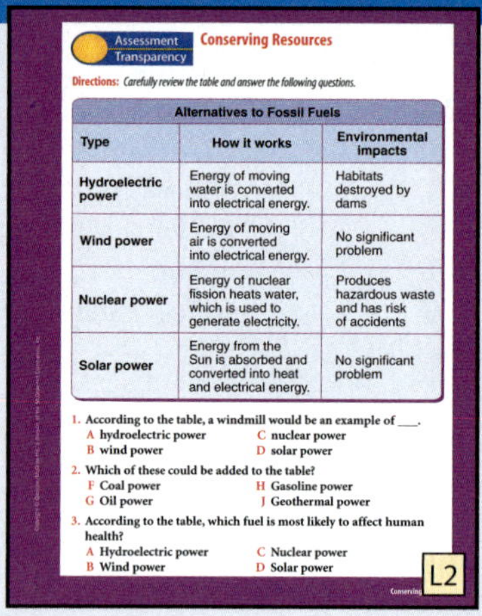
Assessment Transparency **Conserving Resources**

Directions: Carefully review the table and answer the following questions.

Alternatives to Fossil Fuels		
Type	How it works	Environmental impacts
Hydroelectric power	Energy of moving water is converted into electrical energy.	Habitats destroyed by dams
Wind power	Energy of moving air is converted into electrical energy.	No significant problem
Nuclear power	Energy of nuclear fission heats water, which is used to generate electricity.	Produces hazardous waste and has risk of accidents
Solar power	Energy from the Sun is absorbed and converted into heat and electrical energy.	No significant problem

1. According to the table, a windmill would be an example of ___.
   A hydroelectric power C nuclear power
   B wind power D solar power
2. Which of these could be added to the table?
   F Coal power H Gasoline power
   G Oil power J Geothermal power
3. According to the table, which fuel is most likely to affect human health?
   A Hydroelectric power C Nuclear power
   B Wind power D Solar power

L2

Teaching

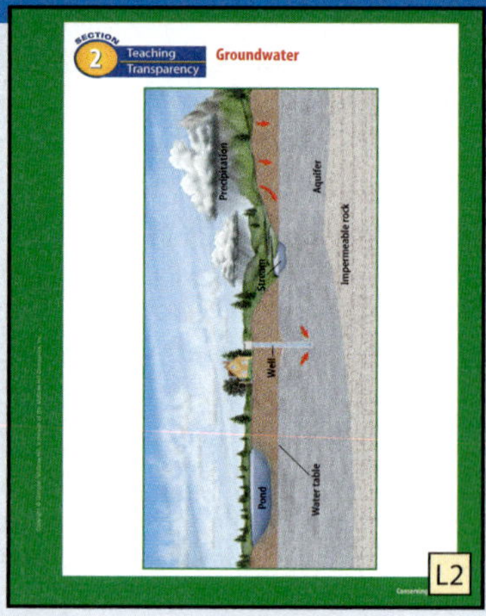
Section 2 Teaching Transparency **Groundwater**

L2

Key to Teaching Strategies

The following designations will help you decide which activities are appropriate for your students.

- **L1** Level 1 activities should be appropriate for students with learning difficulties.
- **L2** Level 2 activities should be within the ability range of all students.
- **L3** Level 3 activities are designed for above-average students.
- **ELL** ELL activities should be within the ability range of English Language Learners.
- **COOP LEARN** Cooperative Learning activities are designed for small group work.
- **LS** Multiple Learning Styles logos, as described on page 12T, are used throughout to indicate strategies that address different learning styles.
- **P** These strategies represent student products that can be placed into a best-work portfolio.
- **PBL** Problem-Based Learning activities apply real-world situations to learning.

Hands-on Activities

Student Text Lab Worksheet

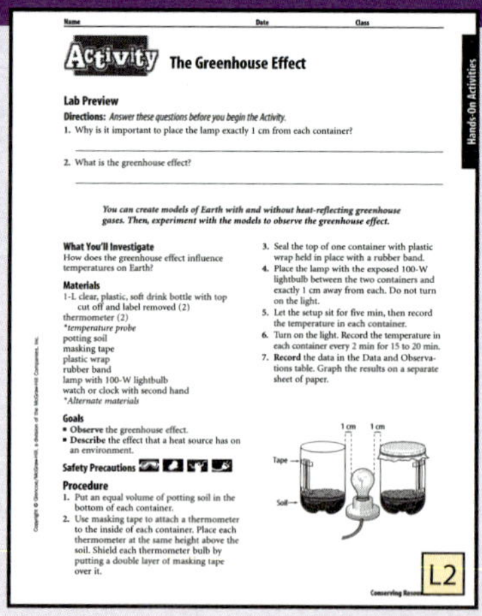
Activity **The Greenhouse Effect**

L2

Laboratory Activities

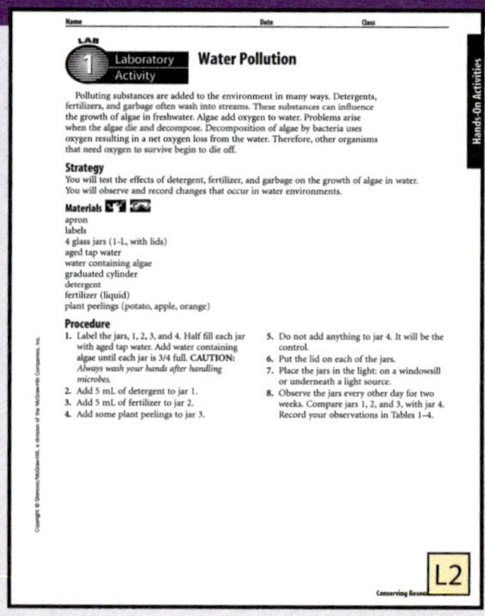
Laboratory Activity 1 **Water Pollution**

L2

Meeting Different Ability Levels

Content Outline

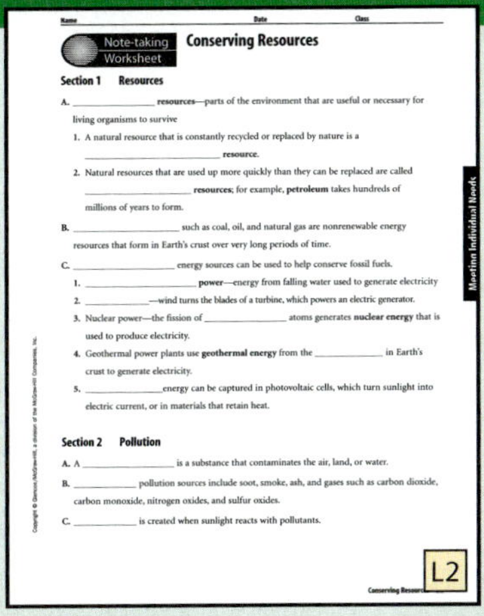

Reinforcement

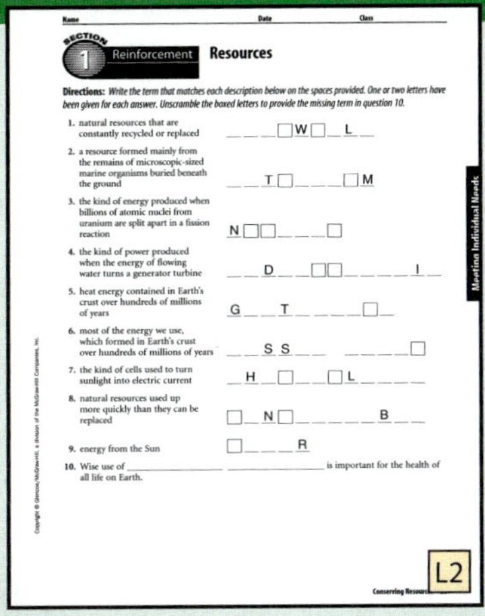

Enrichment

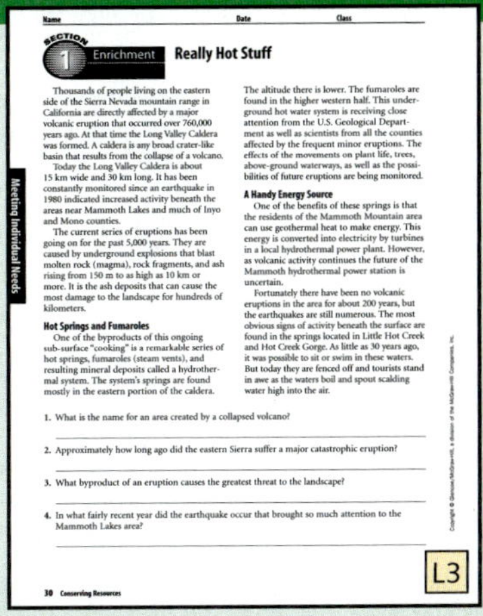

Directed Reading (English/Spanish)

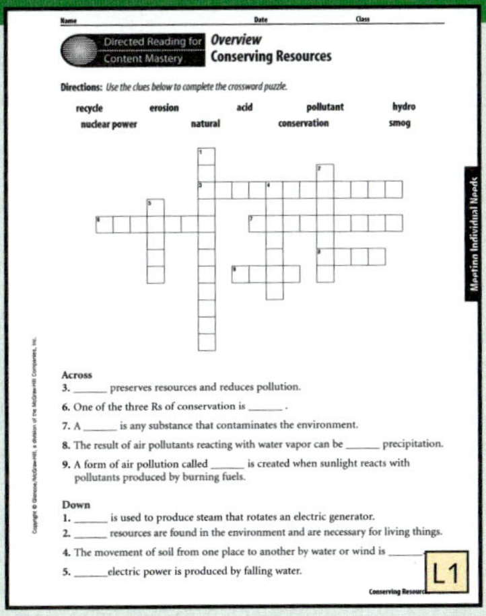

Study Guide

Reading Essentials

Assessment

Test Practice Workbook

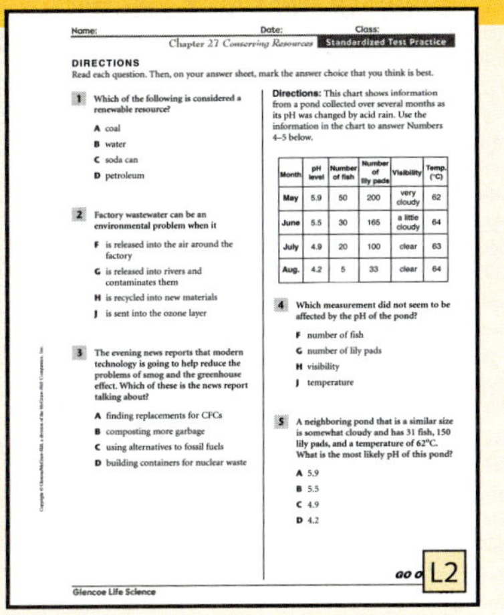

Chapter Review

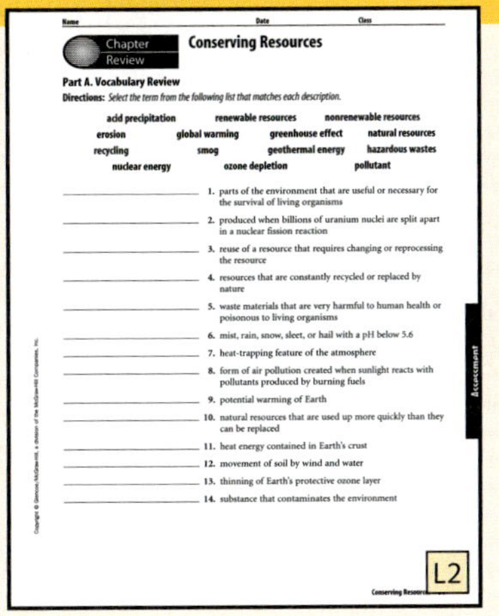

Chapter Tests

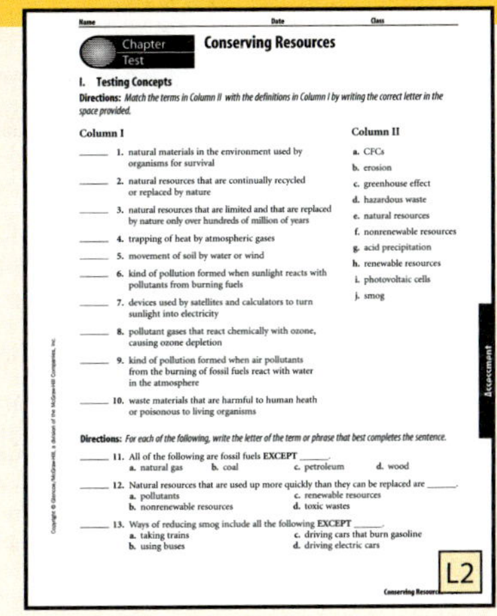

Science Content Background

section 1 Resources

Renewable Resources

Renewable resources include trees in forests; fish in lakes, rivers, and oceans; fertile agricultural soil; and freshwater in lakes and rivers. Nature replaces these resources, and they can be used forever as long as they are not overexploited in the short term. Some resources, like agricultural soil, can be considered renewable or nonrenewable. Soil is constantly replaced but at a very slow rate.

Nonrenewable Resources

Nonrenewable resources, which include minerals and fossil fuels, are present in limited supplies and are depleted by use. They are not replenished by natural processes within a reasonable amount of time. Fossil fuels, for example, take millions of years to form. Recycling nonrenewable resources can help sustain the supply of certain substances. In the future, most products will contain some recycled material.

Nuclear Power

In 1942, Enrico Fermi established the first successful controlled nuclear chain reaction. Nuclear fission reactions release a great deal of energy, several million times the energy of a typical chemical reaction. Nuclear reactors use uranium-235 for the production of power. Once started, the reactions are self-sustaining: the neutrons released by each splitting in turn split other atoms, causing a chain reaction. There is no danger of nuclear chain reactions starting in nature since uranium-235 makes up only 0.71% of the naturally occurring uranium. Most of the remainder is uranium-238, which is not fissible.

Teacher to Teacher

Steve Federman
Loveland Middle School
Loveland, Ohio

"To emphasize the effect of coal mining on our land, I give each student a soft cookie with raisins or chocolate chips and a toothpick. Their objective is to use a toothpick carefully to mine all the chocolate chips (raisins) out of the cookie while leaving the cookie intact."

Steve Federman

section 2 Pollution

Acid Precipitation

Acid precipitation is a worldwide problem. Reports of highly acidic precipitation have come from Canada, England, Germany, France, Scandinavia, and the United States. Precipitation is normally slightly acidic, with a pH between 5.6 and 5.7 that is caused by atmospheric carbonic acid. But acid precipitation sometimes has a concentration of acid 1,000 times higher than normal. The average precipitation in much of the northeastern part of the United States and adjoining parts of Ontario has a pH between 4.0 and 4.5.

Greenhouse Effect

Carbon dioxide is one of the most abundant of the greenhouse gases. It occurs as a natural consequence of respiration. However, carbon dioxide is emitted into the atmosphere as a waste product of energy production. Coal, oil, natural gas, and biomass are burned to provide heat and electricity for industrial processes, home heating, and cooking. These sources are increasing the amount of carbon dioxide in the atmosphere. Some evidence suggests that increased levels of carbon dioxide in the atmosphere may cause Earth's average temperature to increase.

Ozone Depletion

Ozone, a molecule made up of three oxygen atoms, traps excess ultraviolet radiation and prevents it from reaching Earth. In 1985, it was discovered that a significant thinning of the ozone layer over the Antarctic occurred during the southern hemisphere's spring. Some regions of the ozone layer showed 95 percent depletion. Ozone depletion is also occurring farther north. Measurements in arctic regions suggest a thinning of the ozone layer there. These findings have caused countries to stop producing ozone-destroying chemicals.

Soil Loss

Erosion is a natural process that has occurred since Earth was formed. Humans have little control over the erosion caused by glaciers, rivers, and oceans. A number of human activities, mainly farming and logging, have increased erosion. Scientists estimate that between 2 and 3 billion metric tons of soil are lost from farmlands in the United States each year.

Soil Pollution—Solid Wastes

About 90 percent of our solid wastes are disposed of on land, mostly in landfills. Several methods of producing energy from solid waste are used on a small scale and further research is being done in this area. One example is a combustion chamber lined with water lines in which selected wastes are burned and the heat is used to produce steam to power a generator.

Telegraph Colour Library/FPG International

section 3 The Three Rs of Conservation

Recycle

In 1999, the overall recycling rate in the U.S. was 28 percent, with the composting of yard waste accounting for 2/5 of this. The overall recycling rate of plastic soft-drink containers was only 34.9 percent, 26.3 percent for glass containers, and 57.2 percent for steel cans. Paper and paper products are the most voluminous waste products, and they make up about 42 percent of the solid waste produced in the United States.

chapter content resources

Internet Resources

For additional content background, visit **booke.msscience.com** to:

- access your book online
- find references to related articles in popular science magazines
- access Web links with related content background
- access current events with science journal topics

Print Resources

Understanding Garbage and Our Environment, by Andrea Nolan, McGraw-Hill Trade, 1998

Air Pollution (True Books), by Rhonda Lucas Donald, Children's Press, 2001

Global Warming: The Threat of the Earth's Changing Climate, by Laurence Pringle, SeaStar Books, 2003

chapter 4

Chapter Vocabulary

natural resource, p. 94
renewable resource, p. 94
nonrenewable resource, p. 95
petroleum, p. 95
fossil fuel, p. 96
hydroelectric power, p. 97
nuclear energy, p. 98
geothermal energy, p. 99
pollutant, p. 102
acid precipitation, p. 103
greenhouse effect, p. 104
ozone depletion, p. 105
erosion, p. 109
hazardous waste, p. 110
recycling, p. 113

Science Journal Woods, food crops, and soil are examples. Wood can provide shelter and home furnishings, food crops are needed for our own consumption and by other animals, and soil is needed by farmers to produce crops and it filters water.

INTERACTIVE CHALKBOARD with Image Bank

PowerPoint® Presentations

This CD-ROM is an editable Microsoft® PowerPoint® presentation that includes:

- a pre-made presentation for every chapter
- interactive graphics
- animations
- audio clips
- image bank
- all new section and chapter questions
- Standardized Test Practice
- transparencies
- pre-lab questions for all labs
- Foldables directions
- links to booke.msscience.com

chapter 4

Conserving Resources

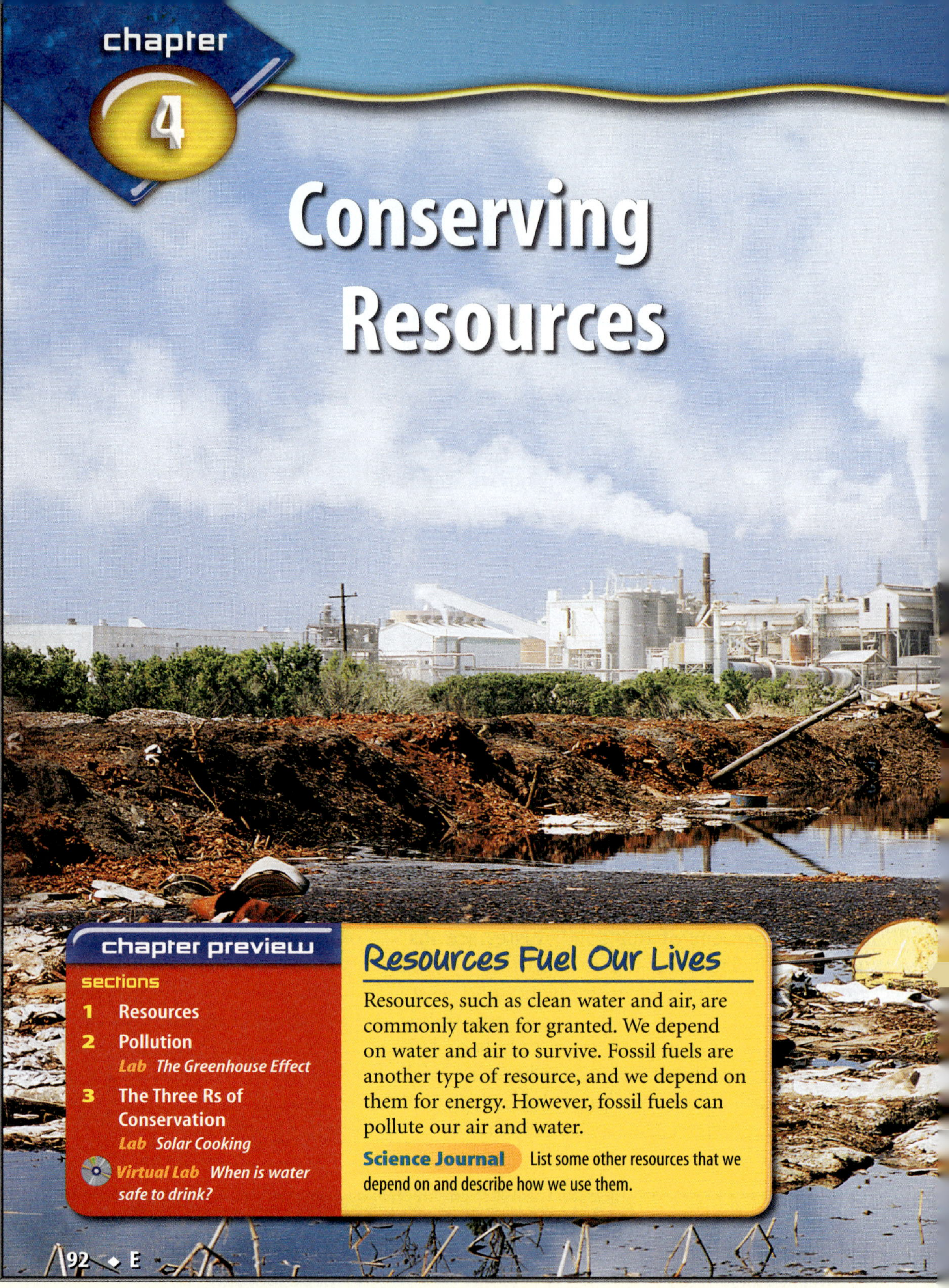

chapter preview

sections

1. Resources
2. Pollution
   Lab *The Greenhouse Effect*
3. The Three Rs of Conservation
   Lab *Solar Cooking*

Virtual Lab *When is water safe to drink?*

Resources Fuel Our Lives

Resources, such as clean water and air, are commonly taken for granted. We depend on water and air to survive. Fossil fuels are another type of resource, and we depend on them for energy. However, fossil fuels can pollute our air and water.

Science Journal List some other resources that we depend on and describe how we use them.

92 ◆ E

Theme Connection

Energy Even though the total amount of energy on Earth remains constant, the sources of usable energy are limited. Energy is often transformed from one form to another, such as fossil fuel energy into electricity and then into light energy.

About the Photo

Pollution Three major sources of pollution are municipal, industrial, and agricultural wastes. It is estimated that 1.5 billion people worldwide lack safe drinking water and at least 5 million people die from waterborne diseases each year.

Start-Up Activities

Launch LAB

What happens when topsoil is left unprotected?

Plants grow in the top, nutrient-rich layer, called topsoil. Plants help keep topsoil in place by protecting it from wind and rain. Try the following experiment to find out what happens when topsoil is left unprotected.

1. Use a mixture of moist sand and potting soil to create a miniature landscape in a plastic basin or aluminum-foil baking pan. Form hills and valleys in your landscape.
2. Use clumps of moss to cover areas of your landscape. Leave some sloping portions without plant cover.
3. Simulate a rainstorm over your landscape by spraying water on it from a spray bottle or by pouring a slow stream of water on it from a beaker.
4. **Think Critically** In your Science Journal, record your observations and describe what happened to the land that was not protected by plant cover.

Science Online | Preview this chapter's content and activities at booke.msscience.com

FOLDABLES Study Organizer

Resources Make the following Foldable to help you organize information and diagram ideas about renewable and nonrenewable resources.

STEP 1 **Fold** a sheet of paper in half lengthwise. Make the back edge about 5 cm longer than the front edge.

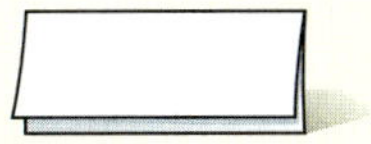

STEP 2 **Turn** the paper so the fold is on the bottom. Then **fold** in half.

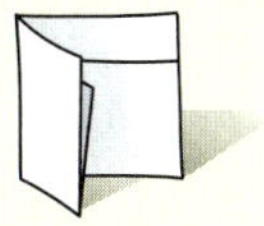

STEP 3 **Unfold and cut** only the top layer along the fold to make two tabs. **Label** the Foldable as shown.

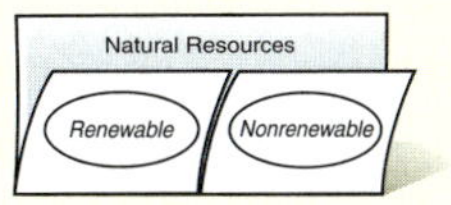

Make a Concept Map Before you read the chapter, list examples of each type of natural resource you already know on the back of the appropriate tabs. As you read the chapter, add to your lists.

Launch LAB

Purpose Use the Launch Lab to show students how plant cover protects topsoil. L2 ELL LS **Kinesthetic**

Preparation Mix one part sand and two parts topsoil.

Materials sand and topsoil mixture, containers, clumps of moss or grass, spray bottles, beaker

Teaching Strategies

- Have students vary the strength of the spray to simulate both light and heavy rainfall and compare the results.
- Point out that water flowing downhill is called "runoff." Have students note the formation of small streams and observe what happens to the land under and around these streams. Students should infer that this is how streams and rivers are formed.

Think Critically

Water washes the uncovered soil down slopes, forming gullies.

Assessment

Performance Have students reuse the materials to demonstrate the effects of terracing. Remove the plants and form two hills, one with a steady incline and one with a series of ridges (terraces) cut into the incline. Observe what happens as rain falls. Soil does not wash away on the incline with ridges. Use **Performance Assessment in the Science Classroom**, p. 89.

FOLDABLES Study Organizer **Dinah Zike Study Fold**

Student preparation materials for this Foldable are available in the **Chapter *FAST FILE* Resources.**

1 Motivate

INTERACTIVE CHALKBOARD
PowerPoint® Presentations

Bellringer

Section Focus Transparencies also are available on the Interactive Chalkboard CD-ROM. L2 ELL

SECTION 1 Section Focus Transparency **The Material World**

Many of the items you use in your daily life are made from materials that come from nature. On the other hand, some items you use are made from manufactured materials.

1. What kinds of materials can you identify in the scene above?
2. Where do you think the materials you identified come from?
3. Can the materials you named be replaced or will they eventually disappear?

Conserving Resources

Tie to Prior Knowledge

Oil Hold up a bottle of corn oil and a quart of motor oil. Ask students to explain how they are similar and different? Both are oil and come from living things. Corn is a renewable resource that grew in one summer. The motor oil is a nonrenewable resource that took millions of years to form from the remains of organisms.

section 1

Resources

as you read

What You'll Learn

- **Compare** renewable and nonrenewable resources.
- **List** uses of fossil fuels.
- **Identify** alternatives to fossil fuel use.

Why It's Important

Wise use of natural resources is important for the health of all life on Earth.

Review Vocabulary

geyser: a spring that emits intermittent jets of heated water and steam

New Vocabulary

- natural resource
- renewable resource
- nonrenewable resource
- petroleum
- fossil fuel
- hydroelectric power
- nuclear energy
- geothermal energy

Natural Resources

An earthworm burrowing in moist soil eats decaying plant material. A robin catches the worm and flies to a tree. The leaves of the tree use sunlight during photosynthesis. Leaves fall to the ground, decay, and perhaps become an earthworm's meal. What do these living things have in common? They rely on Earth's **natural resources**—the parts of the environment that are useful or necessary for the survival of living organisms.

What kinds of natural resources do you use? Like other organisms, you need food, air, and water. You also use resources that are needed to make everything from clothes to cars. Natural resources supply energy for automobiles and power plants. Although some natural resources are plentiful, others are not.

Renewable Resources The Sun, an inexhaustible resource, provides a constant supply of heat and light. Rain fills lakes and streams with water. When plants carry out photosynthesis, they add oxygen to the air. Sunlight, water, air, and the crops shown in **Figure 1** are examples of renewable resources. A **renewable resource** is any natural resource that is recycled or replaced constantly by nature.

Figure 1 Cotton and wood are renewable resources. Cotton cloth is used for rugs, curtains, and clothing. A new crop of cotton can be grown every year. Wood is used for furniture, building materials, and paper. It will take 20 years for these young trees to grow large enough to harvest.

Section 1 Resource Manager

Chapter FAST FILE Resources

- Transparency Activity, p. 44
- Directed Reading for Content Mastery, pp. 19, 20
- Note-taking Worksheets, pp. 33–35
- MiniLAB, p. 3
- Enrichment, p. 30
- Reinforcement, p. 27

Reading and Writing Skill Activities, p. 9
Cultural Diversity, p. 41
Mathematics Skill Activities, p. 9
Life Science Critical Thinking/Problem Solving, p. 23

94 ◆ E CHAPTER 4 Conserving Resources

Supply and Demand Even though renewable resources are recycled or replaced, they are sometimes in short supply. Rain and melted snow replace the water in streams, lakes, and reservoirs. Sometimes, there may not be enough rain or snowmelt to meet all the needs of people, plants, and animals. In some parts of the world, especially desert regions, water and other resources usually are scarce. Other resources can be used instead, as shown in **Figure 2.**

Nonrenewable Resources Natural resources that are used up more quickly than they can be replaced by natural processes are **nonrenewable resources.** Earth's supply of nonrenewable resources is limited. You use nonrenewable resources when you take home groceries in a plastic bag, paint a wall, or travel by car. Plastics, paints, and gasoline are made from an important nonrenewable resource called petroleum, or oil. **Petroleum** is formed mostly from the remains of microscopic marine organisms buried in Earth's crust. It is nonrenewable because it takes hundreds of millions of years for it to form.

Reading Check *What are nonrenewable resources?*

Minerals and metals found in Earth's crust are nonrenewable resources. Petroleum is a mineral. So are diamonds and the graphite in pencil lead. The aluminum used to make soft-drink cans is a metal. Iron, copper, tin, gold, silver, tungsten, and uranium also are metals. Many manufactured items, like the car shown in **Figure 3,** are made from nonrenewable resources.

Figure 2 In parts of Africa, firewood has become scarce. People in this village now use solar energy instead of wood for cooking.

Figure 3 Iron, a nonrenewable resource, is the main ingredient in steel. Steel is used to make cars, trucks, appliances, buildings, bridges, and even tires.
Infer *what other nonrenewable resources are used to build a car.*

2 Teach

Use an Analogy

Savings Account Compare nonrenewable resources to a noninterest-bearing savings account to which nothing is added. The depletion of the savings account depends on how fast withdrawals are made and how large they are. L2

Activity

Daily Resource Use Have pairs work together to make a list of the things they used or relied upon today. Have them categorize the items as food, transportation, clothing, or household items. Finally, have them make check marks by the items they think are renewable. Have groups compare and discuss their answers. L2 COOP LEARN LS **Interpersonal**

Reading Check

Answer resources that are used up more quickly than they can be replaced

IDENTIFYING Misconceptions

Recognize Nonrenewability Some students may not realize that fossil fuels and other mineral resources are nonrenewable. Point out that much of the coal used today was formed during the Carboniferous Period. Have students research how long ago that was. about 345–280 million years ago

Active Reading

News Summary Assign students the job of being television reporters. Allow different students to summarize the information on renewable and nonrenewable resources. Then have students present their summaries to the "television" audience.

Caption Answer

Figure 3 Most of the materials are nonrenewable and include chrome, plastic, and synthetic rubber.

Mini LAB

Purpose to "mine" a model mineral deposit and attempt to restore the site to its original condition L1 LS **Kinesthetic**

Materials chocolate-chip, oatmeal-raisin cookie, or nut filled brownie; paper plate; toothpick

Teaching Strategy Suggest students use drops of water to soften the surface if they have trouble removing the "minerals."

Analysis

1. Students will experience difficulty in the restoration process.
2. The minerals closer to the surface are easier to remove than those found deeper within Earth's crust.
3. An ecosystem could be permanently damaged by mining.

Assessment

Oral Which do you think is more expensive: mining at the surface or deep within Earth's crust? Explain. In general, surface mining is less expensive. It doesn't require as much machinery or time. Use **Performance Assessment in the Science Classroom**, p. 89.

Try at Home

Caption Answer

Figure 4 Coal, oil, and natural gas are fossil fuels that provided 84 percent of the energy used in the U.S. in 1999.

Mini LAB

Observing Mineral Mining Effects

Procedure

1. Place a **chocolate-chip cookie** on a **paper plate.** Pretend the chips are mineral deposits and the rest of the cookie is Earth's crust.
2. Use a **toothpick** to locate and dig up the mineral deposits. Try to disturb the land as little as possible.
3. When mining is completed, try to restore the land to its original condition.

Analysis

1. How well were you able to restore the land?
2. Compare the difficulty of digging for mineral deposits found close to the surface with digging for those found deep in Earth's crust.
3. Describe environmental changes that might result from a mining operation.

Try at Home

Fossil Fuels

Coal, oil, and natural gas are nonrenewable resources that supply energy. Most of the energy you use comes from these fossil fuels, as the graph in **Figure 4** shows. **Fossil fuels** are fuels formed in Earth's crust over hundreds of millions of years. Cars, buses, trains, and airplanes are powered by gasoline, diesel fuel, and jet fuel, which are made from oil. Coal is used in many power plants to produce electricity. Natural gas is used in manufacturing, for heating and cooking, and sometimes as a vehicle fuel.

Fossil Fuel Conservation Billions of people all over the world use fossil fuels every day. Because fossil fuels are nonrenewable, Earth's supply of them is limited. In the future, they may become more expensive and difficult to obtain. Also, the use of fossil fuels can lead to environmental problems. For example, mining coal can require stripping away thick layers of soil and rock, as shown in **Figure 4,** which destroys ecosystems. Another problem is that fossil fuels must be burned to release the energy stored in them. The burning of fossil fuels produces waste gases that cause air pollution, including smog and acid rain. For these reasons, many people suggest reducing the use of fossil fuels and finding other sources of energy.

You can use simple conservation measures to help reduce fossil fuel use. Switch off the light when you leave a room and turn off the television when you're not watching it. These actions reduce your use of electricity, which often is produced in power plants that burn fossil fuels. Hundreds of millions of automobiles are in use in the United States. Riding in a car pool or taking public transportation uses fewer liters of gasoline than driving alone in a car. Walking or riding a bicycle uses even less fossil fuel. Reducing fossil fuel use has an added benefit—the less you use, the more money you save.

Figure 4 Coal is a fossil fuel. It often is obtained by strip mining, which removes all the soil above the coal deposit. The soil is replaced, but it takes many years for the ecosystem to recover.

Identify *the resource that provided 84 percent of the energy used in the United States in 1999.*

Curriculum Connection

Social Studies Have students investigate ways petroleum was used in ancient cultures. Egyptians used petroleum to help preserve mummies. The Chinese used petroleum to heat and light their homes and to cook. Tar pits supplied crude oil that was used for fuel, cooking, and lighting in Latin America. At the same time, petroleum was used in Europe for lubricating wagon wheels and to make ointments. L3 LS **Naturalist**

Figure 5 Most power plants use turbine generators to produce electricity. In fossil fuel plants, burning fuel boils water and produces steam that turns the turbine.

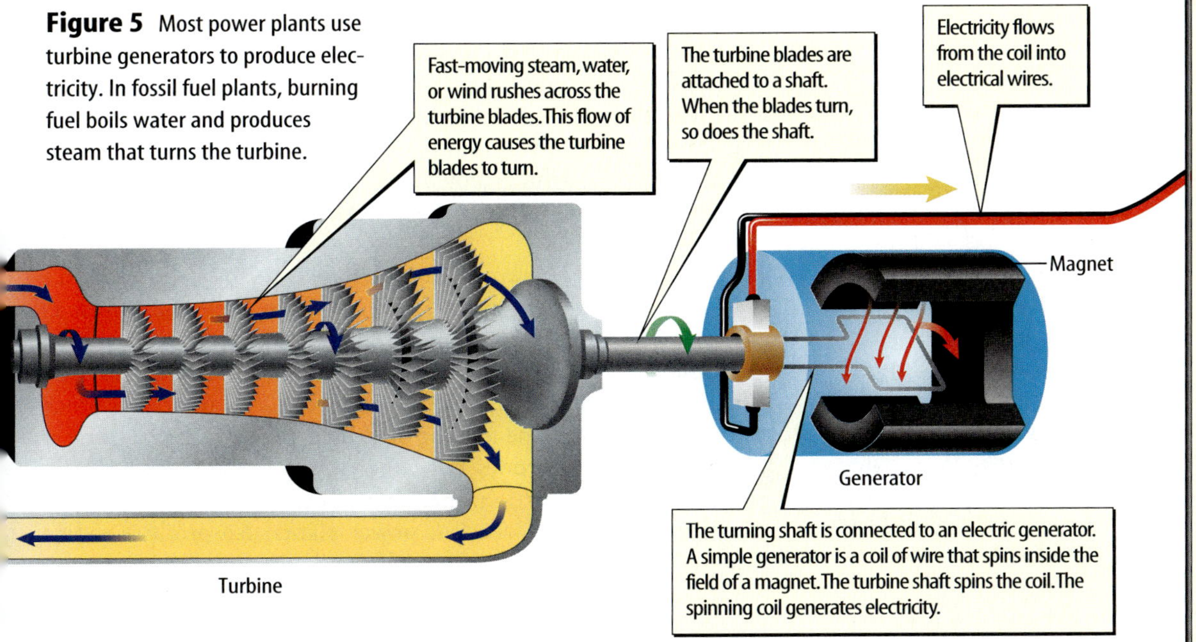

Alternatives to Fossil Fuels

Another approach to reducing fossil fuel use is to develop other sources of energy. Much of the electricity used today comes from power plants that burn fossil fuels. As **Figure 5** shows, electricity is generated when a rotating turbine turns a coil of wires in the magnetic field of an electric generator. Fossil-fuel power plants boil water to produce steam that turns the turbine. Alternative energy sources, including water, wind, and atomic energy can be used instead of fossil fuels to turn turbines. Also, solar cells can produce electricity using only sunlight, with no turbines at all. Some of these alternative energy sources—particularly wind and solar energy—are so plentiful they could be considered inexhaustible resources.

Water Power Water is a renewable energy source that can be used to generate electricity. **Hydroelectric power** is electricity that is produced when the energy of falling water is used to turn the turbines of an electric generator. Hydroelectric power does not contribute to air pollution because no fuel is burned. However, it does present environmental concerns. Building a hydroelectric plant usually involves constructing a dam across a river. The dam raises the water level high enough to produce the energy required for electricity generation. Many acres behind the dam are flooded, destroying land habitats and changing part of the river into a lake.

INTEGRATE Social Studies

Energy Oil and natural gas are used to produce over 60 percent of the energy supply in the United States. Over half of the oil used is imported from other countries. Many scientists suggest that emissions from the burning of fossil fuels are principally responsible for global warming. In your Science Journal, write what you might do to persuade utility companies to increase their use of water, wind, and solar power.

Visual Learning

Figure 5 Have students trace the process of producing electricity using a turbine generator. Then have students identify at which point of the process different energy sources can be used to generate the energy. at the first stage of the process, where the turbine blades are turned L2

Teacher FYI

Niagara Falls Students may wonder why they don't see the water actually turning turbines at waterfalls such as Niagara Falls. The reason is that water diverted before it actually gets to the falls is what goes through the turbines. During the night at Niagara Falls, more water is diverted than in the day in order to keep the daytime visitors happy.

Discussion

Hydroelectricity Hydroelectricity is a clean way to produce electricity, but should dams that flood thousands of acres be built? Have students discuss the advantages and disadvantages of this type of electricity production. Advantages: clean, doesn't use nonrenewable resources, provides inexpensive electricity; disadvantages: loss of natural habitats or human communities

INTEGRATE Social Studies

Energy Letters could be written to legislators asking them to propose or support bills that would reduce the use of fossil fuels and generate funds to support and improve the use of alternative energy.

Research Have students investigate new advances in technology that could promote the use of alternate energy sources such as wind and solar energy.

Differentiated Instruction

Visually Impaired Natural waterfalls are not numerous enough or high enough, so dams are often built to raise the water level. Height is an important determinant of the amount of hydroelectric power produced. Have students feel the effects of water dropped from a cup held one, two, and four feet above their hands. L1 LS **Kinesthetic**

Fun Fact

The movement of coils of wire within a magnetic field is how electricity is produced. This electricity is called alternating current (AC) as opposed to electricity from batteries or photovoltaic cells, which is called direct current (DC).

Make a Model

Turbines Use a pinwheel and a fan to show how wind can be used to turn a turbine. Give some pinwheels to students and ask them to find the spot on the school grounds that would be best for generating electricity using the wind. L1 ELL LS **Naturalist**

IDENTIFYING Misconceptions

Radiation Glows Some students may think that nuclear radiation makes a person glow or turn green. Combat this misconception by asking students if they turn green when they receive dental X rays, which are a form of radiation.

Discussion

Windmills Ask students to compare modern windmills with old-style windmills found in Holland, Spain, or the United States. They both have blades that are turned by the force of the wind. Modern windmills are more streamlined and generate electricity. Older ones were used to grind grain or pump water from underground.

Caption Answer

Figure 6 A fast-moving neutron from the nucleus of a uranium atom collides with another atom. The collision splits the atom, releasing more neutrons and producing heat.

Wind Power Wind power is another renewable energy source that can be used for electricity production. Wind turns the blades of a turbine, which powers an electric generator. When winds blow at least 32 km/h, energy is produced. Wind power does not cause air pollution, but electricity can be produced only when the wind is blowing. So far, wind power accounts for only a small percentage of the electricity used worldwide.

Nuclear Power Another alternative to fossil fuels makes use of the huge amounts of energy in the nuclei of atoms, as shown in **Figure 6. Nuclear energy** is released when billions of atomic nuclei from uranium, a radioactive element, are split apart in a nuclear fission reaction. This energy is used to produce steam that rotates the turbine blades of an electric generator.

Nuclear power does not contribute to air pollution. However, uranium is a nonrenewable resource, and mining it can disrupt ecosystems. Nuclear power plants also produce radioactive wastes that can seriously harm living organisms. Some of these wastes remain radioactive for thousands of years, and their safe disposal is a problem that has not yet been solved. Accidents also are a danger.

Figure 6 Nuclear power plants are designed to withstand the high energy produced by nuclear reactions.
Describe *how heat is produced in a nuclear reactor.*

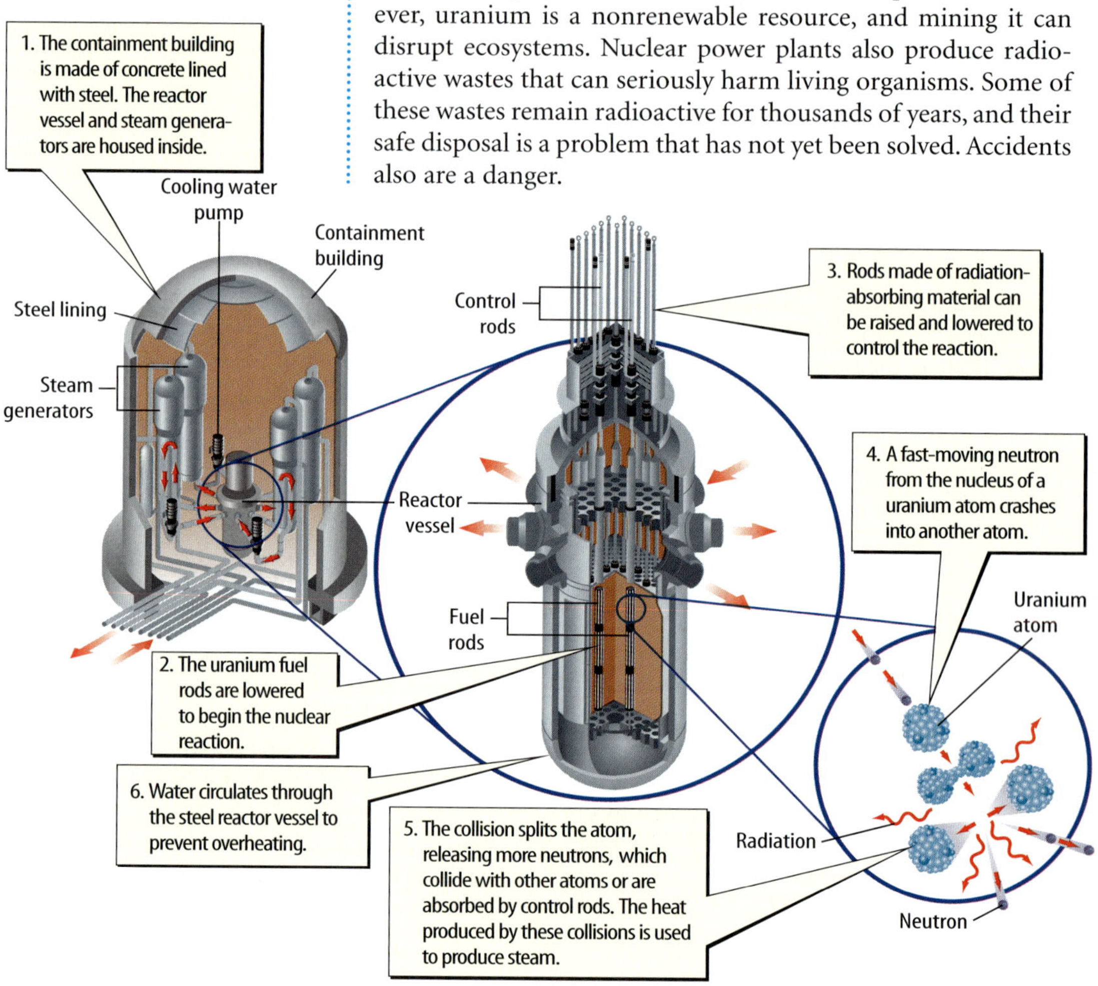

Science Journal

Nuclear Power Plant Sometimes people think something is a good idea but they don't want it in their own backyards. Have students describe in their journals how they might feel and what they might do if they learned of plans to build a nuclear power plant near their neighborhood. L2 LS **Intrapersonal**

Teacher FYI

Yucca Mountain The Yucca Mountain nuclear waste disposal site in Nevada has been under consideration since the late 1980s. To date, it is not operational and some people still are not convinced of the site's safety.

Geothermal Energy The hot, molten rock that lies deep beneath Earth's surface is also a source of energy. You see the effects of this energy when lava and hot gases escape from an erupting volcano or when hot water spews from a geyser. The heat energy contained in Earth's crust is called **geothermal energy.** Most geothermal power plants use this energy to produce steam to generate electricity.

Geothermal energy for power plants is available only where natural geysers or volcanoes are found. A geothermal power plant in California uses steam produced by geysers. The island nation of Iceland was formed by volcanoes, and geothermal energy is plentiful there. Geothermal power plants supply heat and electricity to about 90 percent of the homes in Iceland. Outdoor swimming areas also are heated with geothermal energy, as shown in **Figure 7.**

Reading Check *Where does geothermal energy come from?*

Figure 7 In Iceland, a geothermal power plant pumps hot water out of the ground to heat buildings and generate electricity. Leftover hot water goes into this lake, making it warm enough for swimming even when the ground is covered with snow.

Solar Energy The most inexhaustible source of energy for all life on Earth is the Sun. Solar energy is an alternative to fossil fuels. One use of solar energy is in solar-heated buildings. During winter in the northern hemisphere, the parts of a building that face south receive the most sunlight. Large windows placed on the south side of a building help heat it by allowing warm sunshine into the building during the day. Floors and walls of most solar-heated buildings are made of materials that absorb heat during the day. During the night, the stored heat is released slowly, keeping the building warm. **Figure 8** shows how solar energy can be used.

Figure 8 The Zion National Park Visitor Center in Utah is a solar-heated building designed to save energy. The roof holds solar panels that are used to generate electricity. High windows can be opened to circulate air and help cool the building on hot days. The overhanging roof shades the windows during summer.

Differentiated Instruction

Challenge Have interested students research tidal energy, nuclear fusion, or biomass fuels as alternative energy sources and present their findings in a poster. After students have examined all the posters, have them discuss which energy source they think has the most potential. L3

English-Language Learners Have students sort the following terms into renewable and nonrenewable resources: sisal, palladium, paprika, topaz, papaya, vinegar, diamond. Suggest they use a dictionary or the Internet to look up unfamiliar words. renewable: sisal, paprika, papaya, vinegar; nonrenewable: topaz, palladium, diamond L2

Use Science Words

Word Meaning Have students think about the parts of the word *geothermal. Geo* refers to Earth, and *thermal* refers to heat. Have students list other words that have either of these word parts in them. Possible answers: geology, geography, geomagnetic; thermometer, thermostat, homeotherm L2

Reading Check

Geothermal energy comes from the heat energy contained in Earth's crust.

Quick Demo

Energy

Materials radiometer

Estimated Time two minutes

Procedure Place the radiometer in an area where light will shine on the vanes. Have students speculate what causes the vanes to turn. Block the light, and have them observe the vanes slowing and eventually stopping. Explain the radiant energy is reflected from the polished sides of the vanes and is absorbed by the dark sides of the vanes. This raises the temperature of the dark surfaces. The air near the dark surfaces also is heated and exerts pressure on the dark surfaces, causing the rotor to turn.

Teacher FYI

Solar Cooking Nongovernmental organizations in the United States are promoting solar cooking in regions where firewood has become scarce. In African refugee villages and in many other areas, women and men are trained to make and use simple solar cookers in exchange for training others. Cooker designs are modified to make use of materials that are inexpensive and easy to obtain locally.

SECTION 1 Resources E ◆ 99

3 Assess

DAILY INTERVENTION

Check for Understanding

Logical-Mathematical Using the board, brainstorm with students about the methods we use to obtain energy. Have them determine whether each source of energy is renewable or not and how the energy is obtained. L2

Reteach

Using Resources Have students list reasons why renewable resources should be used wisely and nonrenewable resources should be conserved. L1

Assessment

Oral Have students explain which energy alternative to fossil fuels they think has the most promise to decrease fossil fuel use. Have them justify their answers. Use **Performance Assessment in the Science Classroom**, p. 89. L2

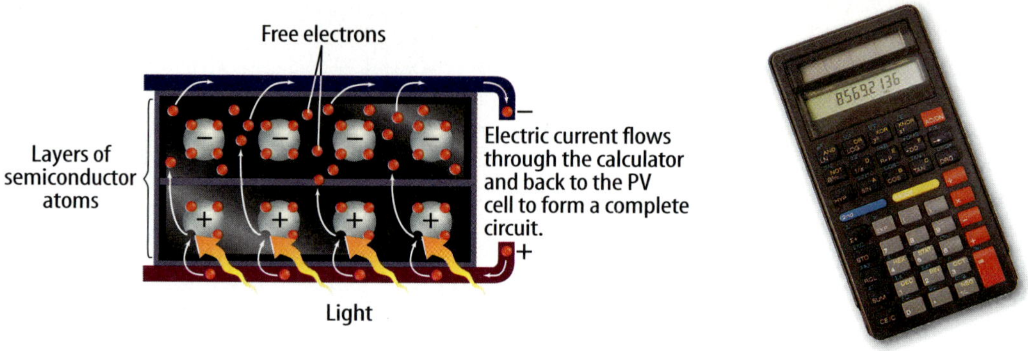

Figure 9 Light energy from the Sun travels in tiny packets of energy called photons. Photons crash into the atoms of PV cells, knocking electrons loose. These electrons create an electric current.

Solar Cells Do you know how a solar-powered calculator works? How do spacecraft use sunlight to generate electricity? These devices use photovoltaic (foh toh vohl TAY ihk) cells to turn sunlight into electric current, as shown in **Figure 9.** Photovoltaic (PV) cells are small and easy to use. However, they produce electricity only in sunlight, so batteries are needed to store electricity for use at night or on cloudy days. Also, PV cells presently are too expensive to use for generating large amounts of electricity. Improvements in this technology continue to be made, and prices probably will go down in the future. As **Figure 10** shows, solar buildings and PV cells are just two of the many ways solar energy can be used to replace fossil fuels.

section 1 review

Summary

Natural Resources

- All living things depend on natural resources to survive.
- Some resources are renewable, while other resources, such as petroleum, are nonrenewable.

Fossil Fuels

- Most of the energy that humans use comes from fossil fuels.
- Fossil fuels must be burned to release the energy stored in them, which causes air pollution.

Alternatives to Fossil Fuels

- Alternatives to fossil fuels include water power, wind power, nuclear power, geothermal energy, and solar energy.
- The Sun provides the most inexhaustible supply of energy for all life on Earth.

Self Check

1. **Summarize** What are natural resources?
2. **Compare and contrast** renewable and nonrenewable resources. Give five examples of each.
3. **Describe** the advantages and disadvantages of using nuclear power.
4. **Describe** two ways solar energy can be used to reduce fossil fuel use.
5. **Think Critically** Explain why the water that is used to cool the reactor vessel of a nuclear power plant is kept separate from the water that is heated to produce steam for the turbine generators.

Applying Math

6. **Solve One-Step Equations** Most cars in the U.S. are driven about 10,000 miles each year. If a car can travel 30 miles on one gallon of gasoline, how many gallons will it use in a year?

Science Online booke.msscience.com/self_check_quiz

section 1 review

1. the raw materials that organisms use for survival or to enhance life
2. renewable: can be replaced quickly by nature, include water, plants, animals, sunlight, air; nonrenewable: cannot be replaced quickly by nature, include petroleum, diamonds, metals, phosphorous, topsoil.
3. advantage—produces huge amounts of energy, does not contribute to air pollution; disadvantage—nonrenewable, produces radioactive wastes that are difficult to dispose of
4. passively, to heat homes (and water) and thus reduce use of fuels for this purpose; directly, to make electricity directly so that fossil fuels don't have to be burned
5. to prevent possible radioactive contamination
6. $10{,}000/30 = 333.33$ gallons per year

NATIONAL GEOGRAPHIC

VISUALIZING SOLAR ENERGY

Figure 10

Sunlight is a renewable energy source that provides an alternative to fossil fuels. Solar technologies use the Sun's energy in many ways—from heating to electricity generation.

▲ POWER PLANTS In California's Mojave Desert, an experimental solar power plant used hundreds of mirrors to focus sunlight on a water-filled tower. The steam produced by this system generates enough electricity to power 2,400 homes.

▼ ELECTRICITY Photovoltaic (PV) cells turn sunlight into electric current. They are commonly used to power small devices, such as calculators. Panels that combine many PV cells provide enough electricity for a home—or an orbiting satellite, such as the International Space Station, below.

▲ WATER HEATING Water is heated as it flows through small pipes in this roof-mounted solar heat collector. The hot water then flows into an insulated tank for storage.

▼ COOKING In hot, sunny weather, a solar oven or panel cooker can be used to cook a pot of rice or heat water. The powerful solar cooker shown below reaches even higher temperatures. It is being used to fry food.

▼ INDOOR HEATING South-facing windows and heat-absorbing construction materials turn a room into a solar collector that can help heat an entire building, such as this Connecticut home.

NATIONAL GEOGRAPHIC

Visualizing Solar Energy

Have students examine the pictures and read the captions. Then ask the following questions.

How are photovoltic cells similar to chloroplasts? How are they different? Possible answer: Both chloroplasts and photovoltic cells capture sunlight and turn it into another form of energy. Chloroplasts convert solar energy to chemical energy, and photovoltic cells convert solar energy into electric energy.

What are some potential disadvantages of using solar energy as the only source of energy for your home? How could you overcome these disadvantages? Possible answer: Solar energy might not be consistent or reliable; a cloudy day could result in no heat or electricity. However, a method of storing solar energy for future use would overcome this problem.

Activity

Capture the Sun's Energy Have students work in pairs to make posters that diagram an innovative way to capture or use the Sun's energy. Encourage creative and different thinking. Have students explain their diagrams to the class. L3 P

Differentiated Instruction

Challenge Have students research the percentage of energy used in the United States that was supplied by solar energy in the past year. Have the students compare this percentage to the percentage in 1950. L3

1 Motivate

INTERACTIVE CHALKBOARD PowerPoint® Presentations

Bellringer

Section Focus Transparencies also are available on the Interactive Chalkboard CD-ROM. L2 ELL

SECTION 2 Section Focus Transparency **Ocean-Sized Ice Cubes**

One of the concerns with rising worldwide temperatures is the melting of ice at Earth's poles. Ice floating in water, as found in the Arctic Ocean, won't raise sea levels if it melts. But ice-sheets on land, as found in Greenland and Antarctica, do contain enough water to raise sea levels if they melt.

1. What is happening in the picture above?
2. What might cause ice sheets to melt?
3. Do you think human activity could affect the amount of ice that melts in Antarctica? Why or why not?

Conserving Resources

Tie to Prior Knowledge

Pollution Most students will be familiar with some forms of pollution. Ask for examples. For each example, have students list what people can do to reduce that form of pollution.

Caption Answer

Figure 11 Smog can be reduced by using public transportation, and using vehicles that produce fewer pollutants.

section 2

Pollution

as you read

What You'll Learn

- **Describe** types of air pollution.
- **Identify** causes of water pollution.
- **Explain** methods that can be used to prevent erosion.

Why It's Important

By understanding the causes of pollution, you can help solve pollution problems.

Review Vocabulary

atmosphere: the whole mass of air surrounding Earth

New Vocabulary

- pollutant
- acid precipitation
- greenhouse effect
- ozone depletion
- erosion
- hazardous waste

Keeping the Environment Healthy

More than six billion people live on Earth. This large human population puts a strain on the environment, but each person can make a difference. You can help safeguard the environment by paying attention to how your use of natural resources affects air, land, and water.

Air Pollution

On a still, sunny day in almost any large city, you might see a dark haze in the air, like that in **Figure 11.** The haze comes from pollutants that form when wood or fuels are burned. A **pollutant** is a substance that contaminates the environment. Air pollutants include soot, smoke, ash, and gases such as carbon dioxide, carbon monoxide, nitrogen oxides, and sulfur oxides. Wherever cars, trucks, airplanes, factories, homes, or power plants are found, air pollution is likely. Air pollution also can be caused by volcanic eruptions, wind-blown dust and sand, forest fires, and the evaporation of paints and other chemicals.

Smog is a form of air pollution created when sunlight reacts with pollutants produced by burning fuels. It can irritate the eyes and make breathing difficult for people with asthma or other lung diseases. Smog can be reduced if people take buses or trains instead of driving or if they use vehicles, such as electric cars, that produce fewer pollutants than gasoline-powered vehicles.

Figure 11 The term *smog* was used for the first time in the early 1900s to describe the mixture of smoke and fog that often covers large cities in the industrial world.
Infer *how smog can be reduced in large cities.*

Section 2 Resource Manager

Chapter FAST FILE Resources
- Transparency Activity, pp. 45, 47–48
- Directed Reading for Content Mastery, p. 21
- MiniLAB, p. 4
- Enrichment, p. 31
- Lab Activity, pp. 9–12
- Reinforcement, p. 28
- Lab Worksheet, pp. 5–6

Reading and Writing Skill Activities, pp. 19, 29

Performance Assessment in the Science Classroom, p. 48

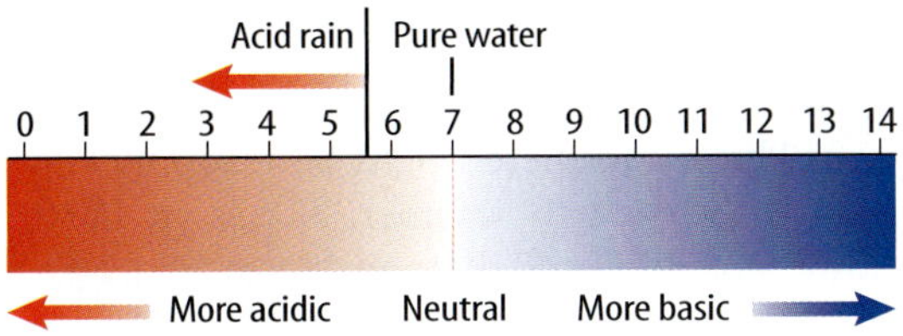

Figure 12 Compare these two photographs of the same statue. The photo on the left was taken before acid rain became a problem. The photo on the right shows acid rain damage. The pH scale, shown below, indicates whether a solution is acidic or basic.

Acid Precipitation

INTEGRATE Chemistry

Water vapor condenses on dust particles in the air to form droplets that combine to create clouds. Eventually, the droplets become large enough to fall to the ground as precipitation—mist, rain, snow, sleet, or hail. Air pollutants from the burning of fossil fuels can react with water in the atmosphere to form strong acids. Acidity is measured by a value called pH, as shown in **Figure 12.** **Acid precipitation** has a pH below 5.6.

Effects of Acid Rain Acid precipitation washes nutrients from the soil, which can lead to the death of trees and other plants. Runoff from acid rain that flows into a lake or pond can lower the pH of the water. If algae and microscopic organisms cannot survive in the acidic water, fish and other organisms that depend on them for food also die.

Preventing Acid Rain Sulfur from burning coal and nitrogen oxides from vehicle exhaust are the pollutants primarily responsible for acid rain. Using low-sulfur fuels, such as natural gas or low-sulfur coal, can help reduce acid precipitation. However, these fuels are less plentiful and more expensive than high-sulfur coal. Smokestacks that remove sulfur dioxide before it enters the atmosphere also help. Reducing automobile use and keeping car engines properly tuned can reduce acid rain caused by nitrogen oxide pollution. The use of electric cars, or hybrid-fuel cars that can run on electricity as well as gasoline, also could help.

Mini LAB

Measuring Acid Rain

Procedure

1. Collect **rainwater** by placing a clean **cup** outdoors. Do not collect rainwater that has been in contact with any object or organism.
2. Dip a piece of **pH indicator paper** into the sample.
3. Compare the color of the paper to the pH chart provided. Record the pH of the rainwater.
4. Use separate pieces of pH paper to test the pH of **tap water** and **distilled water.** Record these results.

Analysis

1. Is the rainwater acidic, basic, or neutral?
2. How does the pH of the rainwater compare with the pH of tap water? With the pH of distilled water?

Curriculum Connection

Social Studies Vinegar is a mild acid that has been used for hundreds of years to preserve foods. Pickles and sauerkraut are examples of foods that were developed to take advantage of vinegar's acidic properties, which kill many microorganisms. Have students research other foods that are prepared in pickling solution. Fish, fruits and vegetables, and pig's feet are pickled, to name a few. L2

2 Teach

Discussion

Automobile Usage Guide students to predict the effects of increased automobile usage on renewable resources such as trees and crops. Have them draw conclusions about the effects of other human activities on Earth's renewable resources.

Mini LAB

Purpose Students collect samples and determine the pH of rainwater. L2 COOP LEARN LS **Kinesthetic**

Materials cups, pH paper, rain gauge (optional), rainwater, tap water, distilled water

Teaching Strategies

- If your school has a rain gauge, this is a convenient place to obtain rainwater. Otherwise, cups can be placed outside to catch rain.
- Have students test the pH of rain on different days throughout the school year to detect variations.

Analysis

1. The pH of most rainwater is acidic and averages 5.6. Student results may range from a pH of 4 to 7.
2. Tap water pH varies, but in most of the U.S. it is about 6; distilled water always has a pH of 7.

Assessment

Performance Set up several liquids of unknown pH. Have students use pH paper to test each and determine whether it is an acid or base. Then have them make a chart showing the pH of each unknown liquid and comparing the liquids with substances of known pH. Use **Performance Assessment in the Science Classroom,** p. 109.

SECTION 2 Pollution E ◆ 103

Use an Analogy

Greenhouse Effect Compare the greenhouse effect to getting into a car with the windows rolled up on a summer day. It is much hotter inside the car than outside. L1

Teacher FYI

CO_2 in the Atmosphere The greenhouse effect has given Earth a sufficiently warm climate for life to flourish. Venus, however, has such a thick layer of CO_2 that its atmosphere is as hot as an oven. Some scientists predict that if people continue to add CO_2 to our atmosphere, Earth's atmospheric temperatures will increase. Other scientists think that if we produce more CO_2, plants will be able to use it to grow faster.

Visual Learning

Figure 13 Ask students what the glass in the greenhouse is analogous to in the atmosphere. carbon dioxide

Science Online

Topic: Global Warming
Visit booke.msscience.com for Web links to information about global warming.

Activity Describe three possible impacts of global warming. Provide one fact that supports global warming and one fact that does not.

Greenhouse Effect

Sunlight travels through the atmosphere to Earth's surface. Some of this sunlight normally is reflected back into space. The rest is trapped by certain atmospheric gases, as shown in **Figure 13.** This heat-trapping feature of the atmosphere is the **greenhouse effect.** Without it, temperatures on Earth probably would be too cold to support life.

Atmospheric gases that trap heat are called greenhouse gases. One of the most important greenhouse gases is carbon dioxide (CO_2). CO_2 is a normal part of the atmosphere. It is also a waste product that forms when fossil fuels are burned. Over the past century, more fossil fuels have been burned than ever before, which is increasing the percentage of CO_2 in the atmosphere. The atmosphere might be trapping more of the Sun's heat, making Earth warmer. A rise in Earth's average temperature, possibly caused by an increase in greenhouse gases, is known as global warming.

Global Warming Temperature data collected from 1895 through 1995 indicate that Earth's average temperature increased about 1°C during that 100-year period. No one is certain whether this rise was caused by human activities or is a natural part of Earth's weather cycle. What kinds of changes might be caused by global warming? Changing rainfall patterns could alter ecosystems and affect the kinds of crops that can be grown in different parts of the world. The number of storms and hurricanes might increase. The polar ice caps might begin to melt, raising sea levels and flooding coastal areas. Warmer weather might allow tropical diseases, such as malaria, to become more widespread. Many people feel that the possibility of global warming is a good reason to reduce fossil fuel use.

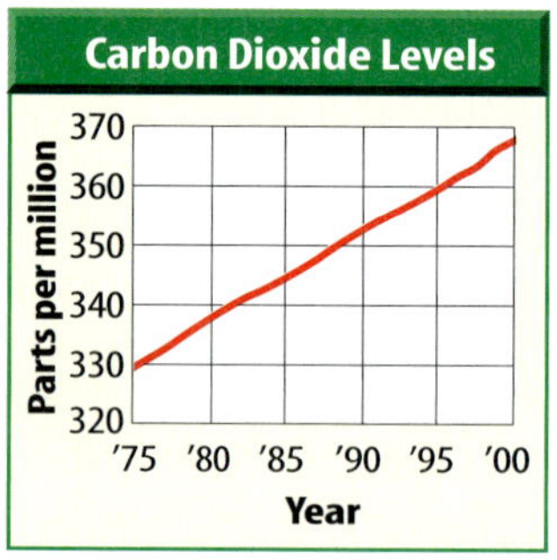

Figure 13 The moment you step inside a greenhouse, you feel the results of the greenhouse effect. Heat trapped by the glass walls warms the air inside. In a similar way, atmospheric greenhouse gases trap heat close to Earth's surface.

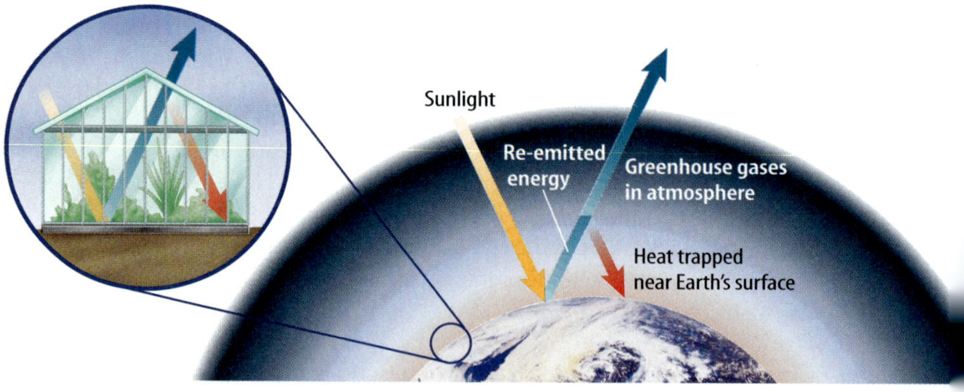

LAB DEMONSTRATION

Purpose to determine relative pH

Materials purple cabbage, hot plate, pot, water, strainer, 4 test tubes, stand, goggles, test materials: club soda, vinegar, ammonia, baking soda water

Preparation Boil the cabbage, cool, and strain. Retain the liquid.

Procedure Explain that purple cabbage juice can be used as a pH indicator. Add 1" juice to each tube. Add a small amount of one test material to each tube. Observe.

Expected Outcome Acids turn cabbage juice purple, bases turn it blue. The darker the color, the stronger the acid or base.

Assessment

Have students sort the test materials into acids. vinegar, club soda and bases ammonia, baking soda Set up another tube. Add lemon juice until the color changes. Have students identify lemon juice as an acid or base. acid

104 ◆ E CHAPTER 4 Conserving Resources

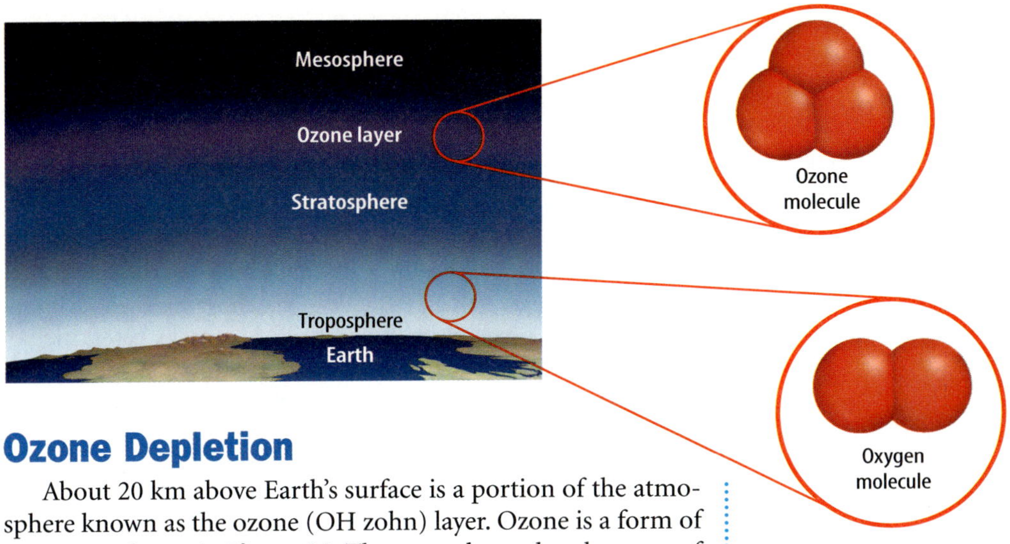

Ozone Depletion

About 20 km above Earth's surface is a portion of the atmosphere known as the ozone (OH zohn) layer. Ozone is a form of oxygen, as shown in **Figure 14.** The ozone layer absorbs some of the Sun's harmful ultraviolet (UV) radiation. UV radiation can damage living cells.

Every year, the ozone layer temporarily becomes thinner over each polar region during its spring season. The thinning of the ozone layer is called **ozone depletion.** This problem is caused by certain pollutant gases, especially chlorofluorocarbons (klor oh FLOR oh kar bunz) (CFCs). CFCs are used in the cooling systems of refrigerators, freezers, and air conditioners. When CFCs leak into the air, they slowly rise into the atmosphere until they arrive at the ozone layer. CFCs react chemically with ozone, breaking apart the ozone molecules.

UV Radiation Because of ozone depletion, the amount of UV radiation that reaches Earth's surface could be increasing. UV radiation could be causing a rise in the number of skin cancer cases in humans. It also might be harming other organisms. The ozone layer is so important to the survival of life on Earth that world governments and industries have agreed to stop making and using CFCs.

Ozone that is high in the upper atmosphere protects life on Earth. Near Earth's surface though, it can be harmful. Ozone is produced when fossil fuels are burned. This ozone stays in the lower atmosphere, where it pollutes the air. Ozone damages the lungs and other sensitive tissues of animals and plants. For example, it can cause the needles of a Ponderosa pine to drop, harming growth.

Reading Check *What is the difference between ozone in the upper atmosphere and ozone in the lower atmosphere?*

Figure 14 The atmosphere's ozone layer absorbs large amounts of UV radiation, preventing it from reaching Earth's surface. Ozone molecules are made of three oxygen atoms. They are formed in a chemical reaction between sunlight and oxygen. The oxygen you breathe has two oxygen atoms in each molecule.
Infer *what will happen if the ozone layer continues to thin.*

Teacher FYI

CFCs Although the production of CFC has halted, there is a reserve supply available for air conditioning and refrigeration units. In many cases, HFCs are being used as coolants to replace CFCs, as they do not lead to ozone depletion.

Quick Demo

Chemical Bonds

Materials none

Estimated Time one minute

Procedure Have two students join hands to represent the normal state of oxygen (O_2) and three students join hands to represent ozone (O_3). L1

Reading Check

Answer Ozone in the upper atmosphere is helpful to life; ozone in the lower atmosphere can be harmful to life as it pollutes the air.

Identifying Misconceptions

Cause and Effect Students sometimes confuse common environmental problems. For example, they may think that not littering will help with the problem of acid rain. Make sure that for each problem, students understand the related causes and effects and what can be done to reduce the problem.

Differentiated Instruction

Challenge Have students research monthly and annual changes in the ozone layer. Have students present graphs showing their results and discuss what environmental factors might explain the changes in the ozone layer. L3

Cultural Diversity

Sun Exposure Australia leads the world in skin cancer because of the many fair-skinned people, the ozone hole over the country, and changes in culture. In the past, wealthy Australians avoided the Sun. After WWI, Australians embraced a tanned look. As a result, skin cancer rates rapidly increased. It has now become a culturally accepted practice to have children wear hats when they go outside.

Caption Answer

Figure 14 The amount of UV radiation that reaches Earth could increase. This could cause an increase in the number of skin cancer cases in humans and might cause harm to other organisms.

Discussion

Indoor Air Pollution Why might better-insulated buildings make indoor air pollution more of a problem? Less air enters and leaves a building. Why do radon levels tend to be higher in wintertime? People are more likely to keep windows tightly closed. Have students discuss the pros and cons of insulating buildings to this extent.

INTEGRATE Health

Air Quality Carbon monoxide has an affinity for hemoglobin in red blood cells, so when it attaches, it doesn't let go. This prevents red blood cells from functioning properly. Because of this, appliances designed for outdoor use should only be used outdoors, as they often are not equipped with ventilation devices for indoor use.

IDENTIFYING Misconceptions

Natural Pollution Some students may think that people are responsible for all pollutants. Although many pollutants can be traced to people, radon is a naturally occurring substance. It is formed from the radioactive decay of uranium present in soil.

Caption Answer

Figure 15 Soils of the northern and northeastern portions of the U.S. produce more radon gas than in most other areas.

INTEGRATE Health

Air Quality Carbon monoxide enters the body through the lungs. It attaches to red blood cells, preventing the cells from absorbing oxygen. In your Science Journal, explain why heaters and barbecues designed for outdoor use never should be used indoors.

Indoor Air Pollution

Air pollution can occur indoors. Today's buildings are better insulated to conserve energy. However, better insulation reduces the flow of air into and out of a building, so air pollutants can build up indoors. For example, burning cigarettes release hazardous particles and gases into the air. Even nonsmokers can suffer ill effects from secondhand cigarette smoke. As a result, smoking no longer is allowed in many public and private buildings. Paints, carpets, glues and adhesives, printers, and photocopy machines also give off dangerous gases, including formaldehyde. Like cigarette smoke, formaldehyde is a carcinogen, which means it can cause cancer.

Carbon Monoxide Carbon monoxide (CO) is a poisonous gas that is produced whenever charcoal, natural gas, kerosene, or other fuels are burned. CO poisoning can cause serious illness or death. Fuel-burning stoves and heaters must be designed to prevent CO from building up indoors. CO is colorless and odorless, so it is difficult to detect. Alarms that provide warning of a dangerous buildup of CO are being used in more and more homes.

Radon Radon is a naturally occurring, radioactive gas that is given off by some types of rock and soil, as shown in **Figure 15.** Radon has no color or odor. It can seep into basements and the lower floors of buildings. Radon exposure is the second leading cause of lung cancer in this country. A radon detector sounds an alarm when levels of the gas in indoor air become too high. If radon is present, increasing a building's ventilation can eliminate any damaging effects.

Figure 15 The map shows the potential for radon exposure in different parts of the United States.
Identify *the area of the country with soils that produce the most radon gas.*

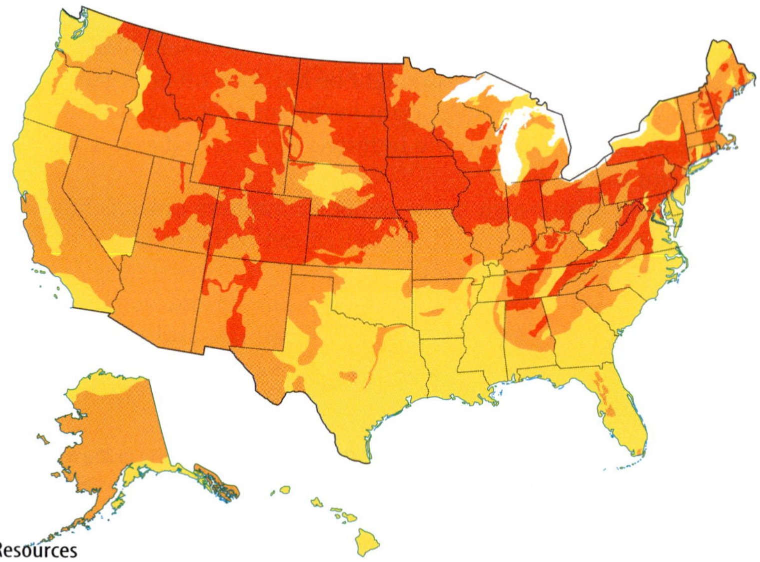

Geologic radon potential
Low | Moderate | High

Teacher FYI

Complete Combustion If combustion were complete, only carbon dioxide and water would be produced. Complete combustion is hard to achieve, especially in a barbecue. This is why barbecues should not be used indoors.

Curriculum Connection

Health Have students research sources of indoor air pollution. Suggest different students study the effects of asbestos, formaldehyde, radon, cigarette smoke, and other substances found in homes. Have students combine their findings and present them as an informative bulletin board. L2 COOP LEARN

When rain falls on roads and parking lots, it can wash oil and grease onto the soil and into nearby streams.

Rain can wash agricultural pesticides and fertilizers into lakes, streams, or oceans.

Industrial wastes are sometimes released directly into surface waters.

Figure 16 Pollution of surface waters can occur in several ways, as shown above.

Water Pollution

Pollutants enter water, too. Air pollutants can drift into water or be washed out of the sky by rain. Rain can wash land pollutants into waterways, as shown in **Figure 16.** Wastewater from factories and sewage-treatment plants often is released into waterways. In the United States and many other countries, laws require that wastewater be treated to remove pollutants before it is released. But, in many parts of the world, wastewater treatment is not always possible. Pollution also enters water when people dump litter or waste materials into rivers, lakes, and oceans.

Surface Water Some water pollutants poison fish and other wildlife, and can be harmful to people who swim in or drink the water. For example, chemical pesticides sprayed on farmland can wash into lakes and streams. These chemicals can harm the insects that fish, turtles, or frogs rely on for food. Shortages of food can lead to deaths among water-dwelling animals. Some pollutants, especially those containing mercury and other metals, can build up in the tissues of fish. Eating contaminated fish and shellfish can transfer these metals to people, birds, and other animals. In some areas, people are advised not to eat fish or shellfish taken from polluted waterways.

Algal blooms are another water pollution problem. Raw sewage and excess fertilizer contain large amounts of nitrogen. If they are washed into a lake or pond, they can cause the rapid growth of algae. When the algae die, they are decomposed by huge numbers of bacteria that use up much of the oxygen in the water. Fish and other organisms can die from a lack of oxygen in the water.

Quick Demo

Pollution Concentration

Materials glass jar or beaker, food coloring

Estimated Time two minutes

Procedure Drop food coloring into a glass jar or beaker full of water. Have students observe how the color spreads out. Explain that dumping a little waste into a lake might not be bad, but as more is added, the concentration increases. Add more drops of food coloring to illustrate this concept. L2

Activity

Algae Obtain aquarium water that has algae growing in it. Have students add varying amounts of commercial fertilizer to test tubes of the water. Place the tubes in sunlight, and add fresh water as the old water evaporates. In a few days, students should observe an algal bloom in one or more of the tubes. Continue the activity so that students can observe the effect of algal decay on the quality of the water. L2 LS **Kinesthetic**

Virtual Labs

Clean Water *When is water safe to drink?*

Teacher FYI

Lung Cancer Radon is the second leading cause of lung cancer in the United States. Cigarette smoking is the leading cause.

Cultural Diversity

Significance of Water Water is so important for civilization that it has become part of human culture. Water and waterscapes have been described in words, drawings, songs, and other art forms by all cultures on Earth. Water also plays an important part in religion. From Japanese Shintoism to Christianity, water symbolizes cleansing in spiritual ceremonies.

Visual Learning

Figure 18 Have students study the illustration. Emphasize that groundwater is an important source of water for many people. Then prepare a beaker with sand or soil on top and gravel on the bottom. Pour a little water on the soil's surface and show students how the water percolates through the layers to the bottom of the beaker. Have students relate this to the movement of groundwater shown in the illustration. L2 LS **Visual-Spatial**

Inquiry Lab

Oil Cleanup

Purpose to explore the process and cleanup of oil spills and discuss waste disposal of materials

Possible Materials cooking oil, clear-plastic bin, cotton balls, aquarium net, sand, liquid detergent, string, plastic spoons, paper towels, garbage bag

Estimated Time 1 class period

Teaching Strategies

- You can have students work in groups and assign each group a bin of water.
- Students can predict and observe what will happen to the oil when it is poured into the water.
- Students can write a procedure using the materials available to clean up the oil spill and then follow their procedure.
- After collecting their data, students can write suggestions for doing the experiment better. PBL
- Have students contact the nearest Hazmat agency and inquire about spills that have occurred locally and how they were handled. Have them compare their methods with Hazmat's to consider better ideas.

For additional inquiry activities, see *Science Inquiry Labs*.

Figure 17 In 1996, the oil tanker *Sea Empress* spilled more than 72 million kg of oil into the sea along the coast of Wales. More than $40 million was spent on the cleanup effort, but thousands of ocean organisms were destroyed, including birds, fish, and shellfish.

Ocean Water Rivers and streams eventually flow into oceans, bringing their pollutants along. Also, polluted water can enter the ocean in coastal areas where factories, sewage-treatment plants, or shipping activities are located. Oil spills are a well-known ocean pollution problem. About 4 billion kg of oil are spilled into ocean waters every year. Much of that oil comes from ships that use ocean water to wash out their fuel tanks. Oil also can come from oil tanker wrecks, as shown in **Figure 17.**

Groundwater Pollution can affect water that seeps underground, as shown in **Figure 18.** Groundwater is water that collects between particles of soil and rock. It comes from precipitation and runoff that soaks into the soil. This water can flow slowly through permeable layers of rock called aquifers. If this water comes into contact with pollutants as it moves through the soil and into an aquifer, the aquifer could become polluted. Polluted groundwater is difficult—and sometimes impossible—to clean. In some parts of the country, chemicals leaking from underground storage tanks have created groundwater pollution problems.

Figure 18 Water from rainfall slowly filters through sand or soil until it is trapped in underground aquifers. Pollutants picked up by the water as it filters through the soil can contaminate water wells.

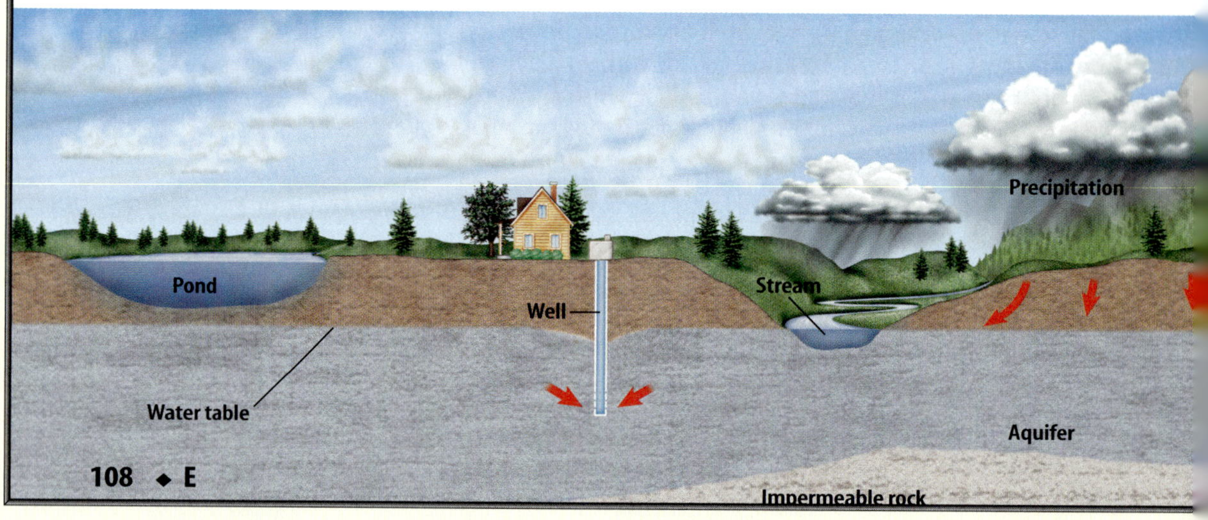

108 ◆ E

Science Journal

Well Water Have students imagine that they depend on a well as their source of water. They find out that pesticides that can injure their health have seeped into the groundwater the well draws from. Have students write in their Science Journals the changes they will make in their use of the well's water. L2 LS **Intrapersonal**

Differentiated Instruction

Hearing Impaired Fill a funnel with soil and place it in a beaker. Add a teaspoon of food coloring to the top of the soil. Slowly pour water through the funnel. At first, water flowing out of the funnel is probably clear, but eventually colored water exits. Explain that soil may act as a filter for a time, but eventually pollutants will work their way through the soil. L1 ELL

Contour plowing reduces the downhill flow of water.

Soil Loss

Fertile topsoil is important to plant growth. New topsoil takes hundreds or thousands of years to form. The Launch Lab at the beginning of this chapter shows that rain washes away loose topsoil. Wind also blows it away. The movement of soil from one place to another is called **erosion** (ih ROH zhun). Eroded soil that washes into a river or stream can block sunlight and slow photosynthesis. It also can harm fish, clams, and other organisms. Erosion is a natural process, but human activities increase it. When a farmer plows a field or a forest is cut down, soil is left bare. Bare soil is more easily carried away by rain and wind. **Figure 19** shows some methods farmers use to reduce soil erosion.

Soil Pollution

Soil can become polluted when air pollutants drift to the ground or when water leaves pollutants behind as it flows through the soil. Soil also can be polluted when people toss litter on the ground or dispose of trash in landfills.

Solid Wastes What happens to the trash you throw out every week? What do people do with old refrigerators, TVs, and toys? Most of this solid waste is dumped in landfills. Most landfills are designed to seal out air and water. This helps prevent pollutants from seeping into surrounding soil, but it slows normal decay processes. Even food scraps and paper, which usually break down quickly, can last for decades in a landfill. In populated areas, landfills fill up quickly. Reducing the amount of trash people generate can reduce the need for new landfills.

Figure 19 The farming methods shown here help prevent soil erosion.
Infer *why soil erosion is a concern for farmers.*

On steep hillsides, flat areas called terraces reduce downhill flow.

In strip cropping, cover crops are planted between rows to reduce wind erosion.

In no-till farming, soil is never left bare.

Caption Answer

Figure 19 Erosion carries away nutrient-rich topsoil that farmers need to grow crops.

Quick Demo

Biodegradable Waste

Materials container of trash including a glass bottle, aluminum can, plastic wrappers, a coffee filter with grounds, vegetable peelings, eggshells, and other household items

Estimated Time three minutes

Procedure Using rubber gloves, remove items from the container of trash. Have students discuss whether the item is biodegradable. Explain that the biodegradable wastes may decay, but the nonbiodegradable wastes will not.

Activity

Cleaning Products Many household cleaners contain toxic chemicals. Numerous new products are available that are just as effective, but do not contain harmful chemicals. Have students obtain an adult's permission to investigate cleaning products in their homes to see what kinds of ingredients they contain. Caution students to take care when handling cleaning products or other chemicals. L2

Fun Fact

Americans represent approximately 5% of the world's population and generate 30% of the world's garbage.

Differentiated Instruction

Challenge Have students calculate the following: If your family produces an average of 1.8 kg of solid waste per person per day, how much waste will your entire family generate in the month of October? Answers will vary depending on the number of family members (f). Students should use the following equation to solve this problem: $(f \times 1.8) \times 31 =$ amount of waste produced by the family. L3

Teacher FYI

Erosion Pollution refers to the degradation of the environment. Erosion fits into this definition because as valuable topsoil is lost, land becomes less fertile. The erosion of land in arid areas may make these places resemble deserts, a process called desertification.

SECTION 2 Pollution E ◆ 109

Reading Check

Answer Hazardous wastes are waste materials that are harmful to human health or poisonous to other living things.

3 Assess

DAILY INTERVENTION

Check for Understanding

Visual-Spatial Present pictures displaying various types of pollution and have students explain the cause and effect of the pollution.

Reteach

Pollution Have students make a chart summarizing the causes and effects of air pollution, soil pollution, and water pollution. L2

Assessment

Content Have student pairs work together to write a song or poem about the dangers of environmental pollution. Use **Performance Assessment in the Science Classroom,** p. 151. L2 P

Figure 20 Leftover paints, batteries, pesticides, drain cleaners, and medicines are hazardous wastes that should not be discarded in the trash. They should never be poured down a drain, onto the ground, or into a storm sewer. Most communities have collection facilities where people can dispose of hazardous materials like these.

Hazardous Wastes Waste materials that are harmful to human health or poisonous to living organisms are **hazardous wastes.** They include dangerous chemicals, such as pesticides, oil, and petroleum-based solvents used in industry. They also include radioactive wastes from nuclear power plants, from hospitals that use radioactive materials to treat disease, and from nuclear weapons production. Many household items also are considered hazardous, such as those shown in **Figure 20.** If these materials are dumped into landfills, they could seep into the soil, surface water, or groundwater over time. Hazardous wastes usually are handled separately from trash. They are treated in ways that prevent environmental pollution.

Reading Check *What are hazardous wastes?*

section 2 review

Summary

Air Pollution and Acid Precipitation

- Vehicles, volcanoes, forest fires, and even wind-blown dust and sand can cause air pollution.
- Acid rain washes nutrients from the soil, which can harm plants.

Greenhouse Effect and Ozone Depletion

- CO_2 is a greenhouse gas that helps warm Earth.
- The ozone layer protects life on Earth.

Indoor Air Pollution, Water Pollution, Soil Loss, and Soil Pollution

- Pollutants can build up inside of buildings.
- There are many sources of water pollutants.
- Wind and rain can erode bare soil.
- Pollutants in soil decay more slowly than in air.

Self Check

1. **List** four ways that air pollution affects the environment.
2. **Explain** how an algal bloom can affect other pond organisms.
3. **Describe** possible causes and effects of ozone depletion.
4. **Think Critically** How could hazardous wastes in landfills eventually affect groundwater?

Applying Math

5. **Solve a One-Step Equation** A solution of pH 4 is 10 times more acidic than one of pH 5, and it is 10 times more acidic than a solution of pH 6. How many times more acidic is the solution of pH 4 than the one of pH 6?

Science Online booke.msscience.com/self_check_quiz

section 2 review

1. smog, acid rain, enhanced greenhouse effect, ozone depletion
2. Algal blooms grow and eventually die. Their decay uses up oxygen, making it difficult for other organisms to survive.
3. CFCs that escape into the air can cause ozone depletion; as ozone is depleted, more UV radiation reached Earth, resulting in more cases of skin cancer.
4. They can leach into the soil and be picked up by groundwater, making the water hazardous for people to drink.
5. A solution with a pH of 4 is 100 times more acidic than a solution with a pH of 6.

LAB

BENCH TESTED

The Greenhouse Effect

You can create models of Earth with and without heat-reflecting green-house gases. Then, experiment with the models to observe the greenhouse effect.

Real-World Question

How does the greenhouse effect influence temperatures on Earth?

Goals

- **Observe** the greenhouse effect.
- **Describe** the effect that a heat source has on an environment.

Materials

1-L clear-plastic, soft-drink bottles with tops cut off and labels removed (2)
thermometers (2)
**temperature probe*
potting soil
masking tape
plastic wrap
rubber band
lamp with 100-W lightbulb
watch or clock with second hand
**Alternate materials*

Safety Precautions

Procedure

1. Copy the data table and use it to record your temperature measurements.
2. Put an equal volume of potting soil in the bottom of each container.
3. Use masking tape to attach a thermometer to the inside of each container. Place each thermometer at the same height above the soil. Shield each thermometer bulb by putting a double layer of masking tape over it.

Changes in Temperature

Time (min)	Open Container Temperature (°C)	Closed Container Temperature (°C)
0		
2		
4		
6		

4. Seal the top of one container with plastic wrap held in place with a rubber band.
5. Place the lamp with the exposed 100-W lightbulb between the two containers and exactly 1 cm away from each. Do not turn on the light.
6. Let the setup sit for 5 min, then record the temperature in each container.
7. Turn on the light. Record the temperature in each container every 2 min for 15 min to 20 min. Graph the results.

Conclude and Apply

1. **Compare and contrast** temperatures in each container at the end of the experiment.
2. **Infer** What does the lightbulb represent in this experimental model? What does the plastic wrap represent?

Communicating Your Data

Average the data obtained in the experiments conducted by all the groups in your class. Prepare a line graph of these data. **For more help, refer to the Science Skill Handbook.**

Real-World Question

Purpose Students make models of the atmosphere and determine what happens when light shines on the model. L2 COOP LEARN IS **Interpersonal**

Process Skills model, observe and infer, interpret data, communicate, compare and contrast, recognize cause and effect, measure in SI, make and use tables, make and use graphs

Time Required 50 minutes

Safety Precautions Caution students to use care in working with thermometers and 100-W lightbulbs. Use only alcohol thermometers.

Teaching Strategies

- Drafts can affect results.
- Make sure the thermometers face outward to make it easier for students to read them.

Troubleshooting If the two thermometers did not read the same temperature when the experiment began, adjust readings so both data sets can be plotted on the same graph.

Conclude and Apply

1. The temperatures of both containers has increased. The container covered with plastic wrap had a greater increase.
2. The light bulb represents the Sun; the plastic wrap represents carbon dioxide in the atmosphere.

Assessment

Oral Have students summarize the information in their graphs in writing. Make sure summaries are clear and concise. Use **Performance Assessment in the Science Classroom,** p. 113.

Communicating Your Data

Have students write the final temperatures for the two conditions on the board. Students should compare data and see if the results were the same for everyone. Discuss how differences may have occurred.

The Three Rs of Conservation

1 Motivate

INTERACTIVE CHALKBOARD
PowerPoint® Presentations

Bellringer

Section Focus Transparencies also are available on the Interactive Chalkboard CD-ROM. L2 ELL

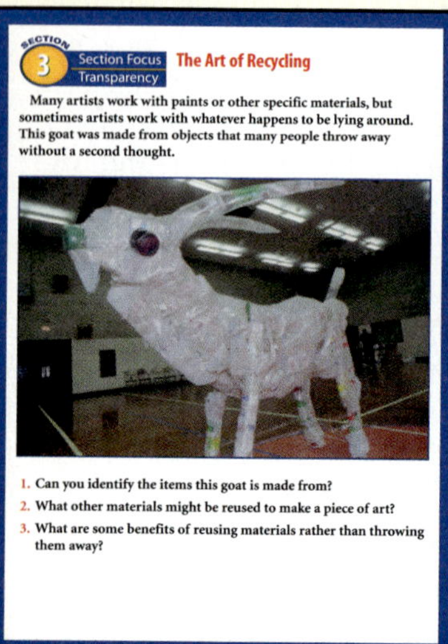
Section 3 Section Focus Transparency **The Art of Recycling**

Many artists work with paints or other specific materials, but sometimes artists work with whatever happens to be lying around. This goat was made from objects that many people throw away without a second thought.

1. Can you identify the items this goat is made from?
2. What other materials might be reused to make a piece of art?
3. What are some benefits of reusing materials rather than throwing them away?

Conserving Resources

Tie to Prior Knowledge

Recycling Most students will be familiar with the idea of recycling. Ask the class to describe some things that are commonly recycled. Ask for reasons why we recycle.

as you read

What **You'll Learn**

- **Recognize** ways you can reduce your use of natural resources.
- **Explain** how you can reuse resources to promote conservation.
- **Describe** how many materials can be recycled.

Why **It's Important**

Conservation preserves resources and reduces pollution.

Review Vocabulary
reprocessing: to subject to a special process or treatment in preparation for reuse

New Vocabulary
- recycling

Conservation

A teacher travels to school in a car pool. In the school cafeteria, students place glass bottles and cans in separate containers from the rest of the garbage. Conservation efforts like these can help prevent shortages of natural resources, slow growth of landfills, reduce pollution levels, and save people money. Every time a new landfill is created, an ecosystem is disturbed. Reducing the need for landfills is a major benefit of conservation. The three Rs of conservation are reduce, reuse, and recycle.

Reduce

You contribute to conservation whenever you reduce your use of natural resources. You use less fossil fuel when you walk or ride a bicycle instead of taking the bus or riding in a car. If you buy a carton of milk, reduce your use of petroleum by telling the clerk you don't need a plastic bag to carry it in.

You also can avoid buying things you don't need. For example, most of the paper, plastic, and cardboard used to package items for display on store shelves is thrown away as soon as the product is brought home. You can look for products with less packaging or with packaging made from recycled materials. What are some other ways you can reduce your use of natural resources?

Figure 21 Worn-out automobile tires can have other useful purposes.

Reuse

Another way to help conserve natural resources is to use items more than once. Reusing an item means using it again without changing it or reprocessing it, as shown in **Figure 21.** Bring reusable canvas bags to the grocery store to carry home your purchases. Donate clothes you've outgrown to charity so that others can reuse them. Take reusable plates and utensils on picnics instead of disposable paper items.

Section 3 Resource Manager

Chapter *FAST FILE* Resources
Transparency Activity, p. 46
Directed Reading for Content Mastery, pp. 21, 22
Lab Worksheet, pp. 7–8
Enrichment, p. 32
Lab Activity, pp. 13–16
Reinforcement, p. 29

Lab Management and Safety, p. 37

Recycle

If you can't avoid using an item, and if you can't reuse it, the next best thing is to recycle it. **Recycling** is a form of reuse that requires changing or reprocessing an item or natural resource. If your city or town has a curbside recycling program, you already separate recyclables from the rest of your garbage. Materials that can be recycled include glass, metals, paper, plastics, and yard and kitchen waste.

Reading Check *How is recycling different from reusing?*

Plastics Plastic is more difficult to recycle than other materials, mainly because several types of plastic are in use. A recycle code marked on every plastic container indicates the type of plastic it is made of. Plastic soft-drink bottles, like the one shown in **Figure 22,** are made of type 1 plastic and are the easiest to recycle. Most plastic bags are made of type 2 or type 4 plastic; they can be reused as well as recycled. Types 6 and 7 can't be recycled at all because they are made of a mixture of different plastics. Each type of plastic must be separated carefully before it is recycled because a single piece of a different type of plastic can ruin an entire batch.

Figure 22 Many soft-drink bottles are made of PETE, which is the most common type of recyclable plastic. It can be melted down and spun into fibers to make carpets, paintbrushes, rope, and clothing.
Identify *other products made out of recycled materials.*

2 Teach

Text Question Answer

Possible answer: Turn off lights in vacant rooms.

Discussion

Over-Packaging Bring in a product that is clearly over-packaged. Have students suggest ways to reduce the amount of material used in the packaging of this item.

Activity

Brainstorm Find an item that is commonly thrown out, such as a plastic soft-drink bottle. Have students brainstorm a list of possible ways to reuse the bottle. L2 ELL IS **Interpersonal**

Reading Check

Answer Reusing does not require reprocessing; recycling does.

Caption Answer

Figure 22 Packaging, paper, glass bottles, furniture, auto parts, and construction and landscaping products are some examples of materials made from recycled material.

Quick Demo

Recycling

Materials cans, cereal boxes, other products with recycling symbol or made from recycled material

Estimated Time one minute

Procedure Lay products out on a table and have students note whether the packaging has a recycle symbol or if the package states that it is made of recycled material. Explain that many products can be recycled and made from recycled materials.

Differentiated Instruction

Challenge Have students find as many products as they can that are recycled and what material they are made up of. Then have them research the types of material that are commonly recycled and present a report of their findings. L3

Visual Learning

Figure 22 The objects in these photographs are made from the same type 1 plastic (PET). Ask students to bring in materials with this symbol on them. When the objects are collected, discuss why recycling is a good idea. Possible answer: It reduces the amount of nonbiodegradable materials going into landfills. L2

Fun Fact

In the United States, each person uses an average of about 650 lbs. of paper each year.

IDENTIFYING Misconceptions

Recycled Paper Many people think that recycled paper is paper that has been used by consumers. Most of it, however, comes from wood or paper scrap in logging or paper manufacturing. Some companies indicate how much recycled paper comes directly from consumers by labeling the percent that is post-consumer content.

Applying Science

Answer

Students' lists should include some of each type of item. Possible exceptions would be yard wastes, depending on the season, and newspapers. The percentages will depend on student data. Tin cans are steel cans coated with tin. They should be included.

Science Online

Topic: Recycling
Visit booke.msscience.com for Web links to information about recycling bottles and cans.

Activity Write one argument in support of a money deposit for bottles and cans and one argument against it. Provide data to support one of your arguments.

Metals The manufacturing industry has been recycling all kinds of metals, especially steel, for decades. At least 25 percent of the steel in cans, appliances, and automobiles is recycled steel. Up to 100 percent of the steel in plates and beams used to build skyscrapers is made from reprocessed steel. About one metric ton of recycled steel saves about 1.1 metric tons of iron ore and 0.5 metric ton of coal. Using recycled steel to make new steel products reduces energy use by 75 percent. Other metals, including iron, copper, aluminum, and lead also can be recycled.

You can conserve metals by recycling food cans, which are mostly steel, and aluminum cans. It takes less energy to make a can from recycled aluminum than from raw materials. Also, remember that recycled cans do not take up space in landfills.

Glass When sterilized, glass bottles and jars can be reused. They also can be melted and re-formed into new bottles, especially those made of clear glass. Most glass bottles already contain at least 25 percent recycled glass. Glass can be recycled again and again. It never needs to be thrown away. Recycling about one metric ton of glass saves more than one metric ton of mineral resources and reduces the energy used to make new glass by 25 percent or more.

Applying Science

What items are you recycling at home?

Many communities have recycling programs. Recyclable items may be picked up at the curbside, taken to a collection site, or the resident may hire a licensed recycling handler to pick them up. What do you recycle in your home?

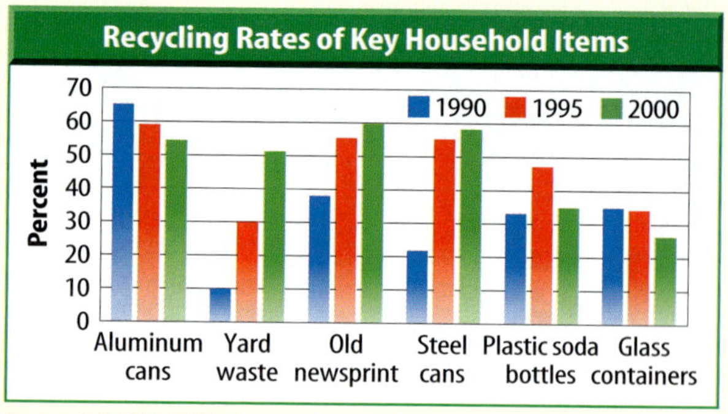

Source: U.S. EPA, 2003

Identifying the Problem

This bar graph shows the recycling rates in the U. S. of six types of household items for the years 1990, 1995, and 2000. What are you and your classmates' recycling rates?

Solving the Problem

For one week, list each glass, plastic, and aluminum item you use. Note which items you throw away and which ones you recycle. Calculate the percentage of glass, plastic, and aluminum you recycled. How do your percentages compare with those on the graph?

Teacher FYI

Compost Composting is considered recycling because the form of the materials is changed (through decay) so they can be used again.

Differentiated Instruction

Challenge Have students investigate the trash generated by your school's cafeteria. Challenge them to devise a plan to reduce the amount of trash that is thrown away. Have them write a report on their investigation and plan. L3 P

114 ◆ E CHAPTER 4 Conserving Resources

Paper Used paper is recycled into paper towels, insulation, newsprint, cardboard, and stationery. Ranchers and dairy farmers sometimes use shredded paper instead of straw for bedding in barns and stables. Used paper can be made into compost. Recycling about one metric ton of paper saves 17 trees, more than 26,000 L of water, close to 1,900 L of oil, and more than 4,000 kW of electric energy. You can do your part by recycling newspapers, notebook and printer paper, cardboard, and junk mail.

Figure 23 Composting is a way of turning plant material you would otherwise throw away into rich garden soil. Dry leaves and weeds, grass clippings, vegetable trimmings, and nonmeat food scraps can be composted.

Reading Check *What nonrenewable resource(s) do you conserve by recycling paper?*

Compost Grass clippings, leaves, and fruit and vegetable scraps that are discarded in a landfill can remain there for decades without breaking down. The same items can be turned into soil-enriching compost in just a few weeks, as shown in **Figure 23.** Many communities distribute compost bins to encourage residents to recycle fruit and vegetable scraps and yard waste.

Buy Recycled People have become so good at recycling that recyclable materials are piling up, just waiting to be put to use. You can help by reading labels when you shop and choosing products that contain recycled materials. What other ways of recycling natural resources can you think of?

section 3 review

Summary

Conservation
- The three Rs of conservation are reduce, reuse, and recycle.

Reduce
- You can contribute to conservation by reducing your use of natural resources.

Reuse
- Some items can be used more than once, such as reusable canvas bags for groceries.

Recycle
- Some items can be recycled, including some plastics, metal, glass, and paper.
- Grass clippings, leaves, and fruit and vegetable scraps can be composted into rich garden soil.

Self Check

1. **Describe** at least three actions you could take to reduce your use of natural resources.
2. **Describe** how you could reuse three items people usually throw away.
3. **Think Critically** Why is reusing something better than recycling it?

Applying Skills

4. **Make and Use Tables** Make a table of data of the number of aluminum cans thrown away in the United States: 2.7 billion in 1970; 11.1 billion in 1974; 21.3 billion in 1978; 22.7 billion in 1982; 35.0 billion in 1986; 33.8 billion in 1990; 38.5 billion in 1994; 45.5 billion in 1998; 50.7 billion in 2001.

Science Online booke.msscience.com/self_check_quiz

Reading Check

Answer oil

Text Question Answer

Answers will vary.

3 Assess

DAILY INTERVENTION

Check for Understanding

Visual-Spatial Place students in small groups and show them various items such as bottles, tin cans, paper, plastic bags, and fruit. Have them decide whether they could reduce their use of the product, reuse it, or recycle it. L2

Reteach

Glass Bottles Ask students to give their opinions about whether a soft-drink company should reuse glass bottles by washing them out or recycle the bottles by melting them and making new ones. L1 LS **Linguistic**

Assessment

Content Hold up a newspaper and ask students to write specific examples of how a community can reduce its use of newspapers, reuse newspapers, and recycle newspapers.

section 3 review

1. Possible answers: turn off lights in vacant rooms, buy materials with less packaging, carpool, buy items made from recycled materials
2. Possible answers: store water in plastic soft-drink bottles, use plastic grocery bags as trash can liners, use margarine tubs to store left-over food.
3. There is less energy used in reusing than in recycling.
4. Check tables for accuracy.

BENCH TESTED

Real-World Question

Purpose Students will design and build a solar cooker.

Process Skills design and make a model, research, compare

Time Required one class period

Discussion Discuss with students the problems some people in the world face just trying to cook a meal. Explain to students that in some places where fuel is scarce, people use solar cookers to prepare food. Why do people cook most food? Possible answers: to improve taste, to kill harmful bacteria Why does cooking food require energy? Energy is needed to raise the temperature of the food.

Alternate Materials Oven bags are available at most supermarkets. Kitchen timers are available at most hardware stores.

Safety Precautions Remind students to use insulated gloves or tongs when handling hot objects.

LAB Model and Invent

Solar Cooking

Goals

- **Research** designs for solar panel cookers or box cookers.
- **Design** a solar cooker that can be used to cook food.
- **Plan** an experiment to measure the effectiveness of your solar cooker.

Possible Materials

poster board
cardboard boxes
aluminum foil
string
wire coat hangers
clear plastic sheets
**oven bags*
black cookware
thermometer
stopwatch
**timer*
glue
tape
scissors
**Alternate materials*

Safety Precautions

WARNING: *Be careful when cutting materials. Your solar cooker will get hot. Use insulated gloves or tongs to handle hot objects.*

Real-World Question

The disappearance of forests in some places on Earth has made firewood extremely difficult and expensive to obtain. People living in these regions often have to travel long distances or sell some of their food to get firewood. This can be a serious problem for people who may not have much food to begin with. Is there a way they could cook food without using firewood? How would you design and build a cooking device that uses the Sun's energy?

Make the Model

1. **Design** a solar cooker. In your Science Journal, explain why you chose this design and draw a picture of it.
2. **Write** a summary explaining how you will measure the effectiveness of your solar cooker. What will you measure? How will you collect and organize your data? How will you present your results?

Fun Fact

The Sun converts approximately 4 million tons of hydrogen gas into energy every second.

Differentiated Instruction

Learning Disabled Assign these students the task of gathering the materials their group decides they want to use. L1

3. **Compare** your solar cooker design to those of other students.
4. Share your experimental plan with students in your class. Discuss the reasoning behind your plan. Be specific about what you intend to test and how you are going to test it.
5. Make sure your teacher approves your plan before you start working on your model.
6. Using all of the information you have gathered, construct a solar cooker that follows your design.

Test the Model

1. **Test** your design to determine how well it works. Try out a classmate's design. How do the two compare?

Analyze Your Data

1. Combine the results for your entire class and decide which type of solar cooker was most effective. How could you design a more effective solar cooker, based on what you learned from this activity?
2. **Infer** Do you think your results might have been different if you tested your solar cooker on a different day? Explain. Why might a solar cooker be more useful in some regions of the world than in others?

Conclude and Apply

1. **Infer** Based on what you've read and the results obtained by you and your classmates, do you think that your solar cooker could boil water? Explain.
2. **Compare** the amount of time needed to cook food with a solar cooker and with more traditional cooking methods. Assuming plenty of sunlight is available, would you prefer to use a solar cooker or a traditional oven? Explain.

Communicating Your Data

Prepare a demonstration showing how to use a solar cooker. Present your demonstration to another class of students or to a group of friends or relatives. **For more help, refer to the Science Skill Handbook.**

Make a Model

Teaching Strategies

- Comparisons of solar cookers will be difficult if students do not use the same method of testing them.
- You might want to have a cook-off with the two best solar ovens to see which one could cook a small red potato in the shortest time. Be sure both potatoes are the same size.
- Students can move the cookers during the testing to keep them in the sunlight.

Test Your Model

Expected Outcome Solar cookers should be able to raise the temperature of a cup of water to at least 75°C. A well-designed cooker might reach temperatures approaching 150°C.

Analyze Your Data

1. Answers will depend on the designs of the solar cookers that are compared.
2. Results would vary in different weather conditions. Solar cookers function best on clear, sunny days. Solar cookers are more useful in areas that have many clear, sunny days.

Conclude and Apply

1. Answers will depend on the success of the design. Some well-designed cookers could boil water.
2. Solar cookers generally take two to three times as long to cook a meal as a traditional oven. However, the energy for a solar cooker is free and the energy for a traditional oven is not.

Assessment

Process Have students use a foam cup of water with a thermometer in it to test whether their solar ovens can boil water. Be sure that each group uses the same amount of water for the test and that students record the time it takes to boil the water. Based on the results, have students rank the solar cooker designs from best to worst. Use **Performance Assessment in the Science Classroom,** p. 97.

Communicating Your Data

Students should prepare a poster that shows the construction of their solar cooker and an explanation of how it works.

Understanding Literature

Cause and Effect Answers will vary but might include that bottle caps litter the picnic area and that exhaust fumes from cars make the snow banks dark and gray.

Respond to the Reading

1. The four verses correspond to the four seasons.
2. Human activity (civilization) has plagiarized beauty.
3. **Linking Science and Writing** Students could review chapter and list conservation methods to include in their poem.

INTEGRATE Life Science **Renewing Resources** Even renewable resources can be rendered nonrenewable. This happens when renewable resources are removed at a rate that prevents their renewal, or if the environment that supports them deteriorates faster than renewal can take place. For example, although forests are considered a renewable resource, if they are not carefully managed they can disappear as a result of excessive logging or clear-cutting. Also, although Earth's supply of air is inexhaustible, the quality of Earth's air can be affected by human activity, and air quality directly affects other renewable resources.

Science and Language Arts

Beauty Plagiarized

by Amitabha Mukerjee

I wandered lonely as a cloud –
Except for a motorboat,
Nary a soul in sight.
Beside the lake beneath the trees,
Next to the barbed wire fence,
There was a picnic table
And beer bottle caps from many years.
A boat ramp to the left,
And the chimney from a power station on the
other side,
A summer haze hung in the air,
And the lazy drone of traffic far away.

Crimson autumn of mists and mellow fruitfulness
Blue plastic covers the swimming pools
The leaves fall so I can see
Dark glass reflections in the building
That came up
where the pine cones crunched underfoot . . .

And then it is snow
White lining on trees and rooftops . . .
And through my windshield wipers
The snow is piled dark and grey . . .
Next to my driveway where I check my mail
Little footprints on fresh snow —
A visiting rabbit.

I knew a bank where the wild thyme blew
Over-canopied with luscious woodbine
It is now a landfill —
Fermentation of civilization
Flowers on TV
Hyacinth rose tulip chrysanthemum
Acres of colour
Wind up wrapped in decorous plastic,
In this landfill where oxlips grew. . .

Understanding Literature

Cause and Effect Recognizing cause-and-effect relationships can help you make sense out of what you read. One event causes another event. The second event is the effect of the first event. In the poem, the author describes the causes and effects of pollution and waste. What effects do pollution and the use of nonrenewable resources have on nature in the poem?

Respond to the Reading

1. To plagiarize is to copy without giving credit to the source. In this poem, who or what has plagiarized beauty?
2. What do the four verses in the poem correspond to?
3. **Linking Science and Writing** Write a poem that shows how conservation methods could restore the beauty in nature.

INTEGRATE Life Science The poet makes a connection between the four seasons of the year and the pollution and waste products created by human activity, or civilization. For example, in the spring, a landfill for dumping garbage replaces a field of wildflowers. Describing four seasons instead of one reinforces the poet's message that the beauty of nature has been stolen, or plagiarized.

Resources for Teachers and Students

A Walk in the Woods: Rediscovering America on the Appalachian Trail, by Bill Bryson, Broadway Books, 1999

The Best American Science & Nature Writing 2000, ed. Burkhard Bilger, Houghton Mifflin Co., 2000

Lost Woods: The Discovered Writing of Rachel Carson, by Rachel Carson, ed. Linda Lear, Beacon Press, 1999

118 ◆ E CHAPTER 4 Conserving Resources

chapter 4 Study Guide

Reviewing Main Ideas

Section 1 Resources

1. Natural resources are the parts of the environment that supply materials needed for the survival of living organisms.
2. Renewable resources are being replaced continually by natural processes.
3. Nonrenewable resources cannot be replaced or are replaced very slowly.
4. Energy sources include fossil fuels, wind, solar energy, geothermal energy, hydroelectric power, and nuclear power.

Section 2 Pollution

1. Most air pollution is made up of waste products from the burning of fossil fuels.
2. The greenhouse effect is the warming of Earth by a blanket of heat-reflecting gases in the atmosphere.
3. Water can be polluted by acid rain and by the spilling of oil or other wastes into waterways.
4. Solid wastes and hazardous wastes dumped on land or disposed of in landfills can pollute the soil. Erosion can cause the loss of fertile topsoil.

Section 3 The Three Rs of Conservation

1. You can reduce your use of natural resources in many ways.
2. Reusing items is an excellent way to practice conservation.
3. In recycling, materials are changed in some way so that they can be used again.
4. Materials that can be recycled include paper, metals, glass, plastics, yard waste, and nonmeat kitchen scraps.

Visualizing Main Ideas

Copy and complete the following concept map using the terms smog, acid precipitation, *and* ozone depletion.

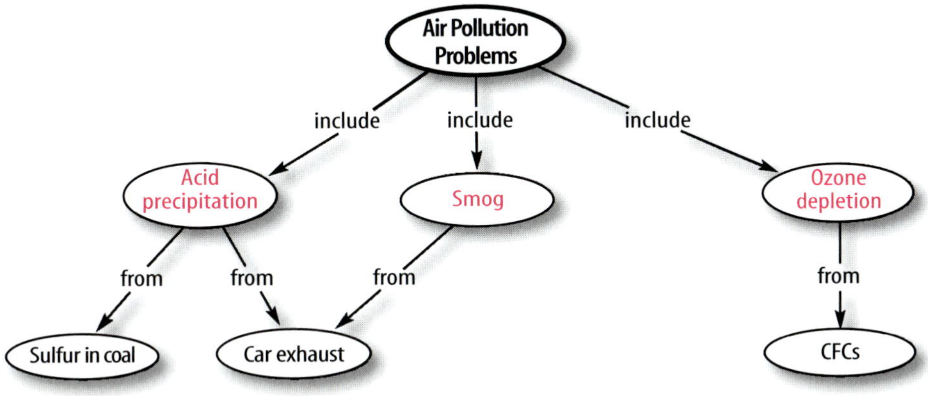

chapter 4 Study Guide

Reviewing Main Ideas

Summary statements can be used by students to review the major concepts of the chapter.

Visualizing Main Ideas

See student page.

Science Online

Visit booke.msscience.com
/self_check_quiz
/interactive_tutor
/vocabulary_puzzlemaker
/chapter_review
/standardized_test

Assessment Transparency

For additional assessment questions, use the *Assessment Transparency* located in the transparency book.

Assessment

Assessment Transparency — **Conserving Resources**

Directions: *Carefully review the table and answer the following questions.*

Alternatives to Fossil Fuels		
Type	**How it works**	**Environmental impacts**
Hydroelectric power	Energy of moving water is converted into electrical energy.	Habitats destroyed by dams
Wind power	Energy of moving air is converted into electrical energy.	No significant problem
Nuclear power	Energy of nuclear fission heats water, which is used to generate electricity.	Produces hazardous waste and has risk of accidents
Solar power	Energy from the Sun is absorbed and converted into heat and electrical energy.	No significant problem

1. According to the table, a windmill would be an example of ___.
   A hydroelectric power
   B wind power
   C nuclear power
   D solar power
2. Which of these could be added to the table?
   F Coal power
   G Oil power
   H Gasoline power
   J Geothermal power
3. According to the table, which fuel is most likely to affect human health?
   A Hydroelectric power
   B Wind power
   C Nuclear power
   D Solar power

Conserving Resources

Using Vocabulary

1. Petroleum is a type of fossil fuel.
2. Erosion produces pollutants and degrades the environment.
3. These are both types of air pollution, but one does not cause the other.
4. The burning of fossil fuels and the release of excess carbon dioxide may enhance the greenhouse effect.
5. The use of nuclear energy generates radioactive materials, which are hazardous wastes.
6. The use of hydroelectric power can decrease the use of fossil fuels.
7. Coal is a fossil fuel. The burning of coal that contains sulfur causes acid precipitation.
8. The pollutant called CFC causes ozone depletion.
9. Recycling is a good way of conserving nonrenewable resources.
10. The use of geothermal energy can decrease the use of fossil fuels.

Checking Concepts

11. B	14. B	17. B
12. C	15. C	18. A
13. C	16. C	

Thinking Critically

19. Most geothermal energy is used to produce steam that rotates turbine blades of an electric generator.
20. Fossil fuels were formed from plants and animals. Wood is from a plant.
21. Desert region; sunshine is more constant there than at the poles.
22. to prevent erosion of topsoil
23. Yes; it is constantly replaced and has many uses.
24. Solar, wind, geothermal, and water energy are renewable resources. Nuclear energy is not a renewable resource, but no severe shortages are expected.
25. Reusing the bag would save energy because it wouldn't need to be changed back into its raw form and then a new bag made.

chapter 4 Review

Using Vocabulary

acid precipitation p. 103
erosion p. 109
fossil fuel p. 96
geothermal energy p. 99
greenhouse effect p. 104
hazardous waste p. 110
hydroelectric power p. 97
natural resource p. 94
nonrenewable resource p. 95
nuclear energy p. 98
ozone depletion p. 105
petroleum p. 95
pollutant p. 102
recycling p. 113
renewable resource p. 94

Explain the differences in the vocabulary words given below. Then explain how the words are related. Use complete sentences in your answers.

1. fossil fuel—petroleum
2. erosion—pollutant
3. ozone depletion—acid precipitation
4. greenhouse effect—fossil fuels
5. hazardous wastes—nuclear energy
6. hydroelectric power—fossil fuels
7. acid precipitation—fossil fuels
8. ozone depletion—pollutant
9. recycle—nonrenewable resources
10. geothermal energy—fossil fuels

Checking Concepts

Choose the word or phrase that best answers the question.

11. An architect wants to design a solar house in the northern hemisphere. For maximum warmth, which side of the house should have the most windows?
A) north **C)** east
B) south **D)** west

12. Of the following, which is considered a renewable resource?
A) coal **C)** sunlight
B) oil **D)** aluminum

Use the photo below to answer question 13.

13. Which energy resource is shown in the photo?
A) solar energy
B) geothermal energy
C) hydroelectric energy
D) photovoltaic energy

14. Which of the following is a fossil fuel?
A) wood **C)** nuclear power
B) oil **D)** photovoltaic cell

15. Which of the following contributes to ozone depletion?
A) carbon dioxide **C)** CFCs
B) radon **D)** carbon monoxide

16. What is a substance that contaminates the environment called?
A) acid rain **C)** pollutant
B) pollution **D)** ozone

17. If there were no greenhouse effect in Earth's atmosphere, which of the following statements would be true?
A) Earth would be much hotter.
B) Earth would be much colder.
C) The temperature of Earth would be the same.
D) The polar ice caps would melt.

18. Which of the following can change solar energy into electricity?
A) photovoltaic cells
B) smog
C) nuclear power plants
D) geothermal power plants

 Science Online booke.msscience.com/vocabulary_puzzlemaker

Thinking Critically

19. **Explain** how geothermal energy is used to produce electricity.
20. **Infer** why burning wood and burning fossil fuels produce similar pollutants.

Use the photos below to answer question 21.

21. **Draw a Conclusion** Which would make a better location for a solar power plant—a polar region (left) or a desert region (right)? Why?
22. **Explain** why it is beneficial to grow a different crop on soil after the major crop has been harvested.
23. **Infer** Is garbage a renewable resource? Why or why not?
24. **Summarize** Solar, nuclear, wind, water, and geothermal energy are alternatives to fossil fuels. Are they all renewable? Why or why not?
25. **Draw Conclusions** Would you save more energy by recycling or reusing a plastic bag?
26. **Recognize Cause and Effect** Forests use large amounts of carbon dioxide during photosynthesis. How might cutting down a large percentage of Earth's forests affect the greenhouse effect?
27. **Form a hypothesis** about why Americans throw away more aluminum cans each year.
28. **Compare and contrast** contour farming, terracing, strip cropping, and no-till farming.

Performance Activities

29. **Poster** Create a poster to illustrate and describe three things students at your school can do to conserve natural resources.

Applying Math

Use the table below to answer questions 30 and 31.

Estimated Recycling Rates

Item	Percent Recycled
Aluminum cans	60
Glass beverage bottles	31
Plastic soft-drink containers	37
Newsprint	56
Magazines	23

30. **Recycling Rates** Make a bar graph of the data above.
31. **Bottle Recycling** For every 1,000 glass beverage bottles that are produced, how many are recycled?
32. **Nonrenewable Resources** 45.8 billion (45,800,000,000) cans were thrown away in 2000. If it takes 33.79 cans to equal one pound and the average scrap value is \$0.58/lb, then what was the total dollar value of the discarded cans?
33. **Ozone Depletion** The thin ozone layer called the "ozone hole" over Antarctica reached nearly 27,000,000 km^2 in 1998. To conceptualize this, the United States has a geographical area of 9,363,130 km^2. How much larger is the "ozone hole" in comparison to the United States?
34. **Increased CO_2 Levels** To determine the effects of increased CO_2 levels in the atmosphere, scientists increased the CO_2 concentration by 70 percent in an enclosed rain forest environment. If the initial CO_2 concentration was 430 parts per million, what was it after the increase?

chapter Review 4

26. Cutting large forests could result in more carbon dioxide in the atmosphere, possibly resulting in an enhanced greenhouse effect and accelerated global warming.
27. Answers will vary. Students should support their answers.
28. Contour farming and terracing are similar in that they are farming methods for hillsides that reduce soil erosion. Strip cropping compares to no-till farming in that they are both methods of reducing soil erosion by leaving soil covered by vegetation. The first two methods contrast to the last two in that they are used on hillsides, while the last two are used on level ground.

Performance Activities

29. Posters should involve reducing, reusing, or recycling. Use **Performance Assessment in the Science Classroom,** p. 145.

Applying Math

National Math Standards

1, 2, 5, 9

30. bars (highest to lowest): aluminum cans, newsprint, plastic soft-drink containers, glass beverage bottles, magazines
31. 310 are recycled
32. \$786, 149, 748
33. 2.88 times larger
34. $430 + (430 \times 0.7) = 731$ ppm

Assessment Resources

Reproducible Masters

Chapter *Fast File* Resources
- Chapter Review, pp. 37–38
- Chapter Tests, pp. 39–42
- Assessment Transparency Activity, p. 49

Glencoe Science Web site
- Chapter Review Test
- Standardized Test Practice

Glencoe Technology

- Assessment Transparency
- Exam*View*® Pro Testmaker
- MindJogger Videoquiz
- Interactive Chalkboard

FAST FILE

Answer Sheet A practice answer sheet can be found at booke.msscience.com/answer_sheet.

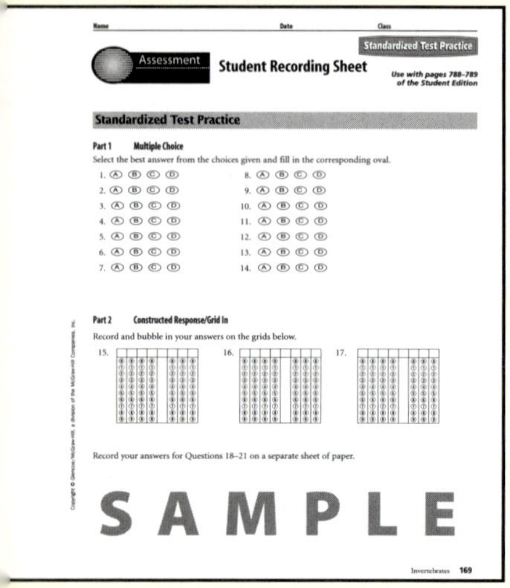
Name Date Class

Standardized Test Practice

Assessment Student Recording Sheet

Use with pages 788–789 of the Student Edition

Standardized Test Practice

Part 1 Multiple Choice

Select the best answer from the choices given and fill in the corresponding oval.

Part 2 Constructed Response/Grid In

Record and bubble in your answers on the grids below.

Record your answers for Questions 18–21 on a separate sheet of paper.

SAMPLE

Part 1 Multiple Choice

1. B	4. C	7. C
2. A	5. B	8. B
3. A	6. D	9. A

Part 2 Short Response

10. Possible answers: renewable source—water, sunlight, wind; nonrenewable source—coal, oil, natural gas
11. solar energy or solar power
12. Benefit—it is renewable, it does not pollute; drawback—you need sunlight to make it work, not as convenient as using fossil fuels.
13. Possible answers: use public transportation, use electric cars, use hand mowers
14. Acid rain has a pH of 5.6 or below. The rain collected by the students is almost neutral on the pH scale. It is not acid rain.
15. Without the greenhouse effect, the heat from Earth would escape into space and the temperatures on Earth would be too cold for us to survive.
16. raw sewage or fertilizers

chapter 4 Standardized Test Practice

Part 1 Multiple Choice

Record your answers on the answer sheet provided by your teacher or on a sheet of paper.

1. From what natural resource are plastics, paints, and gasoline made?
A. coal **C.** iron ore
B. petroleum **D.** natural gas

Use the illustration below to answer questions 2–4.

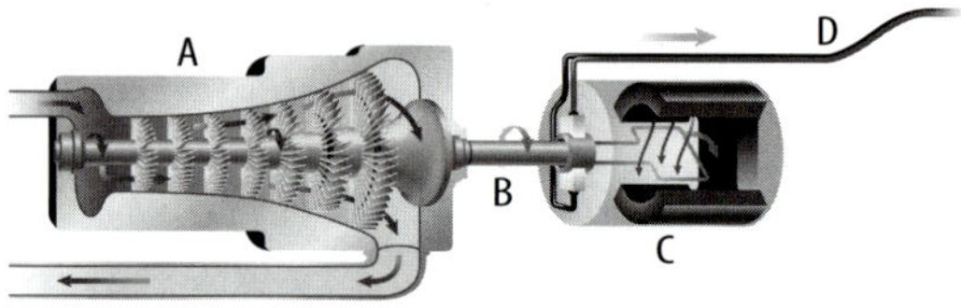

2. What is produced by the mechanism shown in the illustration?
A. electricity **C.** petroleum
B. coal **D.** plastic

3. In which section are the turbine blades found?
A. A **C.** C
B. B **D.** D

4. Which section represents the generator?
A. A **C.** C
B. B **D.** D

5. Which of the following is necessary for the production of hydroelectric power?
A. wind
B. access to a river
C. exposure to sunlight
D. heat from below Earth's crust

6. With which type of alternative energy are photovoltaic cells used?
A. hydroelectric power
B. geothermal energy
C. nuclear energy
D. solar energy

7. Which of the following is a type of air pollution that results when sunlight reacts with pollutants produced by burning fuels?
A. ozones **C.** smog
B. acid rain **D.** UV radiation

Use the photograph below to answer questions 8 and 9.

8. What is the name of the method of farming illustrated above?
A. contour plowing **C.** terracing
B. strip cropping **D.** no-till farming

9. What is the purpose of the method shown in the illustration?
A. to decrease soil erosion from wind
B. to decrease soil erosion from water flow
C. to decrease acid rain production
D. to increase the return of nutrients to the soil

Test-Taking Tip

Qualifying Terms Look for qualifiers in a question. Such questions are not looking for absolute answers. Qualifiers could be words such as most likely, most common, or least common.

Question 18 The qualifier in this question is *possible.* This indicates that there is uncertainty about the effects of global warming.

Part 3 Open Ended

17. Renewable resources are not always available in all areas at all times. Water, for example, may be in short supply if there is a drought. Wind may not be plentiful in a sheltered area. Sunlight may be in short supply in certain locations.
18. As a result of global warming, rainfall patterns may change, which could alter ecosystems; other weather patterns may change; polar ice caps might melt, which would raise sea levels. The cause of global warming is unknown. Burning fossil fuels increases the amount of CO_2 in the atmosphere, which traps more of the Sun's heat, which makes Earth warmer.
19. If there was no sunlight, the family might run out of hot water so it might be difficult to take a warm shower or bath, or wash dishes in warm water.

Part 2 Short Response/Grid In

Record your answers on the answer sheet provided by your teacher or on a sheet of paper.

10. Give one example of a renewable source of energy and one example of a nonrenewable source of energy.

Use the illustration below to answer questions 11 and 12.

11. What type of alternative energy is the girl using in the diagram?

12. Name one benefit and one drawback to using this type of energy for cooking.

13. What are two ways that smog can be reduced?

14. A group of students collects rain outside their classroom, then tests the pH of the collected rain. The pH of the rain is 7.2. Can the students say that their rain is acid rain? Why or why not?

15. Why do we depend on the greenhouse effect for survival?

16. What is the cause of algal blooms in lakes and ponds?

Part 3 Open Ended

Record your answers on a sheet of paper.

17. Are renewable resources always readily available? Explain.

18. What are the possible worldwide effects of global warming? What causes global warming? Why do some people think that using fossil fuels less will decrease global warming?

19. A family lives in a house that uses solar panels to heat the hot water, a wood-burning stove to heat the house, and a windmill for pumping water from a well into a tower where it is stored and then piped into the house as needed. What would be the result if there was no sunlight for two weeks?

Use the illustration below to answer question 20.

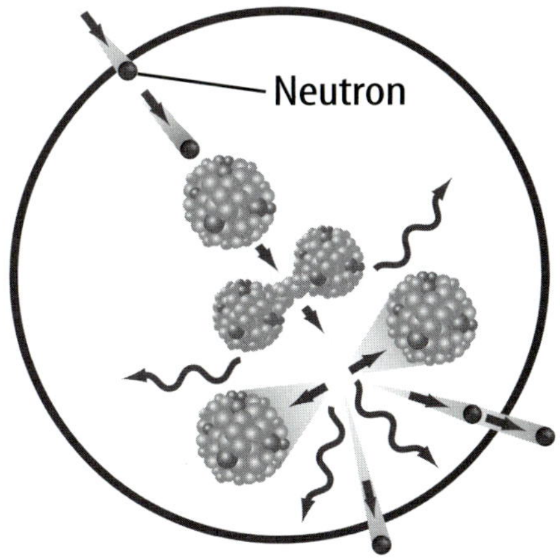

20. What does the diagram represent?

21. Explain how different kinds of plastics are recycled.

20. The diagram shows the source of energy for nuclear power. It is what happens inside the fuel rods of a nuclear reactor.

21. Type 1 plastics (soft-drink bottles) are melted and spun into fibers that are used to make other things. Type 2 or 4 plastics (plastic bags) can be reused or recycled. Type 6 and 7 plastics cannot be recycled.

Rubrics

The following rubrics are sample scoring devices for short response and open-ended questions.

Short Response

Points	Description
2	The student demonstrates a thorough understanding of the science of the task. The response may contain minor flaws that do not detract from the demonstration of a thorough understanding.
1	The student has provided a response that is only partially correct.
0	The student has provided a completely incorrect solution or no response at all.

Open Ended

Points	Description
4	The student demonstrates a thorough understanding of the science of the task. The response may contain minor flaws that do not detract from the demonstration of a thorough understanding.
3	The student demonstrates an understanding of the science of the task. The response is essentially correct and demonstrates an essential but less than thorough understanding of the science.
2	The student demonstrates only a partial understanding of the science of the task. Although the student may have used the correct approach to a solution or may have provided a correct solution, the work lacks an essential understanding of the underlying science concepts.
1	The student demonstrates a very limited understanding of the science of the task. The response is incomplete and exhibits many flaws.
0	The student provides a completely incorrect solution or no response at all.

chapter 5 Organizer

Section/Objectives	Standards		Labs/Features
	National	State/Local	
Chapter Opener	See pp. 9T–10T for a Key to Standards.		**Launch Lab:** Recognize Environmental Differences, p. 125 **Foldables,** p. 125
Section 1 Biodiversity 3 sessions 1.5 blocks 1. **Define** biodiversity. 2. **Explain** why biodiversity is important in an ecosystem. 3. **Identify** factors that limit biodiversity in an ecosystem.	National Content Standards: UCP.2, UCP.3, UCP.5, A.1, A.2, C.1, C.4, C.5, F.2, F.4		**Applying Math:** Measuring Biodiversity, p. 129 **Visualizing Threatened and Endangered Species,** p. 132 **MiniLAB:** Demonstrating Habitat Loss, p. 133 **MiniLAB:** Testing the Effects of Acid Rain, p. 135 **Integrate Health,** p. 136 **Lab:** Oily Birds, p. 137
Section 2 Conservation Biology 3 sessions 1.5 blocks 4. **Identify** several goals of conservation biology. 5. **Recommend** strategies to prevent the extinction of species. 6. **Explain** how an endangered species can be reintroduced into its original habitat.	National Content Standards: UCP.2, UCP.3, UCP.5, A.1, A.2, F.2, F.4		**Science Online,** p. 139 **Integrate Social Studies,** p. 140 **Science Online,** p. 142 **Lab:** Biodiversity and the Health of a Plant Community, pp. 144–145 **Science and Society:** Rain Forest Troubles, p. 146

Glencoe Exclusive! TeacherWorks™ All-In-One Planner and Resource Center

Lab Materials	Reproducible Resources	Section Assessment	Technology
Launch Lab: meterstick, magnifying lens, string or tape, field guides	**Chapter *FAST FILE* Resources** Foldables Worksheet, p. 17 Directed Reading Overview, p. 19 Note-taking Worksheets, pp. 31–32	Glencoe's Assessment Advantage	TeacherWorks includes: • Interactive Teacher Edition • Lesson Planner with calendar • Access to all program blacklines • Correlations to standards • Web links
MiniLAB: banana, jar, pieces of cloth or stocking, rubber band **MiniLAB:** mustard seeds (100), water, plastic cups (2), vinegar, paper towel, self-sealing plastic bags (2) **Lab:** beakers (2); bird feathers; vegetable oil; red, blue, and green food coloring; paper towels, water, cotton balls, cotton swabs, sponge, toothbrush, dish soap	**Chapter *FAST FILE* Resources** Transparency Activity, p. 42 MiniLABs, pp. 3, 4 Enrichment, p. 29 Reinforcement, p. 27 Directed Reading, p. 20 Transparency Activity, pp. 45–47 Lab Activity, pp. 9–12 Lab Worksheet, pp. 5–6 **Cultural Diversity,** p. 19 **Reading and Writing Skill Activities,** pp. 1, 37 **Mathematics Skill Activities,** p. 45	**Portfolio** Differentiated Instruction, p. 127 MiniLAB Assessment, p. 135 Assessment, p. 136 **Performance** Applying Skills, p. 136 **Content** Section Review, p. 136	Section Focus Transparency Teaching Transparency Virtual Labs CD-ROM Guided Reading Audio Program Interactive Chalkboard CD-ROM
Lab: plain white paper, colored paper (red, orange, yellow, green, blue, black), ruler, scissors, pens, die *Need materials?* Contact Science Kit at 1-800-828-7777 or www.sciencekit.com on the Internet.	**Chapter *FAST FILE* Resources** Transparency Activity, p. 43 Enrichment, p. 30 Reinforcement, p. 28 Directed Reading, pp. 21, 22 Lab Activity, pp. 13–16 Lab Worksheet, pp. 7–8 **Lab Management and Safety,** p. 73 **Home and Community Involvement,** p. 33	**Portfolio** Curriculum Connection, p. 141 **Performance** Applying Skills, p. 143 **Content** Section Review, p. 143	Section Focus Transparency Guided Reading Audio Program Interactive Chalkboard CD-ROM Video Lab

End of Chapter Assessment

Glencoe's Assessment Advantage

Blackline Masters	Technology	Professional Series
Chapter *FAST FILE* Resources Chapter Review, pp. 35–36 Chapter Tests, pp. 37–40 **Standardized Test Practice,** pp. 23–26	MindJogger Videoquiz Virtual Labs CD-ROM Exam*View*® Pro Testmaker TeacherWorks CD-ROM Interactive Chalkboard CD-ROM	**Performance Assessment in the Science Classroom (PASC)**

Transparencies

Section Focus

SECTION 1 Section Focus Transparency **Oops!**

This looks like an ordinary fish, right? Well, it's not. It's a coelacanth, a fish thought to be extinct for 70 million years! Occasionally caught in the nets of Madagascar fisherman, a coelacanth came to the attention of western scientists in 1938. Since then, a few coelacanths have been caught each year. The fish has been around, relatively unchanged, for almost 400 million years.

1. What does it mean for a species to be extinct?
2. Why might scientists have thought the coelacanth was extinct? How did they know it had even existed?
3. The coelacanth is now considered endangered. What do you think this means?

L2

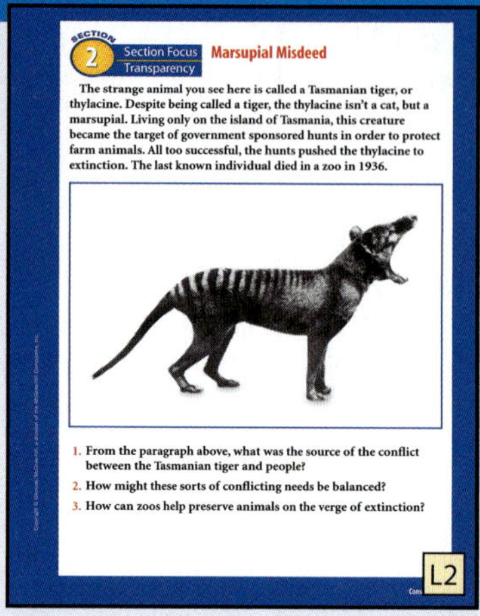
SECTION 2 Section Focus Transparency **Marsupial Misdeed**

The strange animal you see here is called a Tasmanian tiger, or thylacine. Despite being called a tiger, the thylacine isn't a cat, but a marsupial. Living only on the island of Tasmania, this creature became the target of government sponsored hunts in order to protect farm animals. All too successful, the hunts pushed the thylacine to extinction. The last known individual died in a zoo in 1936.

1. From the paragraph above, what was the source of the conflict between the Tasmanian tiger and people?
2. How might these sorts of conflicting needs be balanced?
3. How can zoos help preserve animals on the verge of extinction?

L2

This is a representation of key blackline masters available in the Teacher Classroom Resources. See Resource Manager boxes within the chapter for additional information.

Assessment

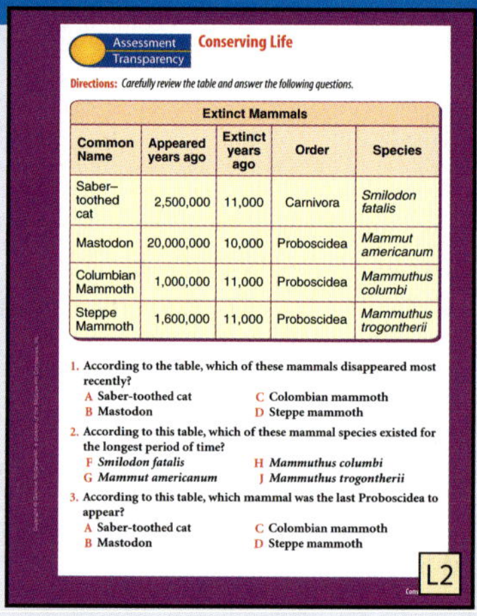
Assessment Transparency **Conserving Life**

Directions: *Carefully review the table and answer the following questions.*

Extinct Mammals				
Common Name	Appeared years ago	Extinct years ago	Order	Species
Saber-toothed cat	2,500,000	11,000	Carnivora	*Smilodon fatalis*
Mastodon	20,000,000	10,000	Proboscidea	*Mammut americanum*
Columbian Mammoth	1,000,000	11,000	Proboscidea	*Mammuthus columbi*
Steppe Mammoth	1,600,000	11,000	Proboscidea	*Mammuthus trogontherii*

1. According to the table, which of these mammals disappeared most recently?
   A Saber-toothed cat
   B Mastodon
   C Colombian mammoth
   D Steppe mammoth
2. According to this table, which of these mammal species existed for the longest period of time?
   F *Smilodon fatalis*
   G *Mammut americanum*
   H *Mammuthus columbi*
   J *Mammuthus trogontherii*
3. According to this table, which mammal was the last Proboscidea to appear?
   A Saber-toothed cat
   B Mastodon
   C Colombian mammoth
   D Steppe mammoth

L2

Teaching

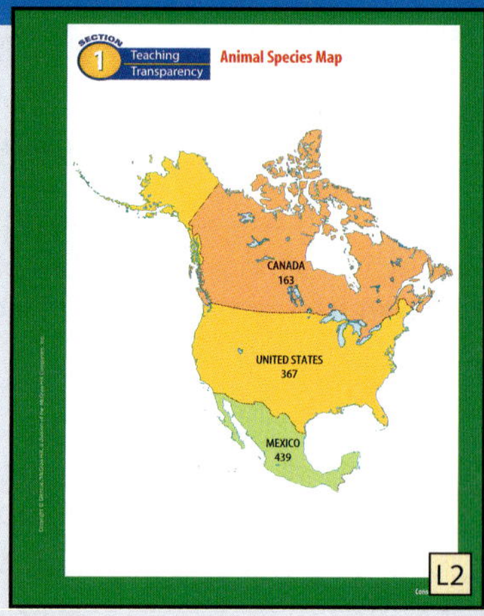

Key to Teaching Strategies

The following designations will help you decide which activities are appropriate for your students.

- **L1** Level 1 activities should be appropriate for students with learning difficulties.
- **L2** Level 2 activities should be within the ability range of all students.
- **L3** Level 3 activities are designed for above-average students.
- **ELL** ELL activities should be within the ability range of English Language Learners.
- **COOP LEARN** Cooperative Learning activities are designed for small group work.
- **LS** Multiple Learning Styles logos, as described on page 6T, are used throughout to indicate strategies that address different learning styles.
- **P** These strategies represent student products that can be placed into a best-work portfolio.
- **PBL** Problem-Based Learning activities apply real-world situations to learning.

Hands-on Activities

Student Text Lab Worksheet

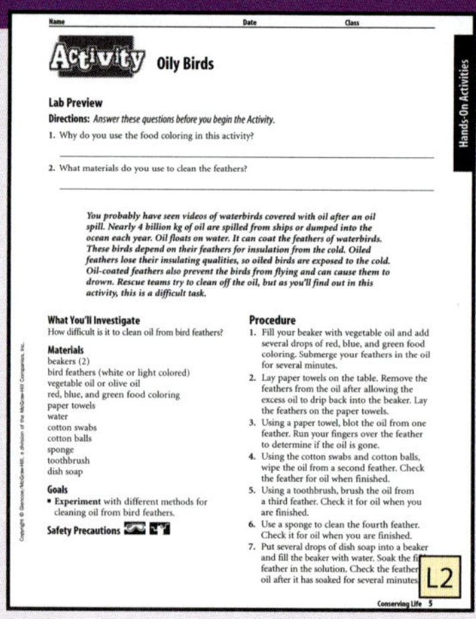
Name Date Class

Activity **Oily Birds**

Hands-On Activities

Lab Preview

Directions: *Answer these questions before you begin the Activity.*

1. Why do you use the food coloring in this activity?
2. What materials do you use to clean the feathers?

You probably have seen videos of waterbirds covered with oil after an oil spill. Nearly 4 billion kg of oil are spilled from ships or dumped into the ocean each year. Oil floats on water. It can coat the feathers of waterbirds. These birds depend on their feathers for insulation from the cold. Oiled feathers lose their insulating qualities, so oiled birds are exposed to the cold. Oil-coated feathers also prevent the birds from flying and can cause them to drown. Rescue teams try to clean off the oil, but as you'll find out in this activity, this is a difficult task.

What You'll Investigate
How difficult is it to clean oil from bird feathers?

Materials
beakers (2)
bird feathers (white or light colored)
vegetable oil or olive oil
red, blue, and green food coloring
paper towels
water
cotton swabs
cotton balls
sponge
toothbrush
dish soap

Goals
- **Experiment** with different methods for cleaning oil from bird feathers.

Safety Precautions

Procedure
1. Fill your beaker with vegetable oil and add several drops of red, blue, and green food coloring. Submerge your feathers in the oil for several minutes.
2. Lay paper towels on the table. Remove the feathers from the oil after allowing the excess oil to drip back into the beaker. Lay the feathers on the paper towels.
3. Using a paper towel, blot the oil from one feather. Run your fingers over the feather to determine if the oil is gone.
4. Using the cotton swabs and cotton balls, wipe the oil from a second feather. Check the feather for oil when finished.
5. Using a toothbrush, brush the oil from a third feather. Check it for oil when you are finished.
6. Use a sponge to clean the fourth feather. Check it for oil when you are finished.
7. Put several drops of dish soap into a beaker and fill the beaker with water. Soak the fi[illegible] feather in the solution. Check the feather[illegible] oil after it has soaked for several minutes.

L2

Conserving Life 5

Laboratory Activities

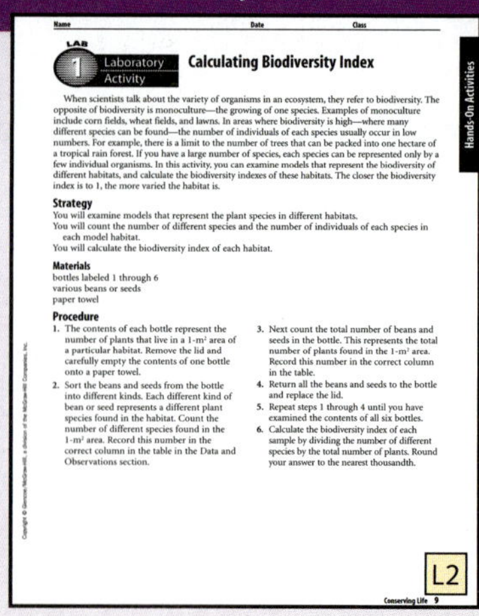
Name Date Class

LAB 1 Laboratory Activity **Calculating Biodiversity Index**

Hands-On Activities

When scientists talk about the variety of organisms in an ecosystem, they refer to biodiversity. The opposite of biodiversity is monoculture—the growing of one species. Examples of monoculture include corn fields, wheat fields, and lawns. In areas where biodiversity is high—where many different species can be found—the number of individuals of each species usually occur in low numbers. For example, there is a limit to the number of trees that can be packed into one hectare of a tropical rain forest. If you have a large number of species, each species can be represented only by a few individual organisms. In this activity, you can examine models that represent the biodiversity of different habitats, and calculate the biodiversity indexes of these habitats. The closer the biodiversity index is to 1, the more varied the habitat is.

Strategy
You will examine models that represent the plant species in different habitats.
You will count the number of different species and the number of individuals of each species in each model habitat.
You will calculate the biodiversity index of each habitat.

Materials
bottles labeled 1 through 6
various beans or seeds
paper towel

Procedure
1. The contents of each bottle represent the number of plants that live in a 1-m^2 area of a particular habitat. Remove the lid and carefully empty the contents of one bottle onto a paper towel.
2. Sort the beans and seeds from the bottle into different kinds. Each different kind of bean or seed represents a different plant species found in the habitat. Count the number of different species found in the 1-m^2 area. Record this number in the correct column in the table in the Data and Observations section.
3. Next count the total number of beans and seeds in the bottle. This represents the total number of plants found in the 1-m^2 area. Record this number in the correct column in the table.
4. Return all the beans and seeds to the bottle and replace the lid.
5. Repeat steps 1 through 4 until you have examined the contents of all six bottles.
6. Calculate the biodiversity index of each sample by dividing the number of different species by the total number of plants. Round your answer to the nearest thousandth.

L2

Conserving Life 9

Meeting Different Ability Levels

Content Outline

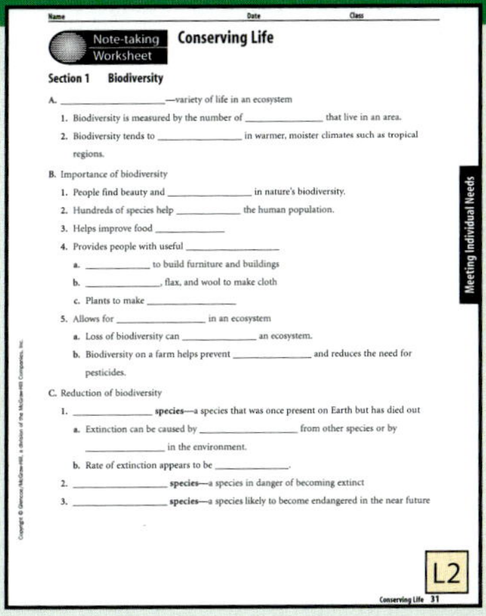

Reinforcement

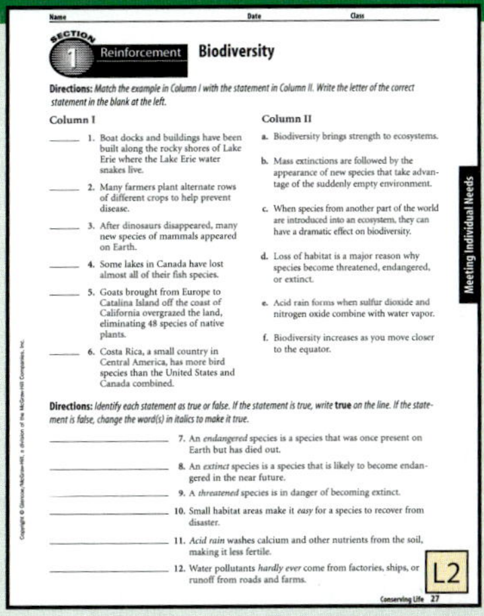

Enrichment

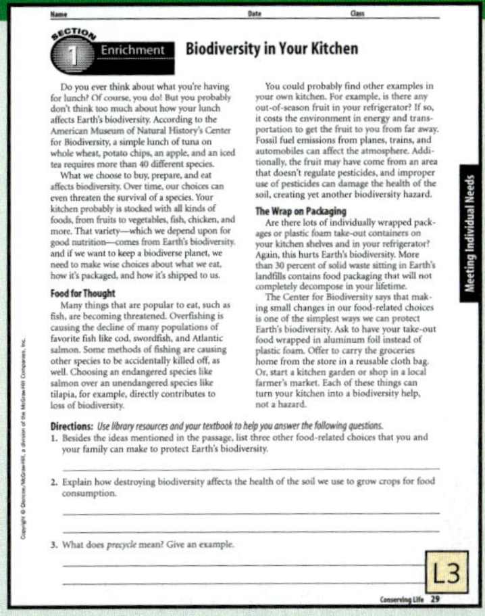

Directed Reading (English/Spanish)

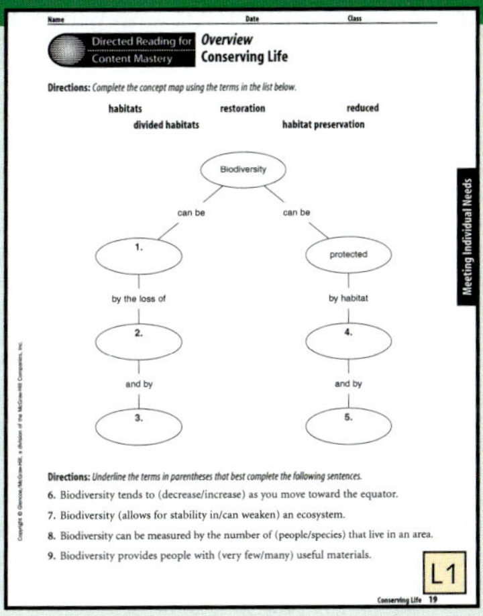

Study Guide

Reading Essentials

Assessment

Test Practice Workbook

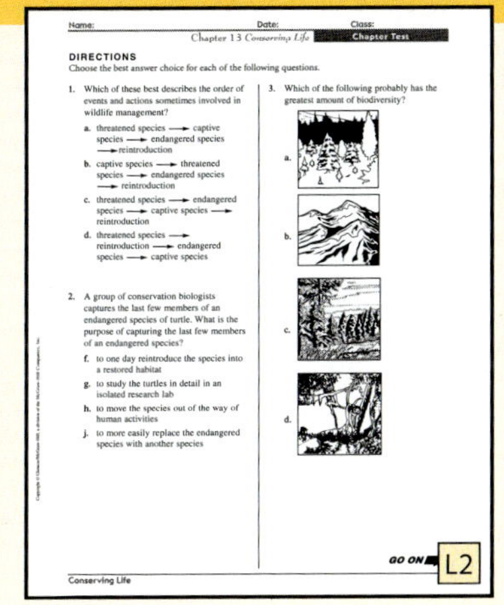

Chapter Review

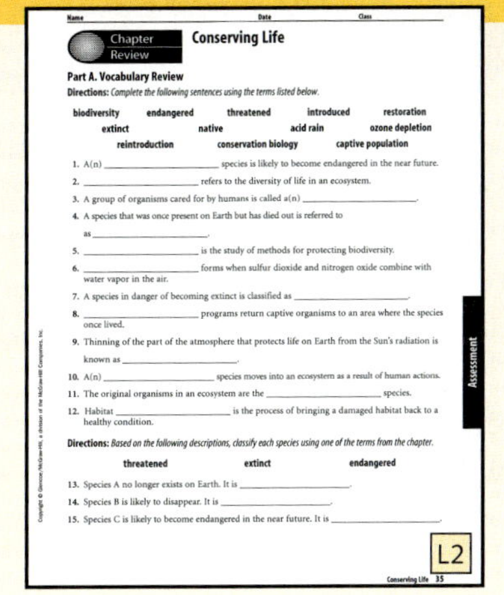

Chapter Tests

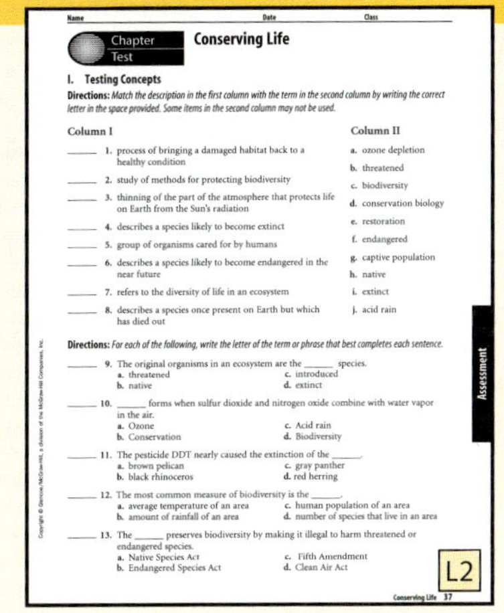

Science Content Background

Gloria H. Chomica/Masterfile

section 1 Biodiversity

Definitions and Understandings

Biodiversity refers to the variety of life forms, the genetic information they carry, and the ecosystems in which the life forms exist. Humans are dependent upon and interact with biotic and abiotic systems for sustenance of life, health, well being, and recreation. Many biologists think that rich biodiversity results in ecosystems having greater resilience and more efficient recovery from natural and human induced stresses.

Threatened, Endangered, and Extinct Species

Threatened species are in danger of becoming extinct because the size of the populations required for breeding is at a critical minimum. Threatened species are different from endangered species—species that are in immediate danger of becoming extinct. Most species are threatened first and then become endangered. Extinct species, such as the passenger pigeon, no longer exist on Earth.

Factors Decreasing Biodiversity—Habitat Loss and Habitat Division

Habitat loss and habitat division, such as that which occurs when cities expand and roads are built, is one of the major reasons why a species becomes threatened or endangered. As habitats decrease, fewer required resources are available and the population decreases. As a result, the biodiversity of the area decreases. Tropical rain forests are examples of habitat loss and a decrease in biodiversity.

chapter content resources

Internet Resources

For additional content background, visit booke.msscience.com to:

- access your book online
- find references to related articles in popular science magazines
- access Web links with related content background
- access current events with science journal topics

Print Resources

A Walk in the Rain Forest, by Kristin Joy Pratt, Dawn Pubns, 1992

Endangered Species: Wild and Rare, by the National Wildlife Federation, McGraw-Hill, 1997

Teacher to Teacher

Steve Federman
Loveland Middle School
Loveland, Ohio

"To create an awareness of the impact an oil spill has on wildlife, I have my students attempt to clean oiled craft feathers with a grease-cutting liquid dish soap and water. We discuss the effect of such disasters as California's 1953 sunken *SS Luckenbach* oil spill, the *Exxon Valdez* in Alaska, or the plight of penguins near Cape Town, South Africa, as well as oiled-wildlife in the Galapagos Islands, or those affected by the Gulf War."

Steve Federman

Stanley Breeden/DRK Photo

Factors Decreasing Biodiversity—Introduced Species and Pollution

Introduced species are nonnative organisms that are brought to an area. Some, like the kudzu vine, are deliberately introduced to an area to solve a specific problem. Others, like the zebra mussel, are accidentally brought to an area as stowaways aboard ships and other transport vessels. However, these species may "take over" an area and force the native plants and animals out of the habitat. In extreme cases, they can cause the extinction of native species. Some of the materials humans use to combat introduced species, such as chemical pesticides, pollute the environment. Polluted air, water, and soil can reduce the biodiversity of an area. Organisms that are unable to relocate to a less polluted habitat or adapt to the changing conditions are in danger of becoming threatened, endangered, or even extinct.

section 2 Conservation Biology

Habitat Preservation

Legal steps have been taken to protect the habitats of endangered and threatened species. The Endangered Species Act of 1973 promotes the conservation of natural habitats and biological processes with the goal of increasing biodiversity. Animal corridors, like the one described in this section, have been constructed to connect divided habitats and reduce encounters between humans and animals. In many areas, habitat that was once destroyed is now being restored so that the species that once lived there can be reintroduced or can return naturally.

Wildlife Management

Wildlife management strategies are directed toward preservation, restoration, and enhancement of natural ecosystems. Practical wildlife management strategies, such as captive breeding programs that result in the reintroduction of an organism, can benefit all species. Strict control of wildlike resources through hunting regulations can help prevent a population from becoming too large for the habitat to support.

chapter 5

Chapter Vocabulary

biodiversity, p. 126
extinct species, p. 130
endangered species, p. 131
threatened species, p. 131
introduced species, p. 134
native species, p. 134
acid rain, p. 135
ozone depletion, p. 136
conservation biology, p. 138
habitat restoration, p. 141
captive populations, p. 142
reintroduction programs, p. 142

Science Journal Answers will vary. Examples: corn because we eat it or a mountain lion because it controls prey populations.

INTERACTIVE CHALKBOARD with Image Bank

PowerPoint® Presentations

This CD-ROM is an editable Microsoft® PowerPoint® presentation that includes:

- a pre-made presentation for every chapter
- interactive graphics
- animations
- audio clips
- image bank
- all new section and chapter questions
- Standardized Test Practice
- transparencies
- pre-lab questions for all labs
- Foldables directions
- links to booke.msscience.com

chapter 5

Conserving Life

chapter preview

sections

1 **Biodiversity**
Lab Oily Birds

2 **Conservation Biology**
Lab Biodiversity and the Health of a Plant Community

Virtual Lab How do introduced species impact the environment?

A Band Helps Preserve Biodiversity

Before formulating a plan to preserve a species, a conservation biologist must first understand what it is that the species needs. Biologists employ a number of techniques to study species, including marking individuals.

Science Journal List an organism that you feel is important for maintaining biodiversity and explain why it's important.

Theme Connection

Systems and Interactions Biodiversity is threatened because human actions alter systems in the biosphere and change interactions. Interventions to help alleviate these threats may help preserve biodiversity.

About the Photo

A Band Helps Preserve Biodiversity The photo shows a bird that was banded with a U.S. Fish and Wildlife aluminum band. These bands are used to identify migration routes, and to estimate survival and other demographic parameters.

Start-Up Activities

Launch LAB

Recognize Environmental Differences

Do human actions affect the number of species present in an ecosystem? What happens to organisms that are pushed out of an area when the environment changes?

1. Using string or tape, mark off a 1-m × 1-m area of lawn, sports field, or sidewalk.
2. Count and record the different species of plants and animals present in the sample area. Don't forget to include insects or birds that fly over the area, or organisms found by gently probing into the soil.
3. Repeat steps 1 and 2 in a partially wooded area, in a weedy area, or at the edge of a pond or stream.
4. **Think Critically** Make notes about your observations in your Science Journal. What kinds of human actions could have affected the environment of each sample area?

FOLDABLES™ Study Organizer

Biodiversity Make the following Foldable to help you identify what you already know, what you want to know, and what you learned about biodiversity.

STEP 1 **Fold** a vertical sheet of paper from side to side. Make the front edge about 1.25 cm shorter than the back edge.

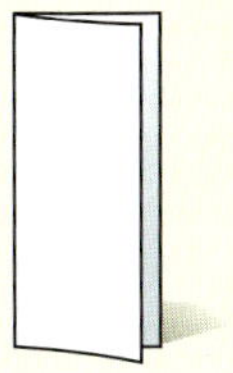

STEP 2 **Turn** lengthwise and **fold** into thirds.

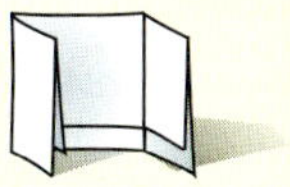

STEP 3 **Unfold and cut** only the top layer along both folds to make three tabs. **Label** each tab as shown.

Ask Questions Before you read the chapter, write what you already know about biodiversity under the left tab of your Foldable, and write questions about what you'd like to know under the center tab. After you read the chapter, list what you learned under the right tab.

Science Online — Preview this chapter's content and activities at booke.msscience.com

Launch LAB

Purpose Use the Launch Lab to allow students to compare biodiversity in two areas, showing the effect of human activities on ecosystems.

Preparation Before going outside, decide on convenient areas to compare.

Materials metersticks, magnifying lens, string or tape

Teaching Strategies

- Be sure students are well-supervised, especially if they are investigating near water.
- Alert students to the appearance of any poisonous plants they might encounter.
- Have students compare their findings. Discuss differences and similarities.

Think Critically

In general, the area that has been converted to use by humans, such as a lawn or sidewalk, will have fewer species than the natural area. Many human actions can affect environment, including cutting down trees, putting down concrete, and so on. Accept all reasonable answers.

Assessment

Performance Give students the following list of places and ask them to sequence them from least to most biodiversity: school lawn, school parking lot, nearby forest, classroom floor. classroom floor, school parking lot, school lawn, nearby forest Use **Performance Assessment in the Science Classroom,** p. 89.

FOLDABLES™ Study Organizer — Dinah Zike Study Fold

Student preparation materials for this Foldable are available in the Chapter *Fast File* Resources.

1 Motivate

INTERACTIVE CHALKBOARD PowerPoint® Presentations

Bellringer

Section Focus Transparencies also are available on the Interactive Chalkboard CD-ROM. L2 ELL

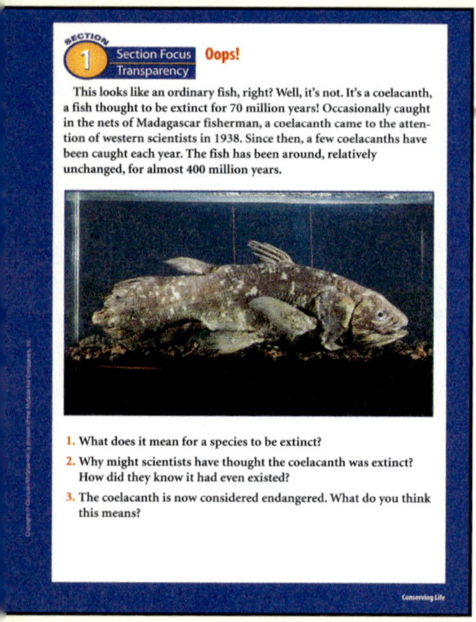

SECTION 1 Section Focus Transparency **Oops!**

This looks like an ordinary fish, right? Well, it's not. It's a coelacanth, a fish thought to be extinct for 70 million years! Occasionally caught in the nets of Madagascar fisherman, a coelacanth came to the attention of western scientists in 1938. Since then, a few coelacanths have been caught each year. The fish has been around, relatively unchanged, for almost 400 million years.

1. What does it mean for a species to be extinct?
2. Why might scientists have thought the coelacanth was extinct? How did they know it had even existed?
3. The coelacanth is now considered endangered. What do you think this means?

Conserving Life

Tie to Prior Knowledge

Endangered Species Have students name endangered or threatened species they know and suggest reasons why the species are endangered or threatened. Have them list any animals they can think of that are now extinct.

section 1

Biodiversity

as you read

What **You'll Learn**

- **Define** biodiversity.
- **Explain** why biodiversity is important in an ecosystem.
- **Identify** factors that limit biodiversity in an ecosystem.

Why **It's Important**

Knowledge of biological diversity can lead to strategies for preventing the loss of species.

Review Vocabulary

mammal: endothermic vertebrate with mammary glands and hair growing from their skin

New Vocabulary

- biodiversity
- extinct species
- endangered species
- threatened species
- introduced species
- native species
- acid rain
- ozone depletion

The Variety of Life

Imagine walking through a forest ecosystem like the one shown in **Figure 1.** Trees, shrubs, and small plants are everywhere. You see and hear squirrels, birds, and insects. You might notice a snake or mushrooms. Hundreds of species live in this forest. Now, imagine walking through a wheat field. You see only a few species—wheat plants, insects, and weeds. The forest contains more species than the wheat field does. The forest has a higher biological diversity, or biodiversity. **Biodiversity** refers to the variety of life in an ecosystem.

Measuring Biodiversity The common measure of biodiversity is the number of species that live in an area. For example, a coral reef can be home to thousands of species including corals, fish, algae, sponges, crabs, and worms. A coral reef has greater biodiversity than the shallow waters that surround it. Before deep-sea exploration, scientists thought that few organisms could live in dark, deep-sea waters. Although the number of organisms living there is likely to be less than the number of organisms on a coral reef, we know that the species biodiversity of deep-sea waters is as great as that of a coral reef.

Figure 1 The forest has more species and is richer in biodiversity than a wheat field.

Section 1 Resource Manager

Chapter *FAST FILE* Resources

Transparency Activity, pp. 42, 45–47
Directed Reading for Content Mastery, pp. 19, 20
Lab Activity, pp. 9–12
Note-taking Worksheets, pp. 31–32
Enrichment, p. 29
MiniLAB, pp. 3, 4
Lab Worksheet, pp. 5–6
Reinforcement, p. 27

Mathematics Skill Activities, p. 45
Life Science Critical Thinking/Problem Solving, pp. 5, 13, 21
Earth Science Critical Thinking/Problem Solving, p. 12
Performance Assessment in the Science Classroom, p. 55

Figure 2 This map shows the number of mammal species found in three North American countries. In general, biodiversity increases as you get closer to the equator.

Mountain lion

Arctic hare

Armadillo

Differences in Biodiversity Biodiversity tends to increase as you move toward the equator because temperatures tend to be warmer. For example, Costa Rica is a Central American country about the size of West Virginia. Yet it is home to as many bird species as there are in the United States and Canada combined. **Figure 2** compares mammal biodiversity in three North American countries. Ecosystems with the highest biodiversity usually have warm, moist climates. In fact, tropical regions contain two-thirds of all of Earth's land species.

Reading Check *What kind of climate usually has a high biodiversity?*

Why is biodiversity important?

Would you rather have a summer picnic on a paved parking lot or on cool grass under a shade tree near a stream? Many people find pleasure in nature's biodiversity. Perhaps you like to visit scenic areas to hike, swim, relax, or enjoy the views. Many painters, writers, and musicians find inspiration for their work by spending time outdoors. However, beauty and pleasure are not the only reasons why biodiversity is important.

2 Teach

Use Science Words

Word Origin Write "biological diversity" on the board. Ask students to suggest why these words were combined to make "biodiversity." Since the words were used so often, they were shortened into one word.

Activity

Local Species Have students make a list of different species that inhabit their local area. L2 ELL COOP LEARN IN **Interpersonal**

Visual Learning

Figure 2 Ask students which North American country has the most mammals. Mexico Is this because of its size? No, Mexico's warm climate is one reason it has greater mammal biodiversity.

Discussion

Biodiversity Why do warmer areas have greater biodiversity? Possible answers: longer growing seasons produce more plant growth and therefore more food; recent ice ages have affected areas near the equator less

Reading Check

Answer warm, moist climates

Differentiated Instruction

English-Language Learners Have students make an alphabet of biodiversity. Use a dictionary and, for each letter of the alphabet, find a species used by people. For "x," it is acceptable if the species name contains an x. Write how the species is used. L2 ELL IN **Linguistic** P

IDENTIFYING Misconceptions

Medicine Beginnings Students may not realize that many medicines began as substances isolated from living organisms. Use **Figure 3** to point this out to students.

Activity

Biodiversity Importance To establish the importance of biodiversity, have students list 20 things they use regularly that come from living organisms. Possible answers: sponges, cotton clothing, glue, paper, foods L2

Reading Check

Answer Biodiversity provides people with food, clothing, building materials, medicines, and natural beauty to enjoy.

Humans Need Biodiversity What foods do you like to eat? If you eat meat or fish, your meals probably include beef, chicken, pork, tuna, shrimp, or clams. Rice, peanuts, green beans, strawberries, or carrots are plant foods you might enjoy. Eating a variety of foods is a good way to stay healthy. Hundreds of species help feed the human population all around the world.

Biodiversity can help improve food crops. For example, plant breeders discovered a small population of wild corn, called maize, growing in Mexico. They crossbred it with domestic corn and developed different, disease-resistant strains of corn. In 1970, a new type of a fungal disease wiped out much of the U.S. corn crop because the strain of corn commonly grown was not resistant to it. Because of biodiversity, farmers could plant other corn strains that proved resistant to this new disease.

Biodiversity provides people with many useful materials. Furniture and buildings are made from wood and bamboo. Fibers from cotton, flax, and wool are woven into clothing. Most of the medicines used today originally came from wild plants, including those shown in **Figure 3.**

Reading Check *Why is biodiversity important to people?*

Maintaining Stability Forests usually contain many different kinds of plants. If one type of plant disappears, the forest still exists. Biodiversity allows for stability in an ecosystem.

Figure 3 Although most medicines are made in factories, many originally came from wild plants.

An extract from the bark of the Pacific yew was traditionally used to treat arthritis and rheumatism. It is now known that the bark contains taxol, which is a treatment for some types of cancer.

The velvet bean plant is the original source of L-dopa, a drug used to treat Parkinson's disease.

The bark of the cinchona tree contains quinine, which has been used to cure malaria in millions of people.

LAB DEMONSTRATION

Purpose to determine the biodiversity of a fish tank

Materials fish tank that includes several species of fish, snails, and plants

Preparation Bring in a community fish tank.

Procedure As a class, count the number of organisms and list the different species contained in the tank. Ask which is the better measure of biodiversity. species count

Expected Outcome The number of organisms will usually be larger than the number of species.

Assessment

Have students look at a paragraph in their textbook. Have them imagine that each different word is a separate species. Have them count the number of species and the number of individuals per species. Answers will vary.

128 ◆ E CHAPTER 5 Conserving Life

A loss of biodiversity can weaken an ecosystem. In a vineyard, as shown in **Figure 4,** vines grow close together. If a disease infects one grapevine, it could move easily from one plant to the next. Soon, the entire vineyard could be infected. Many farmers and gardeners have found that planting alternate rows of different crops can help prevent disease and reduce or eliminate the need for pesticides.

Figure 4 Disease has spread from one grapevine to another in this vineyard.

Applying Math — Solve a One-Step Equation

MEASURING BIODIVERSITY Temperate rain forest ecosystems in Alaska and California are similar. Which one has the higher mammal and bird biodiversity?

Biodiversity

Number of Species	California	Alaska
Mammals	49	38
Birds	94	79

Solution

1 *This is what you know:*

49 = mammal species in California
94 = bird species in California
38 = mammal species in Alaska
79 = bird species in Alaska

2 *This is what you need to find out:*

Which of these ecosystems has the greater bird and mammal biodiversity?

3 *This is the procedure you need to use:*

- Find the total number of bird and mammal species in each ecosystem.
  $49 + 94 = 143$ (species in California)
  $38 + 79 = 117$ (species in Alaska)
- Compare the totals.
  143 is larger than 117, so California has the greater bird and mammal biodiversity.

4 *Check your answer:*

$143 - 94 = 49$ (mammal species in California)
$117 - 79 = 38$ (mammal species in Alaska)

Practice Problems

1. Compare the biodiversity of the coastal temperate rain forest ecosystems of Oregon with 55 mammal species and 99 permanent bird species to the British Columbia ecosystem with 68 mammal species and 116 permanent bird species.
2. Compare the biodiversity of lizards in hot North American deserts that have an average of 6.4 ground-dwelling lizards and 0.9 arboreal lizards with those in Africa that have an average of 9.8 ground-dwelling lizards and 3.5 arboreal lizards.

Science Online — For more practice, visit booke.msscience.com/math_practice

Applying Math

National Math Standards

Correlation to Mathematics Objectives

1, 2, 9

Teaching Strategies

- Use a map to show students the extent of temperate rain forest along the Pacific Coast from Alaska to Northern California. Explain the meaning of temperate rain forest. (temperate = moderate temperatures, rain forest = annual rainfall from 40 to 300 inches).
- Have students hypothesize why Oregon's biodiversity is so much lower than that of British Columbia. Possible answer: There is a greater presence of human actions along Oregon's coast, such as logging, farming, and other development.
- Use a map to show students hot deserts in North America (Sonoran, Chihuahan, Mojave) and Africa (Kalahari, Sahara). Explain the meaning of hot desert. (less than 10 inches of rain per year).
- Have students make two tables like the one shown, the first table using the data for Oregon and British Columbia and the second using data for North American and African deserts.

Answers to Practice Problems

1. British Columbia has 184 species compared to 154 for Oregon. British Columbia shows the higher biodiversity.
2. Africa has 13.3 lizard species compared to 7.3 for North America. Africa shows the higher biodiversity for lizards.

Cultural Diversity

George Washington Carver Carver (1865–1943) is best known for helping southern farmers diversify their crops. He helped them understand that growing only one crop could deplete nutrients in the soil, and encouraged growing crops that would add nitrogen to the soil (i.e. peanuts, peas, soybeans). Have students research the uses Carver identified for the peanut, including peanut butter. L2

Discussion

Keystone Species In ecosystems, many species are often dependent on a few key species. Biologists call these keystone species. For example, many tropical rain forest organisms depend on the fig tree for food, so the fig tree is a keystone species. Have students look up the word keystone and explain why it is an appropriate term. A keystone is the central supporting element of a structure. L2 LS **Linguistic**

Use an Analogy

Airplane Rivets Some scientists use an airplane rivet analogy to explain how living things depend on each other. If a few rivets (metal bolts or pins) off an airplane are lost, the airplane might not be affected. But as more rivets are lost, the chances increase that the plane will eventually fall apart. Rivets are analogous to species and the airplane to the biosphere. LS **Logical-Mathematical**

Caption Answer

Figure 5 Possible answers: prolonged adverse weather, meteorite impact; erupting volcano

Teacher FYI

Extinction The loss of species produces a loss of biodiversity. When the last members of a species die, that species becomes extinct and it is lost from the world forever.

Reading Check

Answer changes in the environment and competition from other species

Figure 5 The passenger pigeon is an extinct species. These North American birds were over-hunted in the 1800s by frontier settlers who used them as a source of food and feathers.
Describe *other causes of extinction.*

What reduces biodiversity?

Flocks of thousands of passenger pigeons used to fly across the skies of North America. Few people today have ever seen one of these birds. The passenger pigeon, shown in **Figure 5,** has been extinct for almost 100 years. An **extinct species** is a species that was once present on Earth but has died out.

INTEGRATE Earth Science Extinction is a normal part of nature. The fossil record shows that many species have become extinct since life appeared on Earth. Extinctions can be caused by competition from other species or by changes in the environment. A mass extinction that occurred about 65 million years ago wiped out almost two-thirds of all species living on Earth, including the dinosaurs. This extinction, shown on the graph in **Figure 6,** occurred in the Mesozoic Era. It might have been caused by a huge meteorite that crashed into Earth's surface. Perhaps the impact filled the atmosphere with dust and ash that blocked sunlight from reaching Earth's surface. This event might have caused climate changes that many species could not survive. Mass extinctions eventually are followed by the appearance of new species that take advantage of the suddenly empty environment. After the dinosaurs disappeared, many new species of mammals appeared on Earth.

Reading Check *What are some causes of extinction?*

Figure 6 This graph shows five mass extinctions in Earth's history. The mass extinctions appear as peaks on the graph.

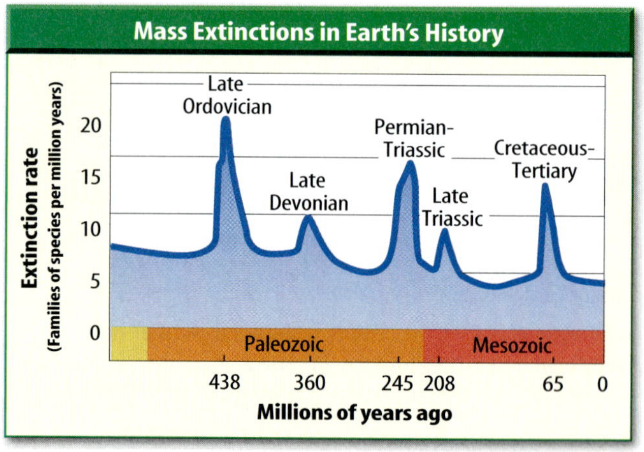

Loss of Species Not everyone agrees about the reason for the extinction of the dinosaurs. One thing is clear—human actions had nothing to do with it. Dinosaurs were extinct millions of years before humans were on Earth. Today is different. The rate of extinctions appears to be rising. From 1980 to 2000, close to 40 species of plants and animals in the United States became extinct. It is estimated that hundreds, if not thousands, of tropical species became extinct during the same 20-year period. Human activities probably contributed to most of these extinctions. As the human population grows, many more species could be lost.

Science Journal

Webbing Loss Have students research food webs. Then, have them draw in their Science Journals realistic food webs in which the loss of one species can affect at least four other species. L2 LS **Visual-Spatial**

Differentiated Instruction

Learning Disabled Draw a line on the board. On the far right, write "extinct," and on the far left write "lots of organisms." Ask students where the word *endangered* should go. Then ask where the word *threatened* should go. Write their responses on the continuum. *Threatened* goes closer to "lots of organisms," and *endangered* goes closer to "extinct." LS **Visual-Spatial**

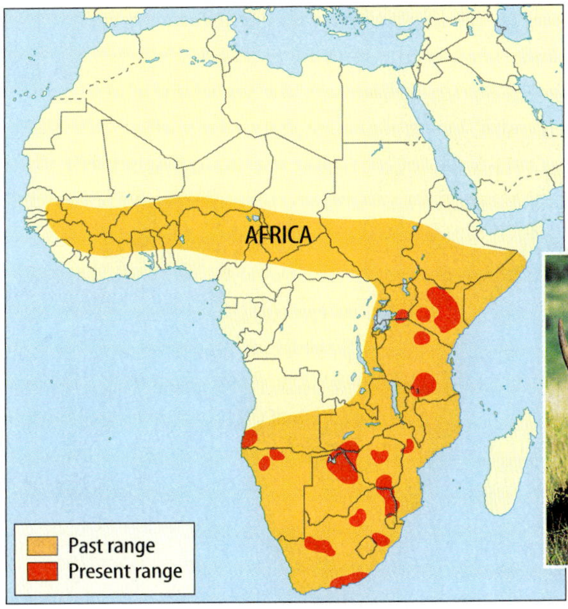

Figure 7 The African black rhinoceros is an endangered species, partly because people continue to kill these animals for their horns. As the map shows, this species once lived throughout southern, eastern, and central Africa.

Endangered Species To help prevent extinctions, it is important to identify species that could soon disappear. A species in danger of becoming extinct is classified as an **endangered species.** The African black rhinoceros, shown in **Figure 7,** is endangered. Rhinoceroses are plant eaters. They use their horns to battle each other for territory and to protect themselves against lions and other predators. For centuries, humans have considered rhinoceros horn to be a rare treasure. It is so valuable that poachers continue to hunt and kill these animals, even though selling rhinoceros horn is against international law. In 1970, there were about 100,000 black rhinoceroses in Africa. By the year 2000, fewer than 3,000 were left.

Figure 8 Koalas are related to kangaroos and opossums. The loss of their habitat threatens the future of wild koala populations.

Threatened Species If a species is likely to become endangered in the near future, it is classified as a **threatened species.** The Australian koala, shown in **Figure 8,** is threatened. People once hunted koalas for their fur. In the 1930s, people realized the koala was in danger. Laws were passed that prohibited the killing of koalas, and the koala populations began to recover. Koalas rely on certain species of Australian eucalyptus trees for food and shelter. After the 1930s laws were passed, Australia's human population grew and the koala's habitat began to disappear. By the year 2000, nearly two-thirds of the koala's habitat had been lost to logging, agriculture, cities, and roads.

Discussion

Elephant Ivory Elephants are threatened because people use their tusks to make ivory objects, including art and jewelry. Most world governments have banned the trade of ivory. Ask students if they feel it was right to ban the trade of ivory or if they feel people should have the right to decide for themselves whether they should buy it or not.

Fun Fact

Blue whales are the largest animal known to live on Earth.

Quick Demo

Loss of Biodiversity

Materials bag of colored candy

Estimated Time ten minutes

Procedure Spread candy out onto a sheet of paper. Tell students that each color will represent a different species. Count how many colors are on the sheet. Divide the habitat in half and remove the candy from one side of the sheet. This will represent the division of habitats. Continue doing this and have students observe how the loss of habitat and removal of species affects the biodiversity. Explain that species are often clumped together or require larger areas of habitat that could eliminate species much quicker than shown in this demonstration.

Science Journal

Going, Going, Gone Have students research and describe in their Science Journals a threatened species, an endangered species, and an extinct species. Make sure they explain the reason why the species is in peril or extinct. L2 ELL LS **Linguistic**

SECTION 1 Biodiversity E ◆ 131

Visualizing Threatened and Endangered Species

Have students examine the pictures and read the captions. Then ask the following questions.

Based on these examples, what seems to be the primary cause of a species becoming threatened or endangered in the United States? Possible answer: habitat destruction and human development.

The vernal pool tadpole shrimp is endangered and the desert tortoise is threatened. What is the difference between being endangered and being threatened? Possible answer: a threatened species is in danger of becoming extinct but there still is time to help the species recover on its own. An endangered species is in immediate danger of becoming extinct and probably will require human intervention to survive.

Activity

Endangered Poster Have students research some other endangered or threatened species and make a poster based on their findings.

Teacher FYI

Deforestation Deforestation is a problem occurring worldwide. The loss of trees drastically changes the habitat. Deforestation of rain forests is becoming a serious problem in South America. Discuss with students what happens to plants and animals as the forests are cleared for farms and ranches or harvested and not replanted.

NATIONAL GEOGRAPHIC VISUALIZING THREATENED AND ENDANGERED SPECIES

Figure 9

The U.S. Fish and Wildlife Service lists more than 1,200 species of animals and plants as being at risk in the United States. Of these, about 20 percent are classified as threatened and 80 percent as endangered.

VERNAL POOL TADPOLE SHRIMP This endangered species lives in the seasonal freshwater ponds of California's Central Valley. Pollution, urban sprawl, and other forces have destroyed 90 percent of the vernal pools in the valley. Loss of habitat makes the survival of these tiny creatures highly uncertain.

HALEAKALA SILVERSWORD One of the world's most spectacular plants, this threatened species is making a recovery. The plant shown here is blooming in Hawaii's Haleakala (ha lee ah kuh LAH) crater.

CALIFORNIA CONDOR The endangered condor came close to extinction at the end of the twentieth century. Some condors have been raised in captivity and successfully released into the wild.

DESERT TORTOISE The future of the threatened desert tortoise is uncertain. Human development is eroding tortoise habitat in the southwestern United States.

SOUTHERN SEA OTTER The threatened sea otter lives in shallow waters along the Pacific coast of the United States. For centuries, sea otters were hunted for their fur.

Differentiated Instruction

Challenge Have students choose an endangered species and write a plan to help the species recover. They should include possible plans to restore the species' habitat, to help the species reproduce (in a controlled environment), and for the species to be reintroduced into the wild. Encourage students to choose plants as well as animals. L3

Habitat Loss In the Launch Lab at the beginning of this chapter, you probably observed that lawns and sidewalks have a lower biodiversity than sunlit woods or weed-covered lots. When people alter an ecosystem, perhaps by replacing a forest or meadow with pavement or a lawn, the habitats of some species may become smaller or disappear completely. If the habitats of many species are lost, biodiversity might be reduced.

Habitat loss is a major reason why species become threatened or endangered, as shown in **Figure 9,** or extinct. The Lake Erie water snake, shown in **Figure 10,** is classified as a threatened species because of habitat loss. These snakes live along the rocky shores of islands in Lake Erie. The islands are popular recreation spots. The development of boat docks and buildings in these areas has seriously reduced the amount of habitat available to the snakes. Also, these snakes are often killed by people who mistakenly think they are poisonous. Adult snakes range from 0.5 m to 1.0 m long, and they feed on fish, frogs, and salamanders. Because they have sharp teeth for capturing prey, they can bite. However, they are not dangerous to humans, especially if left undisturbed.

Conservation strategies for protecting the Lake Erie water snake include preserving its habitat by limiting development in some areas. Also, a public education program has been developed to inform people about this predator species and its importance in the Lake Erie ecosystem.

Mini LAB

Demonstrating Habitat Loss

Procedure

1. Put a small piece of **banana** in an **open jar.** Set the jar indoors near a place where food is prepared or fruit is thrown away.
2. Check the jar every few hours. When at least five fruit flies are in it, place a **piece of cloth or stocking** over the top of the jar and secure it with a **rubber band.**
3. Count and record the number of fruit flies in the jar every two days for three weeks.

Analysis

1. Explain why all the flies in the jar eventually die.
2. Use your results to hypothesize why habitat loss can reduce biodiversity.

Try at Home

Figure 10 The Lake Erie water snake has become a threatened species.

The building of boat docks, roads, and shopping centers is reducing the snake's habitat.

These snakes are not dangerous to humans, but many are killed by people who fear snakes.

Mini LAB

Purpose to observe what happens when a fruit fly habitat is lost LS **Visual-Spatial**

Materials banana, jar, cloth or stocking, rubber band

Teaching Strategy

- Fruit flies are easy to catch if it is warm outside. Some produce sections of supermarkets may also have fruit flies.
- Explain that putting cloth on the jar limits the fruit flies' habitat.
- Each week, have students report how many flies they have.

Troubleshooting Catch fruit flies to give to students who have trouble finding them. If it is winter, you can order nonflying fruit flies from a biological supply company.

Analysis

1. Their food supply was used up and they died.
2. It limits available habitat. In a small area, they may run out of food and perish.

Assessment

Content Have students consider the following scenario: a fruit fly colony begins in a jar with a piece of banana. There are two males and two females. Have students draw a line graph to show possible population growth over an eight-week period. Use **Performance Assessment in the Science Classroom,** p. 111.

Try at Home

Activity

Conserving Resources Have students make a list on the board of ways they can conserve resources by reducing, reusing, or recycling; thereby slowing the loss of habitat. L2 COOP LEARN LS **Interpersonal**

Make a Model

Divided Habitats Use graph paper to show the problem of divided habitats. Have students draw a line to create a square measuring twenty cells by twenty cells. This represents the habitat of a musk ox. One ox needs 200 cells for sufficient habitat to survive. How many musk oxen would this land support? 2 Now divide the land with two roads, one cutting the land vertically and one horizontally. The roads are major highways that the musk oxen cannot cross. Ask if the musk oxen will be able to survive. No, their habitats are now too small.

Teacher FYI

Roads If a road were built through a rain forest, more sunlight and wind would be able to penetrate, making the edges drier. This edge effect—a change in boundary climates—is a problem with divided habitats.

Virtual Labs

Friend or Foe *How do introduced species impact the environment?*

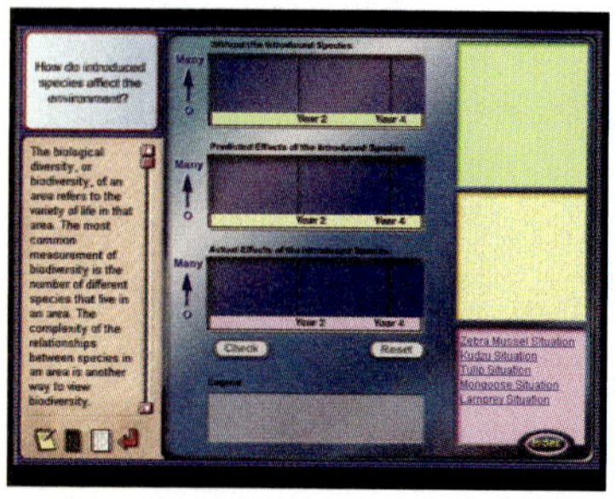

Divided Habitats Biodiversity can be reduced when a habitat is divided by roads, cities, or farms. Small patches of habitat usually have less biodiversity than large areas. One reason for this is that large animals like mountain lions and grizzly bears require hunting territories that cover hundreds of square kilometers. If their habitat becomes divided, they are forced to move elsewhere.

Small habitat areas also make it difficult for species to recover from a disaster. Suppose a fire destroys part of a forest, and all the salamanders living there are destroyed. Later, after new trees have grown, salamanders from the undamaged part of the forest move in to replace those that were lost. But what if fire destroys a grove of trees surrounded by parking lots and buildings? Trees and salamanders perish. Trees might grow back but the salamanders might never return, because none live in the surrounding paved areas.

Introduced Species When species from another part of the world are introduced into an ecosystem, they can have a dramatic effect on biodiversity. An **introduced species** is a species that moves into an ecosystem as a result of human actions. Introduced species often have no competitors or predators in the new area, so their populations grow rapidly. Introduced species can crowd out or consume native species. **Native species** are the original organisms in an ecosystem.

In the early 1800s, European settlers brought goats with them to Santa Catalina Island off the coast of California. The goats overgrazed the land, completely eliminating 48 of the native plant species and exposing patches of bare soil. The bare soil provided a place for hardy, introduced weeds to take root. The weeds crowded out native species. Late in the twentieth century, after goats were removed from the island, some native plant species began to recover. **Figure 11** shows introduced species that have reduced biodiversity in other ecosystems.

Figure 11 Introduced species can reduce or eliminate populations of native species in an ecosystem.

The ruffe was introduced into the Great Lakes by ships from other parts of the world. It quickly takes over habitat and food sources used by native fish.

Purple loosestrife was brought to North America from Europe and Asia in the 1800s. Dense patches of loosestrife grow and crowd out the native plants that some animals need for food and shelter.

Science Journal

Managing Resources The prickly pear cactus is a fast-growing cactus that was brought to Australia because it made a good living fence. However, it quickly spread out of control. Have students investigate and report on how Australians solved this problem. L2 LS **Linguistic**

Differentiated Instruction

Challenge Have students investigate and write a report on problems with introduced species. Possible topics: the brown tree snake, kudzu, sea lamprey, starling, zebra mussel L3 LS **Linguistic**

134 ◆ E CHAPTER 5 Conserving Life

Pollution

Biodiversity also is affected by pollution of land, water, or air. Soil that is contaminated with oil, chemicals, or other pollutants can harm plants or limit plant growth. Because plants provide valuable habitat for many species, any reduction in plant growth can limit biodiversity.

Water Pollution Water-dwelling organisms are easily harmed by pesticides, chemicals, oil, and other pollutants that contaminate the water. Water pollutants often come from factories, ships, or runoff from roads, lawns, and farms. Waterways also can be polluted when people dispose of wastes improperly. For example, excess water from streets and roads runs into storm drains during rainstorms. This water usually flows untreated into nearby waterways. Storm drains should never be used to dispose of used motor oil, paints, solvents, or other liquid wastes. These pollutants can kill aquatic plants, fish, frogs, insects, and the organisms they depend on for food.

Air Pollution A form of water pollution known as acid rain is caused by air pollution. **Acid rain** forms when sulfur dioxide and nitrogen oxide released by industries and automobiles combine with water vapor in the air. As **Figure 12** shows, acid rain can have serious effects on trees. It washes calcium and other nutrients from the soil, making the soil less fertile. One tree species that is particularly vulnerable to acid rain is the sugar maple. Many sugar maple trees in New England and New York have suffered major damage from acid rain. Acid rain also harms fish and other organisms that live in lakes and streams. Some lakes in Canada have become so acidic that they have lost almost all of their fish species. In the United States, 14 eastern states have acid rain levels high enough to harm aquatic life.

Air pollution from factories, power plants, and automobiles can harm sensitive tissues of many organisms. For example, air pollution can damage the leaves or needles of some trees. This can weaken them and make these trees less able to survive diseases, attacks by insects and other pests, or environmental stresses such as drought or flooding.

Figure 12 Acid rain can damage the leaves and other tissues of trees. **Explain** *how this damage can affect other organisms.*

Mini LAB

Modeling the Effects of Acid Rain

Procedure

1. Soak **50 mustard seeds** in **water** and another **50 mustard seeds** in **vinegar** (an acid) for 24 hours.
2. Wrap each group of seeds in a **moist paper towel.**
3. Put them into separate, **self-sealing plastic bags.** Seal and label each bag.
4. After 3 days, open each bag, examine the seeds, and record the number that show evidence of growth.

Analysis

1. Describe the effect of an acid on mustard seed growth.
2. Explain how acid rain could affect plant biodiversity.

Caption Answer

Figure 12 Possible answer: Organisms that depend on these trees for food or shelter also would be affected.

IDENTIFYING Misconceptions

Environmental Terms Students may confuse environmental problems, perhaps associating acid rain with global warming and ozone damage. Write these three terms on the board and ask students to explain each of them.

Mini LAB

Purpose to observe the effects of acid on seedling growth

Materials mustard seeds (100), water, vinegar, plastic cups (2), paper towels, self-sealing plastic bags (2)

Safety Precautions Caution students to wear goggles and aprons when working with acids.

Teaching Strategy Substitute small jars with lids for the open cups to keep the smell of vinegar to a minimum.

Analysis

1. Seeds soaked in vinegar (acid) should show less growth than those soaked in water.
2. Possible answers: Acid rain could stunt plant growth, slowing reproductive cycles to the point that plants die out. Plants could also die from too much acid in the soil.

Assessment

Portfolio Have students draw pictures comparing the growth of the plants over time. Use **Performance Assessment in the Science Classroom**, p. 127. P

Teacher FYI

Acid Rain Precipitation with a pH below 5.6 is called acid rain. Locations in western Europe and the eastern United States have precipitation with pH between 2 and 3. Los Angeles, San Francisco, and Whiteface Mountain in New York, often have fog at least 10 times more acidic than their precipitation.

SECTION 1 Biodiversity E ◆ 135

INTEGRATE Health

UV Radiation Students may suggest limiting the amount of time spent in the Sun or wearing protective hats, sunglasses, and sunblock.

Research Have students investigate seasonal and annual ozone depletion and the effects it has on living organisms.

3 Assess

DAILY INTERVENTION

Check for Understanding

Linguistic Have students divide into groups and brainstorm about causes of biodiversity loss. List answers on the board.

Reteach

Biodiversity Have students explain what biodiversity is, why it is important, and how it is threatened. Have students use construction paper to put together a mini-book containing their answers. L2

Assessment

Performance Have each student make a poster showing an endangered or threatened organism and the problems it is encountering. Use **Performance Assessment in the Science Classroom**, p. 145. P

INTEGRATE Health

UV Radiation Overexposure to the Sun's radiation can cause health problems such as cataracts, which cloud the lens of the eye, and skin cancer. In your Science Journal, describe ways to limit your exposure to the Sun's damaging radiation.

Global Warming Carbon dioxide gas (CO_2) is released into the atmosphere when wood, coal, gas, or any other fuel is burned. People burn large amounts of fuel, and this is contributing to an increase in the percentage of CO_2 in the atmosphere. An increase in CO_2 could raise Earth's average temperature by a few degrees. This average temperature rise, called global warming, might lead to climatic changes that could affect biodiversity. For example, portions of the polar ice caps could melt, causing floods in coastal ecosystems around the world.

Ozone Depletion The atmosphere includes the ozone layer—ozone gas that is about 15 km to 30 km above Earth's surface. It protects life on Earth by preventing damaging amounts of the Sun's ultraviolet (UV) radiation from reaching Earth's surface. Scientists have discovered that the ozone layer is gradually becoming thinner. The thinning of the ozone layer is called **ozone depletion.** This depletion allows increased amounts of UV radiation that can harm living organisms to reach Earth's surface. For humans, it could mean more cases of skin cancer. Ozone depletion occurs over much of Earth. Data collected in the late 1990s indicate that ozone levels over the United States, for example, had decreased by five to ten percent since the 1970s.

section 1 review

Summary

The Variety of Life

- Biodiversity is a measurement of the number of species living in an area.
- In warmer climates, biodiversity usually is greater than in cooler climates.

Why is biodiversity important?

- Biodiversity can improve crops and provide materials for furniture, clothing, and medicine.

What reduces biodiversity?

- Extinction is a normal part of nature that reduces biodiversity.
- Climate changes, habitat loss, and introduced species can decrease biodiversity.

Pollution

- Ozone depletion can harm organisms by increasing their exposure to UV radiation.

Self Check

1. **Describe** how endangered species are different from extinct species.
2. **Explain** why habitat loss and habitat division are threats to biodiversity.
3. **Explain** how introduced species threaten biodiversity.
4. **Think Critically** Aluminum, commonly found in soil, is toxic to fish. It damages the ability of gills to absorb oxygen. Acid rain can wash aluminum out of the soil and into nearby waterways. Explain how acid rain could affect biodiversity in a pond.

Applying Skills

5. **Identify and Manipulate Variables, Constants, and Controls** Lack of calcium in the soil can damage trees. Design an experiment to test the hypothesis that acid rain is removing calcium from forest soil.

Science Online booke.msscience.com/self_check_quiz

section 1 review

1. Extinct species are gone forever; endangered species may soon become extinct.
2. Habitat loss means there is less area in which an organism can obtain what it needs. Habitat division restricts the movement of organisms and may block access to what they need.
3. They can compete with or prey on native organisms, and often have no natural predators in the area.
4. It could result in one or more species dying, reducing biodiversity.
5. Answers will vary, but should involve a control group that receives normal water or rain and a variable group that receives acid water or rain. The levels of calcium in both soils then would be measured.

LAB

Oily Birds

You probably have seen news clips of waterbirds covered with oil after an oil spill. Nearly four billion kilograms of oil are spilled from ships or dumped into the ocean each year. Oil floats on water, and can coat the feathers of waterbirds. These birds depend on their feathers for insulation from the cold. Oiled feathers lose their insulating qualities, so oiled birds are exposed to the cold. Oil-coated feathers also prevent the birds from flying and can even cause them to drown. Rescue teams try to clean off the oil, but as you'll find out in this lab, this is a difficult task.

Real-World Question

How difficult is it to clean oil from bird feathers?

Goals

- **Experiment** with different methods for cleaning oil from bird feathers.

Materials

beakers (2)	water
bird feathers (white or light colored)	cotton balls
vegetable oil or olive oil	cotton swabs
red, blue, and green food coloring	sponge
paper towels	toothbrush
	dish soap

Safety Precautions

Procedure

1. Fill a beaker with vegetable oil and add several drops of red, blue, and green food coloring. Submerge your feathers in the oil for several minutes.
2. Lay paper towels on a table. Remove the feathers from the oil and allow the excess oil to drip back into the beaker. Lay the feathers on the paper towels.
3. Using another paper towel, blot the oil from one feather. Run your fingers over the feather to determine if the oil is gone.
4. Using the cotton swabs and cotton balls, wipe the oil from a second feather. Check the feather for oil when finished.
5. Using a toothbrush, brush the oil from a third feather. Check it for oil when you are finished.
6. Use a sponge to clean the fourth feather. Check it for oil when you are finished.
7. Fill the second beaker with water and add several drops of dish soap. Soak the fifth oil-soaked feather in the solution. Check the feather for oil after it has soaked for several minutes.

Conclude and Apply

1. **Identify** the method or methods that best removed the oil from the feathers.
2. **Describe** the condition of the feathers after you finished cleaning them. How would the condition of these feathers affect birds?
3. **Infer** why rescuers would be hesitant to use soap when cleaning live birds.
4. **Infer** how oil pollution affects water mammals such as otters and seals.

Communicating Your Data

Share your results with your classmates. Decide which methods you would use to try to clean a live bird.

BENCH TESTED

Real-World Question

Purpose Students will experience the difficulty of cleaning oil from bird feathers.

Process Skills observe and infer, describe

Time Required 30 minutes

Procedure

Safety Precautions Instruct students to wash their hands thoroughly after completing the activity.

Teaching Strategies

- In their Science Journals, ask students to write a brief description of the feathers after each cleaning attempt.
- Introduce or conclude the activity with photos or video footage of birds trapped in petroleum and rescue efforts to save them.

Troubleshooting Make certain each pair of students has all supplies at their table before starting the activity to contain messes in one location.

Conclude and Apply

1. The dish soap will work best, but it will not remove all the oil.
2. Answers will vary. It will affect their ability to fly.
3. Soap could prove toxic for birds already in a weakened state.
4. Oil also causes the fur of water mammals to lose its waterproofing ability, exposing the animals to the cold.

Assessment

Content Ask each group to write an international treaty of rules and regulations that would prevent the majority of oil spills around the world from occurring. Use **Performance Assessment in the Science Classroom,** p. 169.

Communicating Your Data

Create a data table on the board listing all the methods students could use for cleaning live birds. Record the student's choices on the table.

LAB E ◆ 137

1 Motivate

INTERACTIVE CHALKBOARD
PowerPoint® Presentations

Bellringer

Section Focus Transparencies also are available on the Interactive Chalkboard CD-ROM. L2 ELL

Section 2 Section Focus Transparency **Marsupial Misdeed**

The strange animal you see here is called a Tasmanian tiger, or thylacine. Despite being called a tiger, the thylacine isn't a cat, but a marsupial. Living only on the island of Tasmania, this creature became the target of government sponsored hunts in order to protect farm animals. All too successful, the hunts pushed the thylacine to extinction. The last known individual died in a zoo in 1936.

1. From the paragraph above, what was the source of the conflict between the Tasmanian tiger and people?
2. How might these sorts of conflicting needs be balanced?
3. How can zoos help preserve animals on the verge of extinction?

Conserving Life

Tie to Prior Knowledge

Parks Most students will have visited some local, state, or national park. Ask about rules or laws enforced by the park rangers to protect these habitats. Answers may include the prohibition of littering, harming or feeding the animals, or taking natural objects from the park.

section 2

Conservation Biology

as you read

What You'll Learn

- **Identify** several goals of conservation biology.
- **Recommend** strategies to prevent the extinction of species.
- **Explain** how an endangered species can be reintroduced into its original habitat.

Why It's Important

Conservation biology provides ways to preserve threatened and endangered species.

Review Vocabulary
habitat: the place or environment where a plant or animal naturally or normally lives and grows

New Vocabulary
- conservation biology
- habitat restoration
- captive population
- reintroduction program

Protecting Biodiversity

The study of methods for protecting biodiversity is called **conservation biology.** Conservation biologists develop strategies to prevent the continuing loss of members of a species. Conservation strategies must be based on a thorough understanding of the principles of ecology. Because human activities are often the reason why a species is at risk, conservation plans also must take into account the needs of the human population. The needs of humans and of other species often conflict. It can be difficult to develop conservation plans that satisfy both. Conservation biologists must consider law, politics, society, and economics, as well as ecology, when they look for ways to conserve Earth's biodiversity.

The Florida manatee, shown in **Figure 13,** is an endangered species. Manatees are plant-eating mammals that live in shallow water along the coasts of Florida and the Carolinas. These animals swim slowly, occasionally rising to the surface to breathe. They swim below the surface, but not deep enough to avoid the propellers of powerboats. Many manatees have been injured or killed by powerboat propellers. Boaters cannot always tell when manatees are nearby, and it is difficult to enforce speed limits to protect the manatees. Also, the manatee's habitat is being affected by Florida's growing human population. More boat docks are being built, which can add to the number of boats in the water. Water pollution from boats, roads, and coastal cities is also a problem.

Reading Check *How do humans affect manatee habitats?*

Figure 13 Conservation biologists are working to protect the endangered Florida manatee by limiting habitat loss, reducing pollution, and encouraging boaters to obey speed limits. Rescued manatees are being rehabilitated at the Columbus Zoo in Ohio.

Section 2 Resource Manager

Chapter *FAST FILE* Resources
- Lab Worksheet, pp. 7–8
- Transparency Activity, p. 43
- Directed Reading for Content Mastery, pp. 21, 22
- Enrichment, p. 30
- Reinforcement, p. 28
- Lab Activity, pp. 13–16

Lab Management and Safety, p. 73

Figure 14 The American bald eagle and American alligator once were listed as endangered. **Infer** *why the American bald eagle and American alligator were taken off the endangered species list.*

Conservation Biology at Work

Almost every conservation plan has two goals. The first goal is to protect a species from harm. The second goal is to protect the species' habitat. Several conservation strategies can be used to meet these goals.

Legal Protections Laws can be passed to help protect a species and its habitat. In the early 1970s, people became concerned about the growing number of species extinctions occurring in the United States. In response to this concern, Congress passed the U.S. Endangered Species Act of 1973. This act makes it illegal to harm, collect, harass, or disturb the habitat of any species on the endangered or threatened species lists. The act also prevents the U.S. government from spending money on projects that would harm these species or their habitats. The Endangered Species Act has helped several species, including those shown in **Figure 14,** recover from near extinction.

The United States and other countries worldwide have agreed to work together to protect endangered or threatened species. In 1975, The Convention on International Trade in Endangered Species of Wild Fauna and Flora, known as CITES, came into existence. One of its goals is to protect certain species by controlling or preventing international trade in these species or any part of them, such as elephant ivory. About 5,000 animal species and 25,000 plant species are protected by this agreement.

Science Online

Topic: Endangered Species

Visit booke.msscience.com for Web links to information about endangered species.

Activity Research an endangered or threatened species. Write a description of the species and explain why it is considered in danger.

2 Teach

Reading Check

Answer Humans pollute the water. They also drive through the water with power boats, which can injure manatees.

Caption Answer

Figure 14 The numbers of American bald eagles and American alligators in the wild increased.

IDENTIFYING Misconceptions

Biologist v. Activist Students might think a conservation biologist is the same thing as an animal rights activist. Conservation biologists are concerned with the protection of species. Animal rights activists are concerned with the protection of individual animals. Sometimes their interests conflict, such as when conservation biologists advocate killing some animals to protect another species.

Active Reading

Reflective Journal In this strategy, students record responses to an activity. Have students divide sheets of paper into columns and record their thoughts under headings such as *What I did*, *What I learned*, *What questions I have*, *What surprises I experienced*, and *Overall response*. Use this strategy for conservation biology.

INTEGRATE Social Studies

Marine Biodiversity Many countries have legislation to protect marine environments. Raising public awareness of the importance of marine environments can help to reduce pollution. Several countries have treaties to monitor, prevent, and clean up marine pollution. Some countries have specialized personnel in pollution clean up.

Quick Demo

Habitat Loss

Materials chessboard

Estimated Time two minutes

Procedure Show students a chessboard and have them imagine that each square is a habitat that can support two pawns. Put two pawns in one square to visually present this idea. How many pawns can the entire chessboard support? $64 \times 2 = 128$ If one whole row of land were removed to create a shopping center, what pawn population could the area support? (Put play buildings on these spaces to visually present this idea.) $56 \times 2 = 112$ Explain that as natural areas are cleared, there is less habitat to support the species that inhabit that area.

Visual Learning

Figure 15 Ask students to use the illustration to explain how wildlife corridors reduce the problem of divided habitats. Organisms pass under highways through the corridors, allowing them the range of habitat they need to survive.

Fun Fact

Adult male grizzly bears usually need at least 647.5 square kilometers of habitat to survive. Female grizzlies require less.

INTEGRATE Social Studies

Marine Biodiversity The marine fish catch has an estimated value of $80 billion per year. Marine environments also provide us with raw materials and medicines. Other uses of marine environments include research, ecotourism, and recreation. Research and then describe in your Science Journal what is being done to protect marine environments.

Habitat Preservation Even if a species is protected by law, it cannot survive unless its habitat also is protected. Conservation biology often focuses on protecting habitats, or even whole ecosystems. One way to do this is to create nature preserves, such as national parks and protected wildlife areas.

The United States established its first national park—Yellowstone National Park—in 1872. At that time, large animals like grizzly bears, elk, and moose ranged over much of North America. These animals roam across large areas of land in search of food. If their habitat becomes too small, they cannot survive. The grizzly bear, for example, requires large quantities of food each day. To feed itself, a grizzly needs a territory of several hundred square kilometers. Without national parks and wildlife areas, some animals would be far fewer in number than they are today.

Wildlife Corridors The successful conservation of a species like the grizzly bear requires enormous amounts of land. However, it is not always possible to create large nature preserves. One alternative is to link smaller parks together with wildlife corridors. Wildlife corridors allow animals to move from one preserve to another without having to cross roads, farms, or other areas inhabited by humans. As **Figure 15** shows, wildlife corridors are part of the strategy for saving the endangered Florida panther. A male panther needs a territory of 712 km^2 or more, which is larger than many of the protected panther habitats.

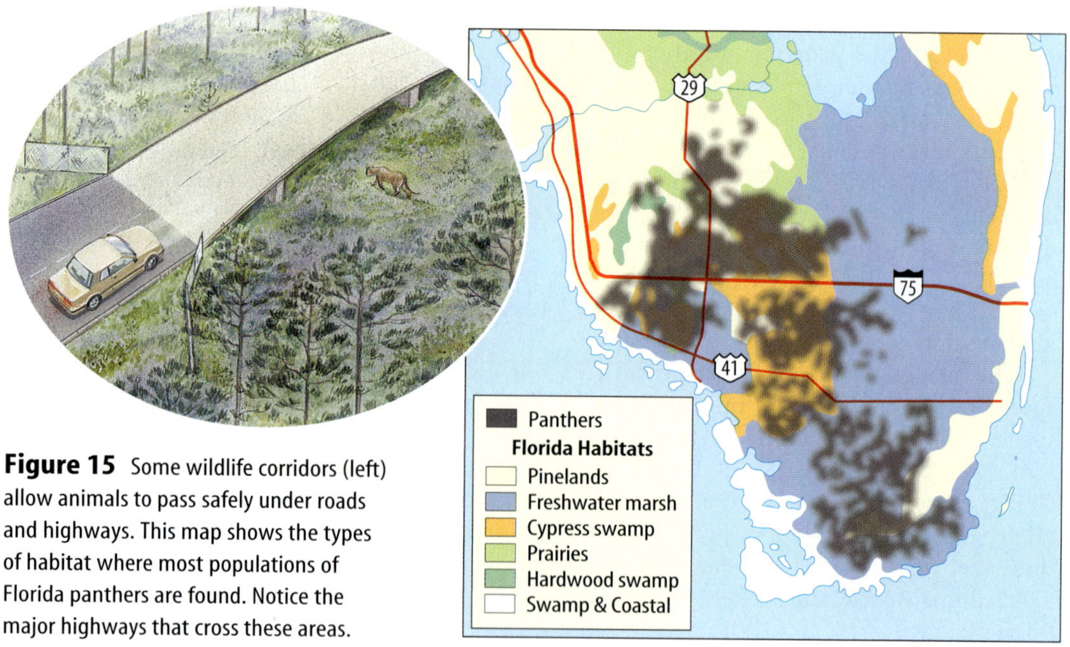

Figure 15 Some wildlife corridors (left) allow animals to pass safely under roads and highways. This map shows the types of habitat where most populations of Florida panthers are found. Notice the major highways that cross these areas.

Teacher FYI

First U.S. National Park Yellowstone National Park was established because of the area's unique geology. In 1872, the wildlife of the park also could be found in many places in the U.S. As the human population grew and took over more land, much of the wildlife that lived outside of the park perished.

Differentiated Instruction

Challenge Ask students to research a national or state park in your local area. Have them find when and why it was created, and any unique features. Have them write a report of their findings. L3

Habitat Restoration Habitats that have been changed or harmed by human activities often can be restored. **Habitat restoration** is the process of taking action to bring a damaged habitat back to a healthy condition.

Rhode Island's Narragansett Bay was once an important fishing area. It supported populations of flounder, bay scallops, blue crabs, and other important food species. Eelgrass, which is an underwater plant that grows in shallow parts of the bay, provides habitat for the young of many of these species. As the human population of Rhode Island grew, most of the bay's eelgrass beds disappeared. Fish and shellfish populations declined. People who lived in the area knew that something needed to be done. Students in Rhode Island schools have helped restore eelgrass habitat by growing thousands of eelgrass seedlings for transplanting into the bay, as shown in **Figure 16.**

Figure 16 By restoring eelgrass habitat, conservation biologists hope to preserve populations of shellfish and fish in Narragansett Bay.
Describe *how the restoration of eelgrass habitat will help preserve populations of shellfish and fish.*

Reading Check *When is habitat restoration used?*

Wildlife Management Preserving or restoring a habitat does not mean that all the species living there are automatically protected from harm. Park rangers, guards, and volunteers often are needed to manage the area. In South Africa, guards patrol wildlife parks to prevent poachers from killing elephants for their tusks. Officials protect the mountain gorillas of Rwanda by limiting the number of visitors allowed to see them. Some wildlife preserves allow no visitors other than biologists who are studying the area.

Hunters and wildlife managers often work together to maintain healthy ecosystems in parks and preserves. People usually are not allowed to hunt or fish in a park unless they purchase a hunting or fishing license. The sale of licenses provides funds for maintaining the wildlife area. It also helps protect populations from overhunting. For example, licenses may limit the number of animals a hunter is allowed to take or may permit hunting only during seasons when a species is not reproducing. Hunting regulations also can help prevent a population from becoming too large for the area.

Activity

Local Pollution Have students research a polluted section of their local area that is now being cleaned up. Have them find information about which species are being affected by the cleanup. L2

Reading Check

Answer when the survival of the species living in an altered or polluted habitat is threatened

Caption Answer

Figure 16 The eelgrass provides a protected habitat for young shellfish and fish, thus, helping to increase their rate of survival.

Inquiry Lab

Conservation Strategies

Purpose to explore what individuals can do to help preserve species and/or their habitat

Possible Materials various

Estimated Time two weeks

Teaching Strategies

- Encourage students to research methods used to protect and restore species and their habitats.
- Students could go to a zoo and ask about reintroduction programs.
- Students could research new laws that are being proposed dealing with species or habitat preservation and write letters to their representatives to find out if they support the law or not.

For additional inquiry activities, see *Science Inquiry Labs*.

Teacher FYI

Banning DDT The banning of DDT led to the recovery of some species, including the bald eagle and the brown pelican. Species higher up the food chain can have problems when they eat organisms that contain a contaminant because the substance accumulates in their bodies.

Curriculum Connection

Art In some tropical rain forest areas of Central and South America, Brazil nuts are harvested. The sale of these nuts brings money to local people, giving them one reason to protect the rain forests. Ask students to design color posters advertising Brazil nuts to the people in their community. L1 ELL LS **Visual-Spatial** P

Figure 17 The Arabian oryx was saved from extinction because a captive population was created before the wild population disappeared.

Captive Populations The Arabian oryx, shown in **Figure 17,** is native to desert lands of the Arabian Peninsula. People hunt the oryx for its horns and hide. Hunting increased during the twentieth century, when four-wheel-drive vehicles became available to hunters. In 1962, to prevent the species from becoming extinct, several oryx were taken to the Phoenix Zoo in Arizona. By 1972, no more wild oryx lived in the desert. A **captive population** is a population of organisms that is cared for by humans. The captive population of oryx in Arizona did well. Their numbers increased, and hundreds of Arabian oryx now live in zoos all over the world.

Keeping endangered or threatened animals in captivity can help preserve biodiversity. It is not ideal, however. It can be expensive to provide proper food, adequate space, and the right kind of care. Also, captive animals sometimes lose their wild behaviors. If that happens, they might not survive if they're returned to their native habitats. The best approach is to preserve the natural habitats of these organisms so they can survive on their own.

Reintroduction Programs In some cases, members of captive populations can be released into the wild to help restore biodiversity. In the 1980s, wildlife managers began reintroducing captive oryx into their desert habitat. **Reintroduction programs** return captive organisms to an area where the species once lived. Programs like this can succeed only if the factors that caused the species to become endangered are removed. In this case, the reintroduced oryx must be protected from illegal hunting.

Plants also can be reintroduced into their original habitats. The island of St. Helena in the South Atlantic was home to the rare St. Helena ebony tree. In the 1500s, European settlers introduced goats to the island. Settlers also cut trees for firewood and timber. Goats overgrazed the land, and native trees were replaced by introduced species. By 1850, the St. Helena ebony was thought to be extinct. In 1980, two small ebony trees were found on the island, high on a cliff. Small branches were cut from these trees and rooted in soil. These cuttings have been used to grow thousands of new ebony trees. The new trees are being replanted all over the island.

Science Online

Topic: Reintroduction Programs

Visit booke.msscience.com for Web links to information about reintroduction programs.

Activity Find out about a reintroduction program and write a summary of the program's progress.

Teacher FYI

Enjoying Nature Allowing local people to benefit from a natural area increases their desire to protect it. For this reason, wildlife areas may allow some harvesting of fruits, nuts, and branches for firewood. Other areas may allow hunting and fishing. The amount of material collected or living things killed is controlled in a process called sustainable development.

Identifying Misconceptions

Reintroduction Programs Students may not realize that reintroduction programs are only successful when the problems affecting the species have been removed from the habitat. For example, if loss of habitat led to the disappearance of jaguars, reintroducing jaguars will probably not be successful unless the lost habitat is restored.

Discussion

Reintroducing Wolves Endangered wolves have been reintroduced into natural areas of some states. Some ranchers object to this because they think the wolves will kill their livestock. Some organizations are reimbursing ranchers for animals killed by the wolves. Have students discuss whether they agree or disagree with reintroduction of wolves.

Seed Banks Throughout the world, seed banks have been created to store the seeds of many endangered plant species. If any of these species become extinct in the wild, the stored seeds can be used to reintroduce them to their original habitats.

Relocation Reintroduction programs do not always involve captive populations. In fact, reintroductions are most successful when wild organisms are transported to a new area of suitable habitat. The brown pelican, shown in **Figure 18,** was once common along the shores of the Gulf of Mexico. Pelicans eat fish that eat aquatic plants. In the mid-twentieth century, DDT—a pesticide banned in 1972 in the U.S.—was used widely to control insect pests. It eventually ended up in the food that pelicans ate. Because of DDT, the pelican's eggshells became so thin that they would break before the chick inside was ready to hatch. Brown pelicans completely disappeared from Louisiana and most of Texas. In 1971, 50 of these birds were taken from Florida to Louisiana. Since then, the population has grown. In the year 2000, more than 7,000 brown pelicans lived in Louisiana and Texas.

Figure 18 Brown pelicans from Florida were relocated successfully to Louisiana.

Reading Check *What two things led to the recovery of the brown pelican?*

section 2 review

Summary

Protecting Biodiversity

- Conservation biology deals with protecting biodiversity.
- Conservation plans must take into account the needs of the human population because human activities often are the reason species are at risk.

Conservation Biology at Work

- Conservation plans are designed to protect a species from harm and protect their habitat.
- Laws, habitat preservation, wildlife corridors, and habitat restoration are techniques used to protect a species and its habitat.
- Sometimes captive or wild populations can be released into a new area to help restore biodiversity.

Self Check

1. **Explain** how the U.S. Endangered Species Act helps protect species.
2. **Describe** what factors can make it difficult to reintroduce captive animals into their natural habitats.
3. **Think Critically** As a conservation biologist, you are helping to restore eelgrass beds in Narragansett Bay. What information will be needed to prepare a conservation plan?

Applying Skills

4. **Recognize Cause and Effect** Imagine creating a new wildlife preserve in your state. Identify uses of the area that would allow people to enjoy it without harming the ecosystem. What kinds of uses might damage the habitat of species that live there?

Science Online booke.msscience.com/self_check_quiz

Reading Check

Answer the banning of DDT and a reintroduction program

3 Assess

DAILY INTERVENTION

Check for Understanding

Visual-Spatial Brainstorm with the class about methods that could be used to preserve an endangered species.

Reteach

Protecting Species Ask students to describe three ways elephants can be protected from possible extinction. L2 LS **Linguistic**

Assessment

Performance Have students write a letter to a newspaper or community leader describing what people can do to prevent a species from becoming extinct. Use **Performance Assessment in the Science Classroom,** p. 139.

section 2 review

1. It makes it illegal to kill endangered or threatened species, to harm their habitats, or use federal funds to build in places that might harm these species.
2. Possible answers: wild behaviors might have been lost; the original problems might not have been solved
3. information on the cause of eelgrass decline; which species depend on the grass; how local people feel about restoring the ecosystem
4. Answers will vary, but uses may include hiking, camping, controlled hunting, and fishing. Destructive uses might include logging, farming, and strip-mining.

BENCH TESTED

Real-World Question

Purpose Students will investigate the relationship between plant biodiversity and the spread of plant diseases. LS **Logical-Mathematical**

Process Skills observe and infer, use numbers, record data, interpret data, use data tables, compare, predict

Time Required 45 minutes

Procedure

Teaching Strategy Once all students have constructed their playing boards, follow the first few steps of round 1 together as a class to make certain everyone understands how the game is played.

Troubleshooting Draw a sample of the field grid for the class to see. Save time by cutting out the colored squares before the activity.

LAB

Biodiversity and the Health of a Plant Community

Goals

- **Record data** on the spread of a plant disease.
- **Compare** the spread of disease in communities with different levels of biodiversity.
- **Predict** a way to increase a crop harvest.

Materials

plain white paper
colored paper (red, orange, yellow, green, blue, black)
ruler
scissors
pens
die

Safety Precautions

Use care when cutting with scissors.

1	2	3	4	5
16				6
15				7
14				8
13	12	11	10	9

Real-World Question

Some plant diseases are carried from plant to plant by specific insects and can spread quickly throughout a garden. If a garden only has plants that are susceptible to one of these diseases, the entire garden can be killed when an infection occurs. One such disease—necrotic leaf spot—can infect impatiens and other garden plants. Could biodiversity help prevent the spread of necrotic leaf spot? A simulation can help you answer this question. How does biodiversity affect the spread of a plant disease?

Procedure

1. On a piece of white paper, draw a square that measures 10 cm on each side. This represents a field. In the square, make a grid with five equal rows and five equal columns, as shown. Number the outer cells from 1 through 16, as shown. The inner cells are not numbered.
2. Cut 1.5-cm × 1.5-cm tiles from colored paper. Cut out 25 black tiles, 25 red tiles, 10 orange tiles, 10 yellow tiles, 5 green tiles, and 5 blue tiles. The black tiles represent a plant that has died from a disease. The other colors represent different plant species. For example, all the red tiles represent one plant species, all the blue tiles are another species, and each remaining color is a different species.
3. There are four rounds in this simulation. For each round, randomly distribute one plant per square as instructed. Roll the die once for each square, proceeding from square 1 to square 2, and so on through square 16.

   Round 1—High Biodiversity
   Distribute five red, five orange, five yellow, five green, and five blue plants in the field.

   Round 2—Moderate Biodiversity
   Distribute ten orange, ten yellow, and five green plants in the field.

   Round 3—Low Biodiversity
   Distribute all 25 of the red tiles in the field.

Differentiated Instruction

Challenge Have students research how top predators and a high biodiversity of herbivores help maintain a higher biodiversity of plants and other organisms for an entire ecosystem. Top predators prey on herbivores preventing them from overgrazing plant species. When there is a large variety of herbivores in an ecosystem, they will eat a wider variety of plants preventing any one type of plant from disappearing. L3

Visually Impaired Ask some of your students to prepare Activity squares to be used by those who are color blind and cannot distinguish between red and green or other color pairs. Instead of using different colored tiles, the tiles can be made to have different shapes. For example, tiles can have one, two, three, or all four corners snipped off.

Round 4—Challenge
You want a harvest of as many red plants as possible. Decide how many red plants you will start with and whether you will plant any other species. Strategically place your tiles on the board. Follow steps 3 through 5 of the procedure.

4. To begin, place a black tile over half of square 1. Roll the die and use the Disease Key to see if the plant in square 1 is infected. If it is, cover it with the black tile. If not, remove the black tile. For example, suppose square 1 has an orange plant on it. Place a black tile over half of the square and roll the die. If you roll a 1, 3, 4, or 5, the plant does not get the disease, so remove the black tile. Proceed through all the squares until you have rolled the die 16 times.
5. If two plants of the same species are next to each other, the disease will spread from one to the other. Suppose squares 2 and 3 contain red plants. If you roll a 1 for square 2, the red plants on squares 2 and 3 die. In this case, the die does not need to be rolled for square 3. Similarly, if the inner square next to square 2 also contains a red plant, it dies and should be covered with a black tile.
6. For each round, record in a data table the number of tiles of each species you started with and the number of each species left alive at the end of the round.

Disease Key

Roll of Die	Infected Color
1	Red
2	Orange
3	Yellow
4	Blue
5	Green
6	All colors

Round ______

Color of Species	Number at Start	Number at End
Red		
Orange		
Yellow		
Blue	Answers will vary.	
Green		
TOTAL		

Analyze Your Data

Identify which round had the lowest number of survivors.

Conclude and Apply

1. **Infer** why a high biodiversity helps a community survive.
2. **Explain** why organic farmers often grow crops of different species in the same field.

Communicating Your Data

Share your results from Round 4 with your classmates. Decide which arrangement and number of red tiles yields the best harvest.

Communicating Your Data

Have students test their set ups several times to determine the best possible arrangements of plant species. Make a display of these arrangements in the classroom.

Analyze Your Data

Expected Outcome Students should realize that fields with high plant biodiversity will resist the spread of infectious diseases much better than fields with a low biodiversity. They will also discover that mixing up plant species within an ecosystem will decrease the spread of disease.

Answers to Questions

Answers may vary depending on the roll of the die, but round 3 should have the fewest survivors.

Error Analysis Inspect the grids made by students before they begin the activity. Make certain they arrange the tiles properly before beginning each round.

Conclude and Apply

1. One disease is not likely to infect more than one species of plant. A high biodiversity of plants isolates a disease to one or a few individual plants preventing it from spreading throughout the population.
2. Varying crop species in a field creates a higher biodiversity and reduces the risk of disease wiping out an entire crop.

Assessment

Process Ask students to infer why exotic species of plants introduced into an ecosystem often choke out native species, lowering the plant biodiversity. Exotic species are often immune to the diseases of the new ecosystem, or there are no predators in the ecosystem to reduce their populations. As a result, they crowd out native species reducing the plant biodiversity. Use **Performance Assessment in the Science Classroom,** p. 89.

Content Background

As rain forests are lost, so are undiscovered compounds in plants that may be of use in the treatment of many diseases.

The heavy rainfall, high humidity and warm temperatures allow the growth of a diversity of trees. These trees support diverse species of vines, mosses, lichens, and orchids. In turn, these plants support communities of insects, spiders, and amphibians. Many of these organisms are host-specific, they depend upon one tree species for their survival.

This diversity of life depends on a rapid recycling of nutrients; so rapid that soil in the tropics is very poor. In a common technique called *slash-and-burn* agriculture, land is cleared and trees are burned to produce nutrient-rich ash. Heavy rains quickly wash away this and other nutrients, resulting in rapid depletion of an area's productivity and requiring further agricultural expansion.

Discussion

Biodiversity Why is biodiversity so high in the tropical rain forest? Possible answers: Environmental conditions are favorable for growth of many species of plants; rapid recycling of leaves and other organic litter; the forest is arranged in layers, and each layer supports many differently adapted organisms.

How much diversity is found in the tropical rain forest? Estimates range from 5 to 30 million different species of plants and animals.

Activity

Forest Layers Have students search discarded nature magazines to obtain pictures of different kinds of rain forest organisms. Organize these pictures according to which layer the organisms are found in the forest. LS **Visual-Spatial**

TIME SCIENCE AND Society

SCIENCE ISSUES THAT AFFECT YOU!

RAIN FOREST Troubles

Large areas of Brazil's rain forests are being burned to make way for homes, farms, and ranches.

Tropical rain forests are located on both sides of the equator. The average temperature in a rain forest is about 25°C and doesn't vary much between day and night. Rain forests receive, on average, between 200 cm and 400 cm of rain each year. Tropical rain forests contain the highest diversity of life on Earth. Scientists estimate that there are about 30 million plant and animal species on this planet—and at least half of them live in rain forests!

Tropical rain forests impact all of our lives every day. Plants in the forest remove carbon dioxide from the atmosphere and give off oxygen. Some of these plants are already used in medicines, and many more are being studied to determine their usefulness.

This tropical rain forest in Costa Rica is untouched by human development.

Humans destroy as many as 20 million hectares of tropical rain forests each year. Farmers who live in tropical areas cut the trees to sell the wood and farm the land. After a few years, the crops use up the nutrients in the soil and more land must be cleared for farming. The logging and mining industries also contribute to the loss of valuable rain forest resources.

Conservationists suggest a number of ways to protect the rain forests.

- Have governments protect rain forest land by establishing parks and conservation areas.
- Reduce the demand for industrial timber and paper.
- Require industries to use different tree-removal methods that cause less damage.
- Teach farmers alternative farming methods to decrease the damage to rain forest land.
- Offer money or lower taxes to farmers and logging companies for not cutting down trees on rain forest land.

Carrying out these plans will be difficult.

The problem extends far beyond the local farmers—the whole world strips rain forest resources in its demand for products such as food ingredients and lumber for building homes.

Present Prepare a multimedia presentation for an elementary class. Teach the class what a rain forest is, where rain forests are located, and how people benefit from rain forests.

Science Online

For more information, visit booke.msscience.com/time

Present Suggest that students prepare their presentations to illustrate how the survival of species is interrelated. Suggest they begin with illustrations of the various trees of the forest and the different habitat layers they form. They can follow these with illustrations of other plants that depend on the trees. Subsequent illustrations may focus on insects that in turn serve as food for birds and mammals.

Resources for Teachers and Students

The Complete National Geographic, 109 Years of National Geographic Magazine on CD-Rom, The National Geographic Society, 1998

Life on Earth: A Natural History Written and Hosted by David Attenborough, 1996, British Broadcasting Corporation and BBC Enterprises, Ltd., available through Warner Home Video

Reviewing Main Ideas

Section 1 Biodiversity

1. A measure of biodiversity is the number of species present in an ecosystem.
2. In general, biodiversity is greater in warm, moist climates than in cold, dry climates.
3. Extinction occurs when the last member of a species dies.
4. Habitat loss, pollution, overhunting, and introduced species can cause a species to become threatened or endangered.
5. Global warming and ozone depletion could affect biodiversity.

Section 2 Conservation Biology

1. Conservation biology is the study of methods for protecting Earth's biodiversity.
2. The Endangered Species Act of 1973 preserves biodiversity by making it illegal to harm threatened or endangered species.
3. Habitat preservation, habitat restoration, and wildlife management strategies can be used to preserve species.
4. Reintroduction programs can be used to restore a species to an area where it once lived.

Visualizing Main Ideas

Copy and complete the following concept map using the following terms: habitat restoration, captive populations, relocation, *and* endangered species.

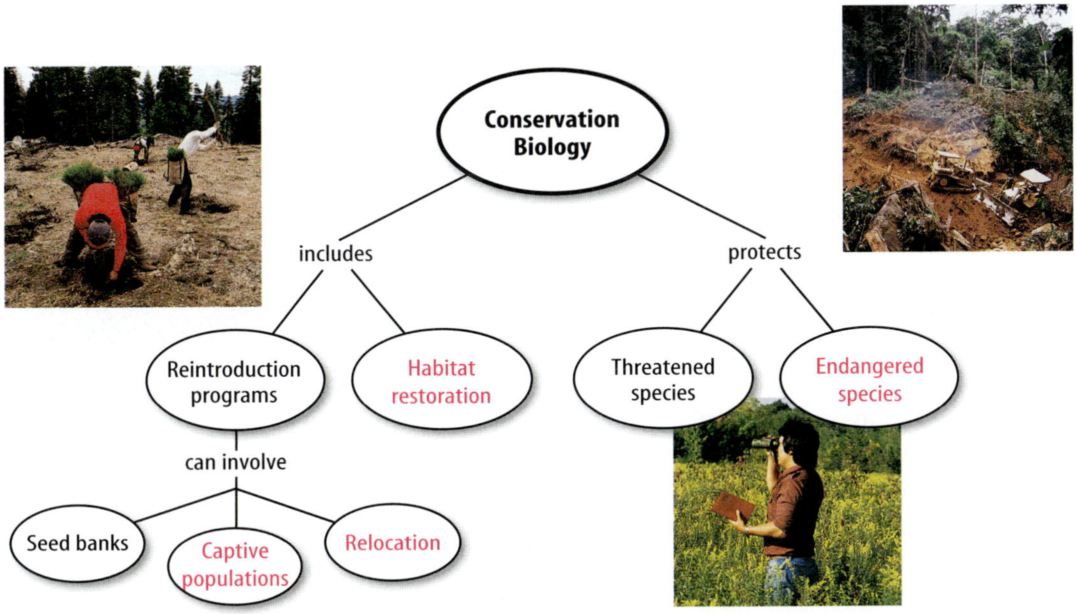

chapter 5 Study Guide

Reviewing Main Ideas

Summary statements can be used by students to review the major concepts of the chapter.

Visualizing Main Ideas

See student page.

Science Online

Visit booke.msscience.com
/self_check_quiz
/interactive_tutor
/vocabulary_puzzlemaker
/chapter_review
/standardized_test

Assessment Transparency

For additional assessment questions, use the *Assessment Transparency* located in the transparency book.

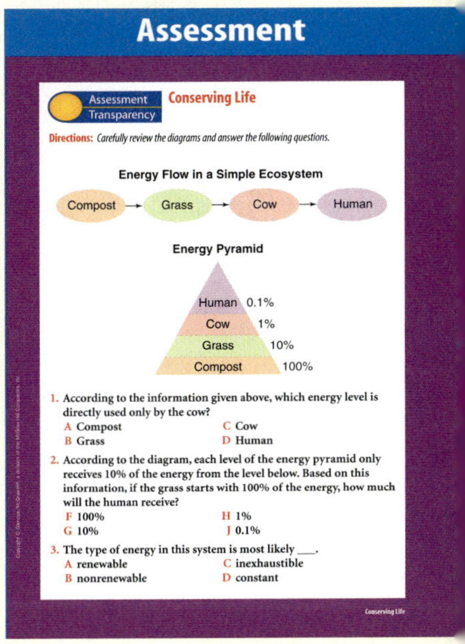

Assessment

Assessment Transparency **Conserving Life**

Directions: *Carefully review the diagrams and answer the following questions.*

1. According to the information given above, which energy level is directly used only by the cow?
   A Compost C Cow
   B Grass D Human
2. According to the diagram, each level of the energy pyramid only receives 10% of the energy from the level below. Based on this information, if the grass starts with 100% of the energy, how much will the human receive?
   F 100% H 1%
   G 10% J 0.1%
3. The type of energy in this system is most likely ___.
   A renewable C inexhaustible
   B nonrenewable D constant

Conserving Life

Using Vocabulary

1. Habitat restoration can help threatened species survive.
2. An endangered species is in jeopardy of becoming an extinct species.
3. Conservation biology studies methods of protecting biodiversity.
4. Captive populations can slow the decline in biodiversity.
5. Acid rain and destruction of the ozone layer are environmental problems that can harm organisms. They are not directly related.
6. Introduced species can create or jeopardize endangered species.
7. Introduced species have no natural predators in the area, therefore they may threaten native species.
8. When the number of extinct species increases, biodiversity declines.
9. Reintroduction programs bring native species back to an area where their population has declined.

Checking Concepts

10. A	14. B
11. B	15. A
12. C	16. B
13. D	17. B

Thinking Critically

18. reducing use of fossil fuels, cutting fewer trees, planting more trees
19. fixing the problem or problems that led to the species' decline in nature and then reintroducing the organism into its natural habitat; captive breeding to maintain population
20. B
21. Answers will vary. Accept all well-reasoned responses.
22. Without sufficient habitat, organisms cannot obtain needed resources.
23. map should show progression from: species → threatened species → endangered species → extinct species

chapter 5 Review

Using Vocabulary

acid rain p. 135
biodiversity p. 126
captive population p. 142
conservation biology p. 138
endangered species p. 131
extinct species p. 130
habitat restoration p. 141
introduced species p. 134
native species p. 134
ozone depletion p. 136
reintroduction program p. 142
threatened species p. 131

Explain the differences in the vocabulary words given below. Then explain how the words are related.

1. habitat restoration—threatened species
2. endangered species—extinct species
3. biodiversity—conservation biology
4. biodiversity—captive population
5. acid rain—ozone depletion
6. introduced species—endangered species
7. native species—introduced species
8. biodiversity—extinct species
9. reintroduction program—native species

Checking Concepts

Choose the word or phrase that best answers the question.

10. Which state probably has the greatest biodiversity?
    - **A)** Florida
    - **B)** Maine
    - **C)** New Hampshire
    - **D)** West Virginia
11. Which of these is the greatest threat to the Lake Erie water snake?
    - **A)** air pollution
    - **B)** habitat loss
    - **C)** ozone
    - **D)** introduced species
12. Which of the following gases might contribute to global warming?
    - **A)** oxygen
    - **B)** nitrogen
    - **C)** carbon dioxide
    - **D)** neon
13. Which of the following can reduce biodiversity?
    - **A)** divided habitats
    - **B)** habitat loss
    - **C)** introduced species
    - **D)** all of these
14. Jen has a tank with 20 guppies. June has a tank with 5 guppies, 5 mollies, 5 swordtails, and 5 platys. Which of the following is true?
    - **A)** June has more fish than Jen.
    - **B)** June's tank has more biodiversity than Jen's tank.
    - **C)** Jen has more fish than June.
    - **D)** Jen's tank has more biodiversity than June's tank.
15. What would you call a species that had never lived on an island until it was brought there by people?
    - **A)** introduced
    - **B)** endangered
    - **C)** native
    - **D)** threatened
16. Which of the following conservation strategies involves passing laws to protect species?
    - **A)** habitat restoration
    - **B)** legal protections
    - **C)** reintroduction programs
    - **D)** captive populations

Use the photo below to answer question 17.

17. The animals above are examples of a(n)
    - **A)** extinct species.
    - **B)** endangered species.
    - **C)** lost species.
    - **D)** introduced species.

Science Online booke.msscience.com/vocabulary_puzzlemaker

Use the Exam*View*® Pro Testmaker CD-ROM to:

- create multiple versions of tests
- create modified tests with one mouse click for inclusion students
- edit existing questions and add your own questions
- build tests aligned with state standards using built-in State Curriculum Tags
- change English tests to Spanish with one mouse click and vice versa

Thinking Critically

18. **Infer** What action(s) could people take to help reduce the types of air pollution that might contribute to global warming?

19. **Describe** what conservation strategies would most likely help preserve a species whose members are found only in zoos.

Use the following diagrams to answer question 20.

a b c d

20. **Draw a Conclusion** Four designs for a road system in a national park are shown above. Which arrangement best would avoid dividing this habitat into pieces? Explain.

21. **Infer** Why might the protection of an endangered species create conflict between local residents and conservation biologists?

22. **Explain** why habitat loss is the most serious threat to biodiversity.

23. **Concept Map** Draw an events-chain concept map using the following terms: *species, extinct species, endangered species,* and *threatened species.*

24. **Form a Hypothesis** In divided wildlife areas, large animals have a greater chance of becoming extinct than other organisms do. Suggest a hypothesis to explain why this is true.

25. **Compare and Contrast** Explain how legal protections and wildlife management are similar and different.

26. **Recognize Cause and Effect** Design a wildlife management plan to allow deer hunting in a state park without damaging the ecosystem.

Performance Activities

27. **Poster** Use images from magazines to create a display about how humans threaten Earth's biodiversity.

Applying Math

Use the following table to answer questions 28–30.

Wildlife Habitat Lost

Country	Area Lost (km^2)	Area Remaining (km^2)
Ethiopia	770,700	330,300
Vietnam	265,680	66,420
Indonesia	708,751	746,860

28. **Habitat Loss** Use the data in the table to make a bar graph.

29. **Original Habitat** Using the table above, calculate the beginning amount of habitat in each country.

30. **Remaining Habitat** Use the table above to calculate the proportion of wildlife habitat area remaining in Ethiopia, Vietnam, and Indonesia.

31. **Extinction** Since 1600, the number of known extinctions includes 29 fish, 2 amphibians, 23 reptiles, 116 birds, and 59 mammals. How many vertebrates are known to have gone extinct since 1600?

32. **Biodiversity** A pond contains 6 species of fish, 3 species of amphibians, and 2 species of reptiles. A new species of fish is released into the pond resulting in the loss of 2 fish species and 2 amphibian species. How many species are in the pond before the new species of fish is released and how many species are in the pond after it is released?

33. **Introduced Species** After the brown tree snake was introduced to the island of Guam, 9 of 11 native forest-dwelling bird species became extinct. What proportion of native forest-dwelling bird species remain?

Science Online booke.msscience.com/chapter_review

Assessment Resources

Reproducible Masters

Chapter *Fast File* Resources
Chapter Review, pp. 35–36
Chapter Tests, pp. 37–40
Assessment Transparency Activity, p. 47

Glencoe Science Web site
Chapter Review Test
Standardized Test Practice

Glencoe Technology

Assessment Transparency
Exam*View*® Pro Testmaker
MindJogger Videoquiz
Interactive Chalkboard

chapter 5 Review

24. Possible answer: In a divided habitat, large animals may be confined to areas that are too small to ensure their survival.

25. Legal protections stop the killing of endangered or threatened organisms. Wildlife management usually affects a specific locale—national or state park—and allows hunting or collection of organisms, but tries to prevent their excessive loss.

26. Answers will vary, but should allow deer hunting when the population rises and prohibit it when populations decline or when species are reproducing.

Performance Activities

27. Possible threats to include: pollution, habitat loss, divided habits, introduced species. Use **PASC,** p. 145.

Applying Math

National Math Standards

1, 5

28. Graphs should show: Area lost (most to least)—Ethiopia, Indonesia, Vietnam; Area remaining (most to least)—Indonesia, Ethiopia, Vietnam

29. Ethiopia 1,101,000; Vietnam 332,100; Indonesia 1,455,611

30. Ethiopia 30%; Vietnam 20%; Indonesia 51%

31. 229

32. before fish released = 11
after fish released = 8

33. 18%

FAST FILE

Answer Sheet A practice answer sheet can be found at booke.msscience.com/answer_sheet.

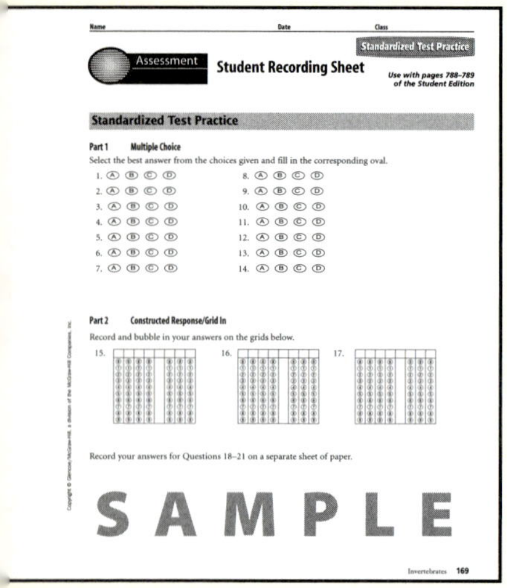
Name Date Class
Standardized Test Practice
Assessment **Student Recording Sheet** *Use with pages 788–789 of the Student Edition*
Standardized Test Practice
Part 1 Multiple Choice
Select the best answer from the choices given and fill in the corresponding oval.
Part 2 Constructed Response/Grid In
Record and bubble in your answers on the grids below.
Record your answers for Questions 18–21 on a separate sheet of paper.
SAMPLE

Part 1 Multiple Choice

1. B
2. D
3. D
4. C
5. B
6. B
7. A

Part 2 Short Response

8. The U.S. Endangered Species Act of 1973 is a law that makes it illegal to harm, collect, harass, or disturb the habitat of threatened species like the America bald eagle or the American alligator.
9. Conservation biology is the study of methods for protecting biodiversity.
10. protect a species from harm; protect the species' habitat
11. sulfur dioxide and nitrogen oxide
12. five mass extinctions
13. No, these mass extinctions occurred before humans were on Earth.
14. because many medicines originally came from wild plants
15. a reintroduction program

chapter 5 Standardized Test Practice

Part 1 Multiple Choice

Record your answers on the answer sheet provided by your teacher or on a sheet of paper.

Use the illustration below to answer questions 1 and 2.

Plan A

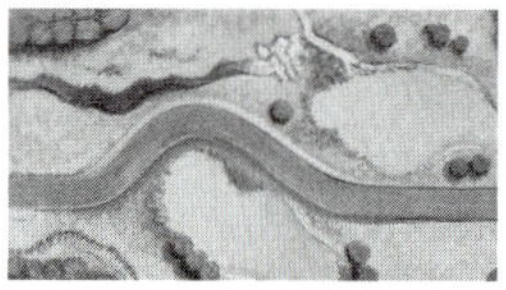

Plan B

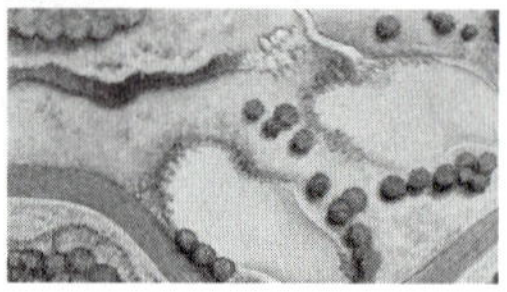

1. A road is being built through an area near two lakes and a construction team must chose between Plan A and Plan B. A conservation biologist evaluates how each plan would affect biodiversity in the ecosystem that includes the lakes and urges the team to follow Plan A. Why?
   - **A.** It introduces new species.
   - **B.** It does not divide the aquatic habitat into small areas.
   - **C.** It makes the road easier to drive along and view animals.
   - **D.** It helps keep more water in both of the lakes.

2. Which species is least likely to be affected if the construction team follows Plan B?
   - **A.** water snakes
   - **B.** frogs
   - **C.** algae
   - **D.** birds

3. Where would you expect to find more biodiversity?
   - **A.** wheat field
   - **B.** polluted stream
   - **C.** lawn
   - **D.** coral reef

Test-Taking Tip

Test Day Preparation Avoid rushing on test day. Prepare your clothes and test supplies the night before. Wake up early and arrive at school on time on test day.

4. What is used to link small parks together so animals can move easily from one to another?
   - **A.** habitat restoration
   - **B.** relocation
   - **C.** wildlife corridors
   - **D.** seed banks

5. How does Earth's ozone layer protect living things?
   - **A.** keeps Earth's temperature constant
   - **B.** blocks much of the Sun's UV radiation
   - **C.** stops polar ice caps from melting
   - **D.** prevents acid rain from falling to Earth

Use the illustration below to answer questions 6 and 7.

6. The small mollusks shown above are zebra mussels, a native species in the Caspian Sea in Asia. They were accidentally brought to the Great Lakes in the ballast water of boats. How would zebra mussels be described in North America?
   - **A.** reintroduced species
   - **B.** introduced species
   - **C.** endangered species
   - **D.** native species

7. Zebra mussels grow more rapidly than mussels native to the Great Lakes and cause the native mussels to die out. Which of the following best describes the problem caused by zebra mussels?
   - **A.** They are decreasing biodiversity.
   - **B.** They are causing divided habitats.
   - **C.** They are polluting the water.
   - **D.** They are causing global warming.

Part 3 Open Ended

16. People are not allowed to hunt or fish in a park without a license. The fees for the licenses provide funds for managing a wildlife area. The license also can limit the number of animals a hunter is allowed to take or can limit hunting only to seasons when the species is not reproducing. Hunting also can help prevent a population from becoming too large for an area.
17. Both global warming and ozone depletion involve changes to gases in Earth's atmosphere. Increases in carbon dioxide might increase Earth's average temperature. This increase in temperature is called global warming and might lead to climate changes that would affect biodiversity. If portions of the polar ice caps melted, floods in coastal ecosystems might occur. The ozone layer is about 15 km to 30 km above Earth's surface and protects life on Earth from harmful amounts

Part 2 | Short Response/Grid In

Record your answers on the answer sheet provided by your teacher or on a sheet of paper.

8. What law makes it illegal to harm, collect, harass, or disturb the habitat of the American bald eagle and the American alligator?

9. What is the study of the methods for protecting biodiversity called?

10. What are two goals of most conservation plans?

11. What are two gases released from automobiles and industries that can combine with water vapor in the air to cause acid rain?

Use the illustration below to answer questions 12 and 13.

Mass Extinctions in Earth's History

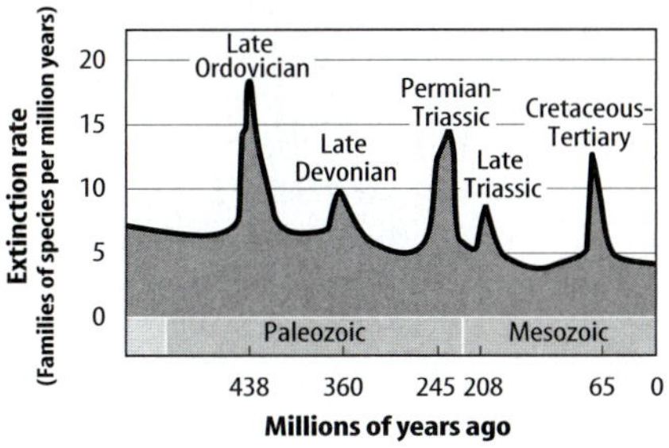

12. The graph shows mass extinctions that have occurred in Earth's history. How many mass extinctions have occurred?

13. Did human activities contribute to any of these extinctions? Explain your answer.

14. Why is biodiversity important to the field of medicine?

15. Brown pelicans once completely disappeared from Louisiana and most of Texas. Today, thousands of them live in these areas. What type of conservation program caused this change?

Part 3 | Open Ended

Record your answers on a sheet of paper.

16. How can hunters and wildlife managers work together to maintain ecosystems in parks and preserves?

17. Global warming and ozone depletion both involve changes to Earth's atmosphere. Compare and contrast the problems these two processes could cause to the environment.

18. Explain why just protecting a species like the grizzly bear with a law may not be enough to conserve this species.

19. Describe what people do that causes the release of large amounts of carbon dioxide (CO_2) into the atmosphere. What problems might an increase in CO_2 cause for people?

Use the photo below to answer questions 20 and 21.

20. Explain why the small round sign placed beside this storm drain says "No Dumping."

21. Describe other ways that pollutants can enter water. How is biodiversity affected by pollutants?

22. Why are captive populations not an ideal way to protect endangered species?

23. Compare and contrast legal protection, habitat restoration, and seed banks as programs used to prevent the loss of species from an ecosystem.

of ultraviolet (UV) radiation. The thinning of the ozone layer means that more UV radiation reaches Earth's surface and can harm living things.

18. Although laws protecting grizzly bears from being harmed or disturbed are important, they cannot survive unless their habitat also is protected. A grizzly bear needs a territory of several hundred square kilometers to feed itself. National parks and wildlife areas preserve the habitats necessary for these bears.

19. People burn large amounts of fuel, like wood, coal, and gasoline. When this fuel burns, carbon dioxide (CO_2) gas is released into the atmosphere. This contributes to an increase in the percentage of CO_2 in Earth's atmosphere and could increase Earth's temperature and cause climate changes.

20. This water usually flows untreated into nearby waterways. Because of this, storm drains should never be used to dispose of used motor oils, paints, or other liquid wastes.

21. Pollutants also can enter water from sources other than storm drains. Pollutants can come from factories, ships, or runoff from lawns and farms. Because pollutants can kill aquatic organisms, water pollution decreases biodiversity in aquatic ecosystems.

22. Although captive populations can help preserve biodiversity, it is not ideal. It can be expensive to provide proper food, space, and care for these populations. Also, captive animals sometimes lose their wild behavior and might not survive if they are released back into their native habitats.

23. Conservation biologists might identify a species as endangered and have it protected legally by the U.S. Endangered Species Act. Other international laws also might be used to protect the species. If the habitat of the species has changed or been harmed by human activities, it might be restored to a healthy condition. Seed banks are used to store the seed of endangered plant species. If a plant species becomes extinct in the wild, the stored seeds can be grown and the species reintroduced to its original habitat.

Rubrics

For more help evaluating open-ended assessment questions, see the rubric on p. 4T.

Student Resources

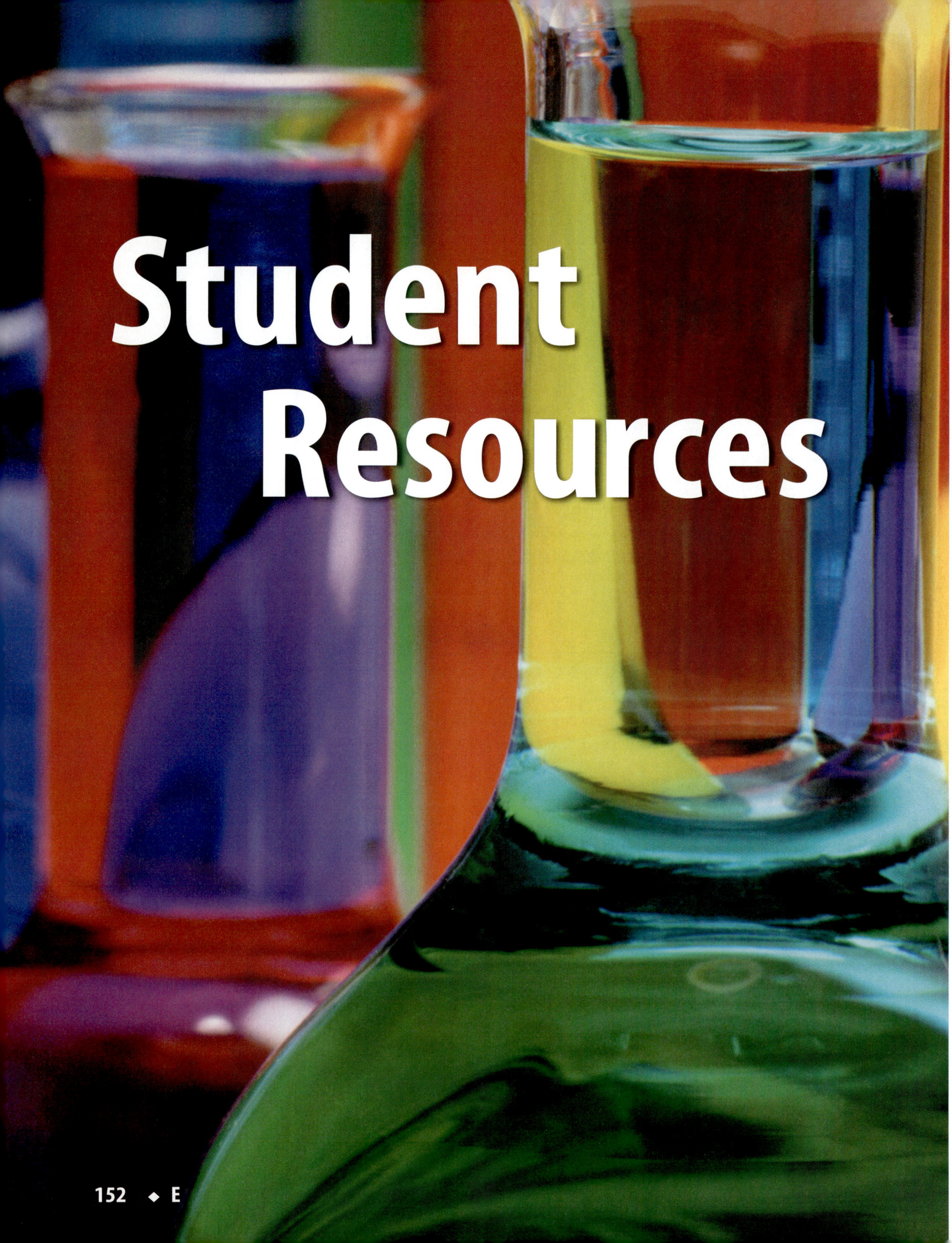

CONTENTS

Scientific Methods

Scientists use an orderly approach called the scientific method to solve problems. This includes organizing and recording data so others can understand them. Scientists use many variations in this method when they solve problems.

Identify a Question

The first step in a scientific investigation or experiment is to identify a question to be answered or a problem to be solved. For example, you might ask which gasoline is the most efficient.

Gather and Organize Information

After you have identified your question, begin gathering and organizing information. There are many ways to gather information, such as researching in a library, interviewing those knowledgeable about the subject, testing and working in the laboratory and field. Fieldwork is investigations and observations done outside of a laboratory.

Researching Information Before moving in a new direction, it is important to gather the information that already is known about the subject. Start by asking yourself questions to determine exactly what you need to know. Then you will look for the information in various reference sources, like the student is doing in **Figure 1.** Some sources may include textbooks, encyclopedias, government documents, professional journals, science magazines, and the Internet. Always list the sources of your information.

Figure 1 The Internet can be a valuable research tool.

Evaluate Sources of Information Not all sources of information are reliable. You should evaluate all of your sources of information, and use only those you know to be dependable. For example, if you are researching ways to make homes more energy efficient, a site written by the U.S. Department of Energy would be more reliable than a site written by a company that is trying to sell a new type of weatherproofing material. Also, remember that research always is changing. Consult the most current resources available to you. For example, a 1985 resource about saving energy would not reflect the most recent findings.

Sometimes scientists use data that they did not collect themselves, or conclusions drawn by other researchers. This data must be evaluated carefully. Ask questions about how the data were obtained, if the investigation was carried out properly, and if it has been duplicated exactly with the same results. Would you reach the same conclusion from the data? Only when you have confidence in the data can you believe it is true and feel comfortable using it.

Interpret Scientific Illustrations As you research a topic in science, you will see drawings, diagrams, and photographs to help you understand what you read. Some illustrations are included to help you understand an idea that you can't see easily by yourself, like the tiny particles in an atom in **Figure 2.** A drawing helps many people to remember details more easily and provides examples that clarify difficult concepts or give additional information about the topic you are studying. Most illustrations have labels or a caption to identify or to provide more information.

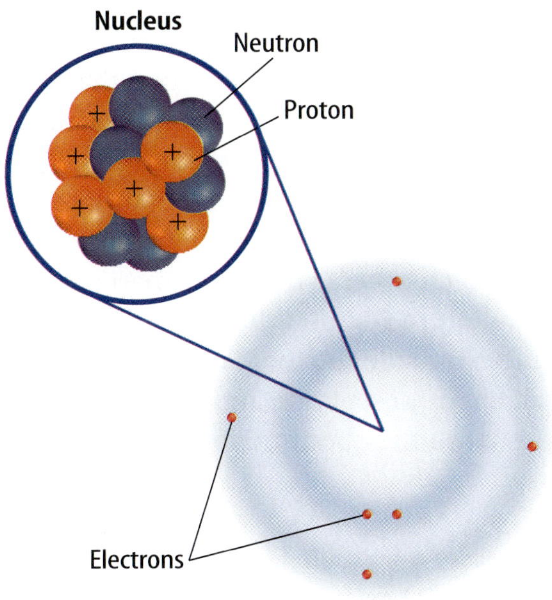

Figure 2 This drawing shows an atom of carbon with its six protons, six neutrons, and six electrons.

Concept Maps One way to organize data is to draw a diagram that shows relationships among ideas (or concepts). A concept map can help make the meanings of ideas and terms more clear, and help you understand and remember what you are studying. Concept maps are useful for breaking large concepts down into smaller parts, making learning easier.

Network Tree A type of concept map that not only shows a relationship, but how the concepts are related is a network tree, shown in **Figure 3.** In a network tree, the words are written in the ovals, while the description of the type of relationship is written across the connecting lines.

When constructing a network tree, write down the topic and all major topics on separate pieces of paper or notecards. Then arrange them in order from general to specific. Branch the related concepts from the major concept and describe the relationship on the connecting line. Continue to more specific concepts until finished.

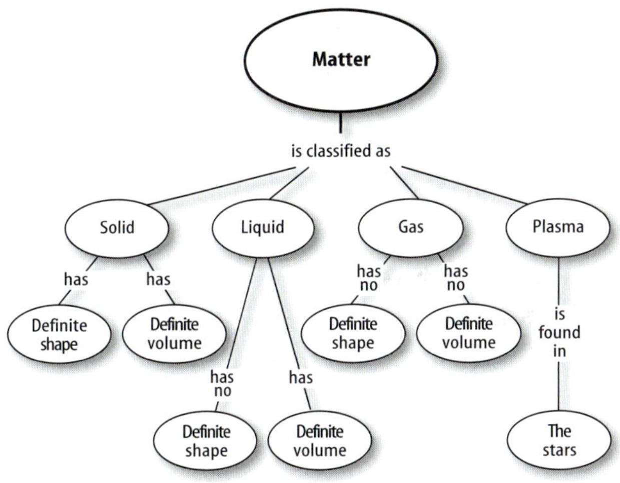

Figure 3 A network tree shows how concepts or objects are related.

Events Chain Another type of concept map is an events chain. Sometimes called a flow chart, it models the order or sequence of items. An events chain can be used to describe a sequence of events, the steps in a procedure, or the stages of a process.

When making an events chain, first find the one event that starts the chain. This event is called the initiating event. Then, find the next event and continue until the outcome is reached, as shown in **Figure 4.**

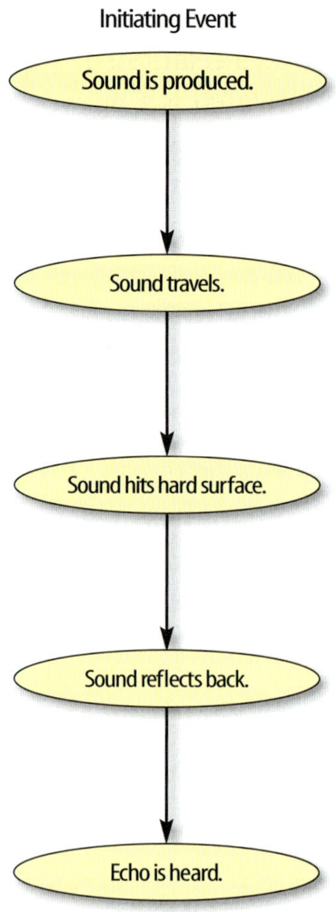

Figure 4 Events-chain concept maps show the order of steps in a process or event. This concept map shows how a sound makes an echo.

Cycle Map A specific type of events chain is a cycle map. It is used when the series of events do not produce a final outcome, but instead relate back to the beginning event, such as in **Figure 5.** Therefore, the cycle repeats itself.

To make a cycle map, first decide what event is the beginning event. This is also called the initiating event. Then list the next events in the order that they occur, with the last event relating back to the initiating event. Words can be written between the events that describe what happens from one event to the next. The number of events in a cycle map can vary, but usually contain three or more events.

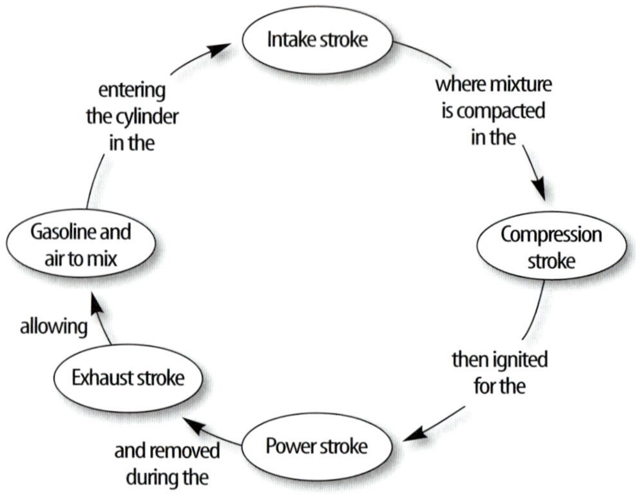

Figure 5 A cycle map shows events that occur in a cycle.

Spider Map A type of concept map that you can use for brainstorming is the spider map. When you have a central idea, you might find that you have a jumble of ideas that relate to it but are not necessarily clearly related to each other. The spider map on sound in **Figure 6** shows that if you write these ideas outside the main concept, then you can begin to separate and group unrelated terms so they become more useful.

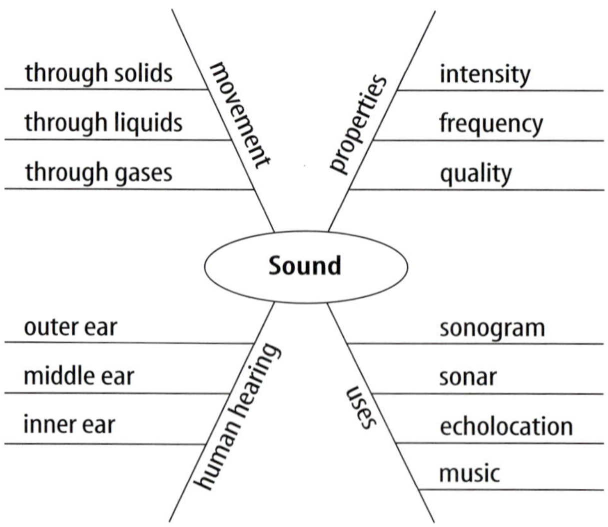

Figure 6 A spider map allows you to list ideas that relate to a central topic but not necessarily to one another.

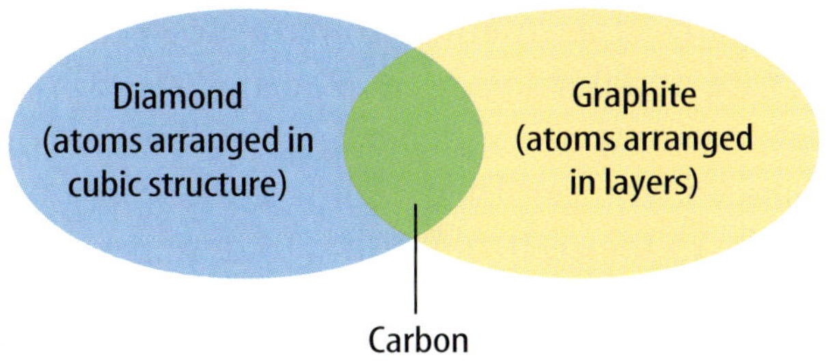

Figure 7 This Venn diagram compares and contrasts two substances made from carbon.

Venn Diagram To illustrate how two subjects compare and contrast you can use a Venn diagram. You can see the characteristics that the subjects have in common and those that they do not, shown in **Figure 7.**

To create a Venn diagram, draw two overlapping ovals that that are big enough to write in. List the characteristics unique to one subject in one oval, and the characteristics of the other subject in the other oval. The characteristics in common are listed in the overlapping section.

Make and Use Tables One way to organize information so it is easier to understand is to use a table. Tables can contain numbers, words, or both.

To make a table, list the items to be compared in the first column and the characteristics to be compared in the first row. The title should clearly indicate the content of the table, and the column or row heads should be clear. Notice that in **Table 1** the units are included.

Table 1 Recyclables Collected During Week

Day of Week	Paper (kg)	Aluminum (kg)	Glass (kg)
Monday	5.0	4.0	12.0
Wednesday	4.0	1.0	10.0
Friday	2.5	2.0	10.0

Make a Model One way to help you better understand the parts of a structure, the way a process works, or to show things too large or small for viewing is to make a model. For example, an atomic model made of a plastic-ball nucleus and pipe-cleaner electron shells can help you visualize how the parts of an atom relate to each other. Other types of models can by devised on a computer or represented by equations.

Form a Hypothesis

A possible explanation based on previous knowledge and observations is called a hypothesis. After researching gasoline types and recalling previous experiences in your family's car you form a hypothesis—our car runs more efficiently because we use premium gasoline. To be valid, a hypothesis has to be something you can test by using an investigation.

Predict When you apply a hypothesis to a specific situation, you predict something about that situation. A prediction makes a statement in advance, based on prior observation, experience, or scientific reasoning. People use predictions to make everyday decisions. Scientists test predictions by performing investigations. Based on previous observations and experiences, you might form a prediction that cars are more efficient with premium gasoline. The prediction can be tested in an investigation.

Design an Experiment A scientist needs to make many decisions before beginning an investigation. Some of these include: how to carry out the investigation, what steps to follow, how to record the data, and how the investigation will answer the question. It also is important to address any safety concerns.

Science Skill Handbook

Test the Hypothesis

Now that you have formed your hypothesis, you need to test it. Using an investigation, you will make observations and collect data, or information. This data might either support or not support your hypothesis. Scientists collect and organize data as numbers and descriptions.

Follow a Procedure In order to know what materials to use, as well as how and in what order to use them, you must follow a procedure. **Figure 8** shows a procedure you might follow to test your hypothesis.

Procedure

1. Use regular gasoline for two weeks.
2. Record the number of kilometers between fill-ups and the amount of gasoline used.
3. Switch to premium gasoline for two weeks.
4. Record the number of kilometers between fill-ups and the amount of gasoline used.

Figure 8 A procedure tells you what to do step by step.

Identify and Manipulate Variables and Controls In any experiment, it is important to keep everything the same except for the item you are testing. The one factor you change is called the independent variable. The change that results is the dependent variable. Make sure you have only one independent variable, to assure yourself of the cause of the changes you observe in the dependent variable. For example, in your gasoline experiment the type of fuel is the independent variable. The dependent variable is the efficiency.

Many experiments also have a control—an individual instance or experimental subject for which the independent variable is not changed. You can then compare the test results to the control results. To design a control you can have two cars of the same type. The control car uses regular gasoline for four weeks. After you are done with the test, you can compare the experimental results to the control results.

Collect Data

Whether you are carrying out an investigation or a short observational experiment, you will collect data, as shown in **Figure 9.** Scientists collect data as numbers and descriptions and organize it in specific ways.

Observe Scientists observe items and events, then record what they see. When they use only words to describe an observation, it is called qualitative data. Scientists' observations also can describe how much there is of something. These observations use numbers, as well as words, in the description and are called quantitative data. For example, if a sample of the element gold is described as being "shiny and very dense" the data are qualitative. Quantitative data on this sample of gold might include "a mass of 30 g and a density of 19.3 g/cm^3."

Figure 9 Collecting data is one way to gather information directly.

Figure 10 Record data neatly and clearly so it is easy to understand.

When you make observations you should examine the entire object or situation first, and then look carefully for details. It is important to record observations accurately and completely. Always record your notes immediately as you make them, so you do not miss details or make a mistake when recording results from memory. Never put unidentified observations on scraps of paper. Instead they should be recorded in a notebook, like the one in **Figure 10.** Write your data neatly so you can easily read it later. At each point in the experiment, record your observations and label them. That way, you will not have to determine what the figures mean when you look at your notes later. Set up any tables that you will need to use ahead of time, so you can record any observations right away. Remember to avoid bias when collecting data by not including personal thoughts when you record observations. Record only what you observe.

Estimate Scientific work also involves estimating. To estimate is to make a judgment about the size or the number of something without measuring or counting. This is important when the number or size of an object or population is too large or too difficult to accurately count or measure.

Sample Scientists may use a sample or a portion of the total number as a type of estimation. To sample is to take a small, representative portion of the objects or organisms of a population for research. By making careful observations or manipulating variables within that portion of the group, information is discovered and conclusions are drawn that might apply to the whole population. A poorly chosen sample can be unrepresentative of the whole. If you were trying to determine the rainfall in an area, it would not be best to take a rainfall sample from under a tree.

Measure You use measurements everyday. Scientists also take measurements when collecting data. When taking measurements, it is important to know how to use measuring tools properly. Accuracy also is important.

Length To measure length, the distance between two points, scientists use meters. Smaller measurements might be measured in centimeters or millimeters.

Length is measured using a metric ruler or meter stick. When using a metric ruler, line up the 0-cm mark with the end of the object being measured and read the number of the unit where the object ends. Look at the metric ruler shown in **Figure 11.** The centimeter lines are the long, numbered lines, and the shorter lines are millimeter lines. In this instance, the length would be 4.50 cm.

Figure 11 This metric ruler has centimeter and millimeter divisions.

Mass The SI unit for mass is the kilogram (kg). Scientists can measure mass using units formed by adding metric prefixes to the unit gram (g), such as milligram (mg). To measure mass, you might use a triple-beam balance similar to the one shown in **Figure 12.** The balance has a pan on one side and a set of beams on the other side. Each beam has a rider that slides on the beam.

When using a triple-beam balance, place an object on the pan. Slide the largest rider along its beam until the pointer drops below zero. Then move it back one notch. Repeat the process for each rider proceeding from the larger to smaller until the pointer swings an equal distance above and below the zero point. Sum the masses on each beam to find the mass of the object. Move all riders back to zero when finished.

Instead of putting materials directly on the balance, scientists often take a tare of a container. A tare is the mass of a container into which objects or substances are placed for measuring their masses. To mass objects or substances, find the mass of a clean container. Remove the container from the pan, and place the object or substances in the container. Find the mass of the container with the materials in it. Subtract the mass of the empty container from the mass of the filled container to find the mass of the materials you are using.

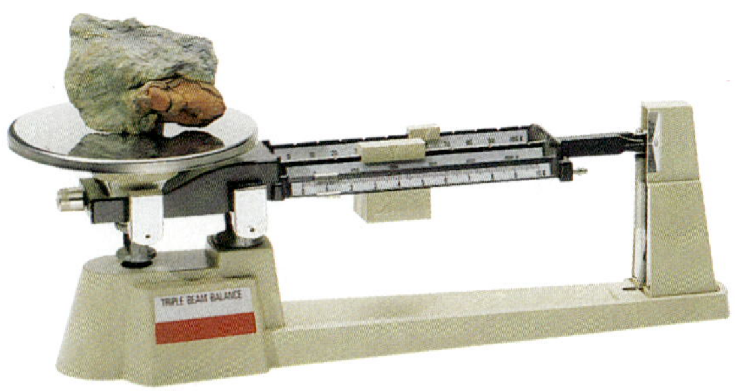

Figure 12 A triple-beam balance is used to determine the mass of an object.

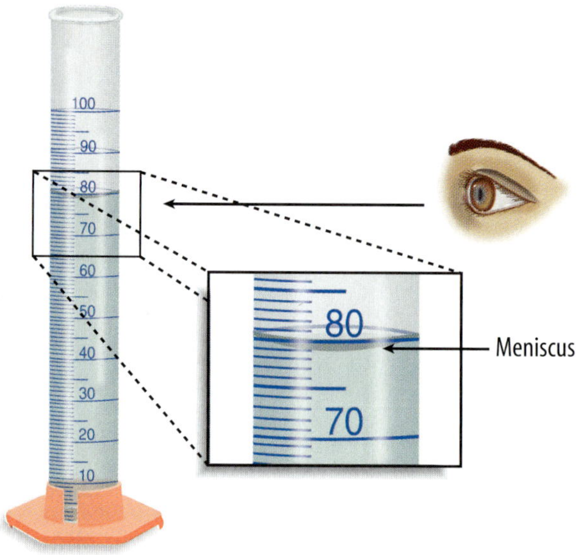

Figure 13 Graduated cylinders measure liquid volume.

Liquid Volume To measure liquids, the unit used is the liter. When a smaller unit is needed, scientists might use a milliliter. Because a milliliter takes up the volume of a cube measuring 1 cm on each side it also can be called a cubic centimeter ($cm^3 = cm \times cm \times cm$).

You can use beakers and graduated cylinders to measure liquid volume. A graduated cylinder, shown in **Figure 13,** is marked from bottom to top in milliliters. In lab, you might use a 10-mL graduated cylinder or a 100-mL graduated cylinder. When measuring liquids, notice that the liquid has a curved surface. Look at the surface at eye level, and measure the bottom of the curve. This is called the meniscus. The graduated cylinder in **Figure 13** contains 79.0 mL, or 79.0 cm^3, of a liquid.

Temperature Scientists often measure temperature using the Celsius scale. Pure water has a freezing point of 0°C and boiling point of 100°C. The unit of measurement is degrees Celsius. Two other scales often used are the Fahrenheit and Kelvin scales.

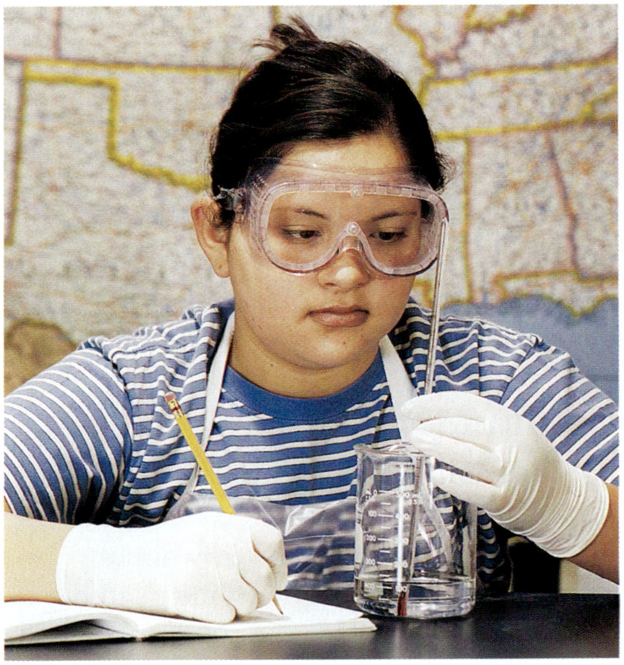

Figure 14 A thermometer measures the temperature of an object.

Scientists use a thermometer to measure temperature. Most thermometers in a laboratory are glass tubes with a bulb at the bottom end containing a liquid such as colored alcohol. The liquid rises or falls with a change in temperature. To read a glass thermometer like the thermometer in **Figure 14,** rotate it slowly until a red line appears. Read the temperature where the red line ends.

Form Operational Definitions An operational definition defines an object by how it functions, works, or behaves. For example, when you are playing hide and seek and a tree is home base, you have created an operational definition for a tree.

Objects can have more than one operational definition. For example, a ruler can be defined as a tool that measures the length of an object (how it is used). It can also be a tool with a series of marks used as a standard when measuring (how it works).

Analyze the Data

To determine the meaning of your observations and investigation results, you will need to look for patterns in the data. Then you must think critically to determine what the data mean. Scientists use several approaches when they analyze the data they have collected and recorded. Each approach is useful for identifying specific patterns.

Interpret Data The word *interpret* means "to explain the meaning of something." When analyzing data from an experiement, try to find out what the data show. Identify the control group and the test group to see whether or not changes in the independent variable have had an effect. Look for differences in the dependent variable between the control and test groups.

Classify Sorting objects or events into groups based on common features is called classifying. When classifying, first observe the objects or events to be classified. Then select one feature that is shared by some members in the group, but not by all. Place those members that share that feature in a subgroup. You can classify members into smaller and smaller subgroups based on characteristics. Remember that when you classify, you are grouping objects or events for a purpose. Keep your purpose in mind as you select the features to form groups and subgroups.

Compare and Contrast Observations can be analyzed by noting the similarities and differences between two more objects or events that you observe. When you look at objects or events to see how they are similar, you are comparing them. Contrasting is looking for differences in objects or events.

Recognize Cause and Effect A cause is a reason for an action or condition. The effect is that action or condition. When two events happen together, it is not necessarily true that one event caused the other. Scientists must design a controlled investigation to recognize the exact cause and effect.

Draw Conclusions

When scientists have analyzed the data they collected, they proceed to draw conclusions about the data. These conclusions are sometimes stated in words similar to the hypothesis that you formed earlier. They may confirm a hypothesis, or lead you to a new hypothesis.

Infer Scientists often make inferences based on their observations. An inference is an attempt to explain observations or to indicate a cause. An inference is not a fact, but a logical conclusion that needs further investigation. For example, you may infer that a fire has caused smoke. Until you investigate, however, you do not know for sure.

Apply When you draw a conclusion, you must apply those conclusions to determine whether the data supports the hypothesis. If your data do not support your hypothesis, it does not mean that the hypothesis is wrong. It means only that the result of the investigation did not support the hypothesis. Maybe the experiment needs to be redesigned, or some of the initial observations on which the hypothesis was based were incomplete or biased. Perhaps more observation or research is needed to refine your hypothesis. A successful investigation does not always come out the way you originally predicted.

Avoid Bias Sometimes a scientific investigation involves making judgments. When you make a judgment, you form an opinion. It is important to be honest and not to allow any expectations of results to bias your judgments. This is important throughout the entire investigation, from researching to collecting data to drawing conclusions.

Communicate

The communication of ideas is an important part of the work of scientists. A discovery that is not reported will not advance the scientific community's understanding or knowledge. Communication among scientists also is important as a way of improving their investigations.

Scientists communicate in many ways, from writing articles in journals and magazines that explain their investigations and experiments, to announcing important discoveries on television and radio. Scientists also share ideas with colleagues on the Internet or present them as lectures, like the student is doing in **Figure 15.**

Figure 15 A student communicates to his peers about his investigation.

Science Skill Handbook

SAFETY SYMBOLS	HAZARD	EXAMPLES	PRECAUTION	REMEDY
DISPOSAL	Special disposal procedures need to be followed.	certain chemicals, living organisms	Do not dispose of these materials in the sink or trash can.	Dispose of wastes as directed by your teacher.
BIOLOGICAL	Organisms or other biological materials that might be harmful to humans	bacteria, fungi, blood, unpreserved tissues, plant materials	Avoid skin contact with these materials. Wear mask or gloves.	Notify your teacher if you suspect contact with material. Wash hands thoroughly.
EXTREME TEMPERATURE	Objects that can burn skin by being too cold or too hot	boiling liquids, hot plates, dry ice, liquid nitrogen	Use proper protection when handling.	Go to your teacher for first aid.
SHARP OBJECT	Use of tools or glassware that can easily puncture or slice skin	razor blades, pins, scalpels, pointed tools, dissecting probes, broken glass	Practice common-sense behavior and follow guidelines for use of the tool.	Go to your teacher for first aid.
FUME	Possible danger to respiratory tract from fumes	ammonia, acetone, nail polish remover, heated sulfur, moth balls	Make sure there is good ventilation. Never smell fumes directly. Wear a mask.	Leave foul area and notify your teacher immediately.
ELECTRICAL	Possible danger from electrical shock or burn	improper grounding, liquid spills, short circuits, exposed wires	Double-check setup with teacher. Check condition of wires and apparatus.	Do not attempt to fix electrical problems. Notify your teacher immediately.
IRRITANT	Substances that can irritate the skin or mucous membranes of the respiratory tract	pollen, moth balls, steel wool, fiberglass, potassium permanganate	Wear dust mask and gloves. Practice extra care when handling these materials.	Go to your teacher for first aid.
CHEMICAL	Chemicals can react with and destroy tissue and other materials	bleaches such as hydrogen peroxide; acids such as sulfuric acid, hydrochloric acid; bases such as ammonia, sodium hydroxide	Wear goggles, gloves, and an apron.	Immediately flush the affected area with water and notify your teacher.
TOXIC	Substance may be poisonous if touched, inhaled, or swallowed.	mercury, many metal compounds, iodine, poinsettia plant parts	Follow your teacher's instructions.	Always wash hands thoroughly after use. Go to your teacher for first aid.
FLAMMABLE	Flammable chemicals may be ignited by open flame, spark, or exposed heat.	alcohol, kerosene, potassium permanganate	Avoid open flames and heat when using flammable chemicals.	Notify your teacher immediately. Use fire safety equipment if applicable.
OPEN FLAME	Open flame in use, may cause fire.	hair, clothing, paper, synthetic materials	Tie back hair and loose clothing. Follow teacher's instruction on lighting and extinguishing flames.	Notify your teacher immediately. Use fire safety equipment if applicable.

Eye Safety
Proper eye protection should be worn at all times by anyone performing or observing science activities.

Clothing Protection
This symbol appears when substances could stain or burn clothing.

Animal Safety
This symbol appears when safety of animals and students must be ensured.

Handwashing
After the lab, wash hands with soap and water before removing goggles.

Safety in the Science Laboratory

The science laboratory is a safe place to work if you follow standard safety procedures. Being responsible for your own safety helps to make the entire laboratory a safer place for everyone. When performing any lab, read and apply the caution statements and safety symbol listed at the beginning of the lab.

General Safety Rules

1. Obtain your teacher's permission to begin all investigations and use laboratory equipment.
2. Study the procedure. Ask your teacher any questions. Be sure you understand safety symbols shown on the page.
3. Notify your teacher about allergies or other health conditions which can affect your participation in a lab.
4. Learn and follow use and safety procedures for your equipment. If unsure, ask your teacher.

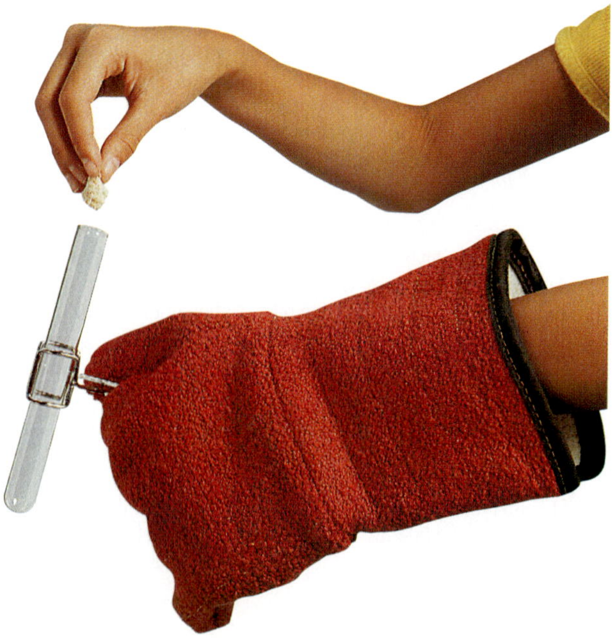

5. Never eat, drink, chew gum, apply cosmetics, or do any personal grooming in the lab. Never use lab glassware as food or drink containers. Keep your hands away from your face and mouth.
6. Know the location and proper use of the safety shower, eye wash, fire blanket, and fire alarm.

Prevent Accidents

1. Use the safety equipment provided to you. Goggles and a safety apron should be worn during investigations.
2. Do NOT use hair spray, mousse, or other flammable hair products. Tie back long hair and tie down loose clothing.
3. Do NOT wear sandals or other open-toed shoes in the lab.
4. Remove jewelry on hands and wrists. Loose jewelry, such as chains and long necklaces, should be removed to prevent them from getting caught in equipment.
5. Do not taste any substances or draw any material into a tube with your mouth.
6. Proper behavior is expected in the lab. Practical jokes and fooling around can lead to accidents and injury.
7. Keep your work area uncluttered.

Laboratory Work

1. Collect and carry all equipment and materials to your work area before beginning a lab.
2. Remain in your own work area unless given permission by your teacher to leave it.

3. Always slant test tubes away from yourself and others when heating them, adding substances to them, or rinsing them.
4. If instructed to smell a substance in a container, hold the container a short distance away and fan vapors towards your nose.
5. Do NOT substitute other chemicals/substances for those in the materials list unless instructed to do so by your teacher.
6. Do NOT take any materials or chemicals outside of the laboratory.
7. Stay out of storage areas unless instructed to be there and supervised by your teacher.

Laboratory Cleanup

1. Turn off all burners, water, and gas, and disconnect all electrical devices.
2. Clean all pieces of equipment and return all materials to their proper places.
3. Dispose of chemicals and other materials as directed by your teacher. Place broken glass and solid substances in the proper containers. Never discard materials in the sink.
4. Clean your work area.
5. Wash your hands with soap and water thoroughly BEFORE removing your goggles.

Emergencies

1. Report any fire, electrical shock, glassware breakage, spill, or injury, no matter how small, to your teacher immediately. Follow his or her instructions.
2. If your clothing should catch fire, STOP, DROP, and ROLL. If possible, smother it with the fire blanket or get under a safety shower. NEVER RUN.
3. If a fire should occur, turn off all gas and leave the room according to established procedures.
4. In most instances, your teacher will clean up spills. Do NOT attempt to clean up spills unless you are given permission and instructions to do so.
5. If chemicals come into contact with your eyes or skin, notify your teacher immediately. Use the eyewash or flush your skin or eyes with large quantities of water.
6. The fire extinguisher and first-aid kit should only be used by your teacher unless it is an extreme emergency and you have been given permission.
7. If someone is injured or becomes ill, only a professional medical provider or someone certified in first aid should perform first-aid procedures.

1 Rock Creatures

Time Required 45 minutes

Materials Organisms are easier to see in white trays than in trays of other colors.

Safety Precautions
- Students should thoroughly wash their hands after handling organisms.
- Warn students that insects called backswimmers have a painful bite and large crayfish have painful pincers.
- Students should avoid deep water and flooded streams after heavy rains.

Teaching Strategy Conduct your own search of a stream near your school to survey the types of organisms your class will find.

Conclude and Apply
1. Answers may include crayfish, caddisfly larvae, mayfly nymphs, scuds, planaria, leeches, snails, and dragonfly nymphs.
2. Stream rocks provide organisms protection against predators, currents, and ultraviolet sunlight.

EXTRA Try at Home Labs

From Your Kitchen, Junk Drawer, or Yard

1 Rock Creatures

Real-World Question
What types of organisms live under stream rocks?

Possible Materials
- waterproof boots
- ice cube tray (white)
- aquarium net
- bucket
- collecting jars
- guidebook to pond life

Procedure
1. With permission, search under the rocks of a local stream. Look for aquatic organisms under the rocks and leaves of the stream. Compare what you find in fast- and slow-moving water.
2. With permission, carefully pull organisms you find off the rocks and put them into separate compartments of your ice cube tray. Take care not to injure the creatures you find.
3. Use your net and bucket to collect larger organisms.
4. Use your guidebook to pond life to identify the organisms you find.
5. Release the organisms back into the stream once you identify them.

Conclude and Apply
1. Identify and list the organisms you found under the stream rocks.
2. Infer why so many aquatic organisms make their habitats beneath stream rocks.

2 A Light in the Forest

Real-World Question
Does the amount of sunlight vary in a forest?

Possible Materials
- empty toilet paper or paper towel roll
- Science Journal

Procedure:
1. Copy the data table into your Science Journal.
2. Go with an adult to a nearby forest or large grove of trees.
3. Stand near the edge of the forest and look straight up through your cardboard tube. Estimate the percentage of blue sky and clouds you can see in the circle. This percentage is the amount of sunlight reaching the forest floor.
4. Record your location and estimated percentage of sunlight in your data table.
5. Test several other locations in the forest. Choose places where the trees completely cover the forest floor and where sunlight is partially coming through.

Data Table

Location	% of Sunlight

Conclude and Apply
1. Explain how the amount of sunlight reaching the forest floor changed.
2. Infer why it is important for leaves and branches to stop sunlight from reaching much of the forest floor.

Adult supervision required for all labs.

These labs are available at booke.msscience.com.

2 A Light in the Forest

Time required 30 minutes to test three sites

Materials Cutting a paper towel roll in half will make it easier to use.

Safety Precautions
- Caution students to look for poisonous plants. Students should avoid forest regions with heavy undergrowth.
- Proper clothing includes long pants, long sleeves, light-colored clothing, long socks tucked into pants, and hats.

Teaching Strategy Check out forest sites before the lab and tack colorful sheets of paper to trees to mark the spots where students could take measurements.

Conclude and Apply
1. The percentages of sunlight should be higher near the edge of the forest.
2. Canopies shade the forest floor to prevent the soil from drying out, and they moderate forest temperatures throughout the year.

Extra Try at Home Labs

3 Echinoderm Hold

Real-World Question
How do echinoderms living in intertidal ecosystems hold on to rocks?

Possible Materials
- plastic suction cup
- water
- paper towel or sponge

Procedure
1. Moisten a paper towel or sponge with water.
2. Press a plastic suction cup on the moist towel or sponge until the entire bottom surface of the cup is wet.
3. Firmly press the suction cup down on a kitchen counter for 10 s.
4. Grab the top handle of the suction cup and try removing the cup from the counter by pulling it straight up.

Conclude and Apply
1. Describe what happened when you tried to remove the cup from the counter.
2. Infer how echinoderms living in intertidal ecosystems withstand the constant pull of ocean waves and currents.

4 UV Watch

Real-World Question
How can you find out about the risks of ultraviolet radiation each day?

Possible Materials
- daily newspaper with weekly weather forecasts
- graph paper

Procedure
1. Use the local newspaper or another resource to get the weather forecast for the day.
2. Check the UV (ultraviolet light) index for the day. If it provides an hourly UV index level, record the level for 1:00 P.M.
3. Find a legend or do research to discover what the numbers of the UV index mean.
4. Record the UV index everyday for ten days and graph your results on graph paper.

Conclude and Apply
1. Explain how the UV index system works.
2. Research several ways you can protect yourself from too much ultraviolet light exposure.

Adult supervision required for all labs.

Extra Try at Home Labs

3 Echinoderm Hold

Time Required ten minutes

Materials
- Plastic suction cups measuring two-to-three centimeters in diameter work best for this lab.
- Remove any wire or string from the suction cups' handles.

Safety Precaution Caution students not to place the suction cups on surfaces where they would leave a permanent mark.

Teaching Strategy Students should not place their suction cups on wood surfaces. The cups do not adhere to wood surfaces as firmly as they do to nonporous surfaces.

Conclude and Apply
1. Students will probably not be able to remove the suction cup by pulling it straight up.
2. Echinoderms have tube feet that they use to create a powerful suction on rocks to prevent them from being washed away by waves and currents.

4 UV Watch

Time Required one-to-two minute sessions for ten consecutive days

Materials
- Students can use online resources and UV updates from news channels to do this lab.
- If you complete this lab in class, supply students with ten daily newspapers from ten consecutive days.

Safety Precaution Warn students not to spend time in the Sun unprotected.

Teaching Strategy This lab works best during warm weather seasons.

Conclude and Apply
1. Ultraviolet light levels range from 0–10. Levels of 0–3 are low levels of UV radiation; 4–6 are moderate UV levels, and 7–10 are high levels of UV exposure.
2. When in the Sun, wear a wide brim hat, sunscreen with an SPF of 15, long sleeves, and UV protection sunglasses. Time in the Sun between 10:00 A.M. and 3:00 P.M. should be limited.

5 Biodiversity

Time Required two class sessions

Safety Precaution For the wild area, find a place where you can keep an eye on the students for safety, or consider asking a parent to come along to help supervise.

Teaching Strategy If the season is not suitable for outside work, or you do not have a good place to take the students, you can do this exercise as a thought experiment. A poster, slide, or picture from a book could be used as the wild plot.

Conclude and Apply

1. The lawn has fewer species than the wild. It might have had fertilizer added to increase some species and pesticides to kill off other plant and insect species. Therefore, there is less food for animals, and fewer types of animals to be found.
2. There would be less biodiversity in the flower garden because the gardener has removed pest plant and insect species and maintains only the desirable species.
3. In the wild there are more types of plants, insects, and other animals for food.

Extra Try at Home Labs

Extra Try at Home Labs

5 Biodiversity

Real-World Question

How does biodiversity vary around your home?

Possible Materials

- wooden ground stakes or nails
- string
- poster paper
- measuring tape, yard stick, or ruler

Procedure

1. With an adult, visit a wild area near your home or school. Mark off an area that is 1 m by 1 m. Put stakes at the four corners and string between them to make a fence around your plot.
2. Count or estimate how many of each animal or plant species are in your plot.
3. Make a drawing on your poster paper and a list of the species found. If you cannot identify a species by name, just describe and sketch it.
4. Repeat the procedure on a lawn. Make your sketch and list on the opposite side of the poster paper.

Conclude and Apply

1. Describe the difference in biodiversity between the lawn plot and the wild plot.
2. Think about a prize flower garden. What is the biodiversity like in the flower garden, compared to the wild?
3. Would you find more bird species in the wild, or near the lawn or flower garden? Why?

Adult supervision required for all labs.

These labs are available at booke.msscience.com.

Computer Skills

People who study science rely on computers, like the one in **Figure 16,** to record and store data and to analyze results from investigations. Whether you work in a laboratory or just need to write a lab report with tables, good computer skills are a necessity.

Using the computer comes with responsibility. Issues of ownership, security, and privacy can arise. Remember, if you did not author the information you are using, you must provide a source for your information. Also, anything on a computer can be accessed by others. Do not put anything on the computer that you would not want everyone to know. To add more security to your work, use a password.

Use a Word Processing Program

A computer program that allows you to type your information, change it as many times as you need to, and then print it out is called a word processing program. Word processing programs also can be used to make tables.

Figure 16 A computer will make reports neater and more professional looking.

Learn the Skill To start your word processing program, a blank document, sometimes called "Document 1," appears on the screen. To begin, start typing. To create a new document, click the *New* button on the standard tool bar. These tips will help you format the document.

- The program will automatically move to the next line; press *Enter* if you wish to start a new paragraph.
- Symbols, called non-printing characters, can be hidden by clicking the *Show/Hide* button on your toolbar.
- To insert text, move the cursor to the point where you want the insertion to go, click on the mouse once, and type the text.
- To move several lines of text, select the text and click the *Cut* button on your toolbar. Then position your cursor in the location that you want to move the cut text and click *Paste.* If you move to the wrong place, click *Undo.*
- The spell check feature does not catch words that are misspelled to look like other words, like "cold" instead of "gold." Always reread your document to catch all spelling mistakes.
- To learn about other word processing methods, read the user's manual or click on the *Help* button.
- You can integrate databases, graphics, and spreadsheets into documents by copying from another program and pasting it into your document, or by using desktop publishing (DTP). DTP software allows you to put text and graphics together to finish your document with a professional look. This software varies in how it is used and its capabilities.

Use a Database

A collection of facts stored in a computer and sorted into different fields is called a database. A database can be reorganized in any way that suits your needs.

Learn the Skill A computer program that allows you to create your own database is a database management system (DBMS). It allows you to add, delete, or change information. Take time to get to know the features of your database software.

- Determine what facts you would like to include and research to collect your information.
- Determine how you want to organize the information.
- Follow the instructions for your particular DBMS to set up fields. Then enter each item of data in the appropriate field.
- Follow the instructions to sort the information in order of importance.
- Evaluate the information in your database, and add, delete, or change as necessary.

Use the Internet

The Internet is a global network of computers where information is stored and shared. To use the Internet, like the students in **Figure 17,** you need a modem to connect your computer to a phone line and an Internet Service Provider account.

Learn the Skill To access internet sites and information, use a "Web browser," which lets you view and explore pages on the World Wide Web. Each page is its own site, and each site has its own address, called a URL. Once you have found a Web browser, follow these steps for a search (this also is how you search a database).

Figure 17 The Internet allows you to search a global network for a variety of information.

- Be as specific as possible. If you know you want to research "gold," don't type in "elements." Keep narrowing your search until you find what you want.
- Web sites that end in *.com* are commercial Web sites; *.org*, *.edu*, and *.gov* are nonprofit, educational, or government Web sites.
- Electronic encyclopedias, almanacs, indexes, and catalogs will help locate and select relevant information.
- Develop a "home page" with relative ease. When developing a Web site, NEVER post pictures or disclose personal information such as location, names, or phone numbers. Your school or community usually can host your Web site. A basic understanding of HTML (hypertext mark-up language), the language of Web sites, is necessary. Software that creates HTML code is called authoring software, and can be downloaded free from many Web sites. This software allows text and pictures to be arranged as the software is writing the HTML code.

Use a Spreadsheet

A spreadsheet, shown in **Figure 18,** can perform mathematical functions with any data arranged in columns and rows. By entering a simple equation into a cell, the program can perform operations in specific cells, rows, or columns.

Learn the Skill Each column (vertical) is assigned a letter, and each row (horizontal) is assigned a number. Each point where a row and column intersect is called a cell, and is labeled according to where it is located—Column A, Row 1 (A1).

- Decide how to organize the data, and enter it in the correct row or column.
- Spreadsheets can use standard formulas or formulas can be customized to calculate cells.
- To make a change, click on a cell to make it activate, and enter the edited data or formula.
- Spreadsheets also can display your results in graphs. Choose the style of graph that best represents the data.

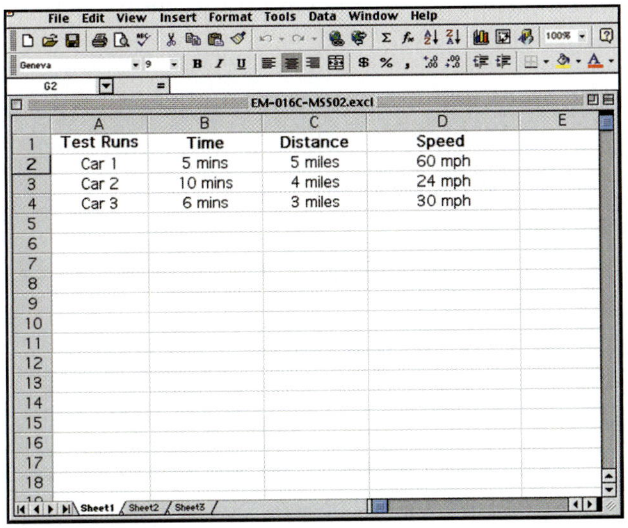

	A	B	C	D	E
1	Test Runs	Time	Distance	Speed	
2	Car 1	5 mins	5 miles	60 mph	
3	Car 2	10 mins	4 miles	24 mph	
4	Car 3	6 mins	3 miles	30 mph	

Figure 18 A spreadsheet allows you to perform mathematical operations on your data.

Use Graphics Software

Adding pictures, called graphics, to your documents is one way to make your documents more meaningful and exciting. This software adds, edits, and even constructs graphics. There is a variety of graphics software programs. The tools used for drawing can be a mouse, keyboard, or other specialized devices. Some graphics programs are simple. Others are complicated, called computer-aided design (CAD) software.

Learn the Skill It is important to have an understanding of the graphics software being used before starting. The better the software is understood, the better the results. The graphics can be placed in a word-processing document.

- Clip art can be found on a variety of internet sites, and on CDs. These images can be copied and pasted into your document.
- When beginning, try editing existing drawings, then work up to creating drawings.
- The images are made of tiny rectangles of color called pixels. Each pixel can be altered.
- Digital photography is another way to add images. The photographs in the memory of a digital camera can be downloaded into a computer, then edited and added to the document.
- Graphics software also can allow animation. The software allows drawings to have the appearance of movement by connecting basic drawings automatically. This is called in-betweening, or tweening.
- Remember to save often.

Presentation Skills

Develop Multimedia Presentations

Most presentations are more dynamic if they include diagrams, photographs, videos, or sound recordings, like the one shown in **Figure 19.** A multimedia presentation involves using stereos, overhead projectors, televisions, computers, and more.

Learn the Skill Decide the main points of your presentation, and what types of media would best illustrate those points.

- Make sure you know how to use the equipment you are working with.
- Practice the presentation using the equipment several times.
- Enlist the help of a classmate to push play or turn lights out for you. Be sure to practice your presentation with him or her.
- If possible, set up all of the equipment ahead of time, and make sure everything is working properly.

Figure 19 These students are engaging the audience using a variety of tools.

Computer Presentations

There are many different interactive computer programs that you can use to enhance your presentation. Most computers have a compact disc (CD) drive that can play both CDs and digital video discs (DVDs). Also, there is hardware to connect a regular CD, DVD, or VCR. These tools will enhance your presentation.

Another method of using the computer to aid in your presentation is to develop a slide show using a computer program. This can allow movement of visuals at the presenter's pace, and can allow for visuals to build on one another.

Learn the Skill In order to create multimedia presentations on a computer, you need to have certain tools. These may include traditional graphic tools and drawing programs, animation programs, and authoring systems that tie everything together. Your computer will tell you which tools it supports. The most important step is to learn about the tools that you will be using.

- Often, color and strong images will convey a point better than words alone. Use the best methods available to convey your point.
- As with other presentations, practice many times.
- Practice your presentation with the tools you and any assistants will be using.
- Maintain eye contact with the audience. The purpose of using the computer is not to prompt the presenter, but to help the audience understand the points of the presentation.

Math Review

Use Fractions

A fraction compares a part to a whole. In the fraction $\frac{2}{3}$, the 2 represents the part and is the numerator. The 3 represents the whole and is the denominator.

Reduce Fractions To reduce a fraction, you must find the largest factor that is common to both the numerator and the denominator, the greatest common factor (GCF). Divide both numbers by the GCF. The fraction has then been reduced, or it is in its simplest form.

Example Twelve of the 20 chemicals in the science lab are in powder form. What fraction of the chemicals used in the lab are in powder form?

Step 1 Write the fraction.

$$\frac{\text{part}}{\text{whole}} = \frac{12}{20}$$

Step 2 To find the GCF of the numerator and denominator, list all of the factors of each number.

Factors of 12: 1, 2, 3, 4, 6, 12 (the numbers that divide evenly into 12)

Factors of 20: 1, 2, 4, 5, 10, 20 (the numbers that divide evenly into 20)

Step 3 List the common factors.

1, 2, 4.

Step 4 Choose the greatest factor in the list.

The GCF of 12 and 20 is 4.

Step 5 Divide the numerator and denominator by the GCF.

$$\frac{12 \div 4}{20 \div 4} = \frac{3}{5}$$

In the lab, $\frac{3}{5}$ of the chemicals are in powder form.

Practice Problem At an amusement park, 66 of 90 rides have a height restriction. What fraction of the rides, in its simplest form, has a height restriction?

Add and Subtract Fractions To add or subtract fractions with the same denominator, add or subtract the numerators and write the sum or difference over the denominator. After finding the sum or difference, find the simplest form for your fraction.

Example 1 In the forest outside your house, $\frac{1}{8}$ of the animals are rabbits, $\frac{3}{8}$ are squirrels, and the remainder are birds and insects. How many are mammals?

Step 1 Add the numerators.

$$\frac{1}{8} + \frac{3}{8} = \frac{(1 + 3)}{8} = \frac{4}{8}$$

Step 2 Find the GCF.

$$\frac{4}{8} \text{ (GCF, 4)}$$

Step 3 Divide the numerator and denominator by the GCF.

$$\frac{4}{4} = 1, \ \frac{8}{4} = 2$$

$\frac{1}{2}$ of the animals are mammals.

Example 2 If $\frac{7}{16}$ of the Earth is covered by freshwater, and $\frac{1}{16}$ of that is in glaciers, how much freshwater is not frozen?

Step 1 Subtract the numerators.

$$\frac{7}{16} - \frac{1}{16} = \frac{(7 - 1)}{16} = \frac{6}{16}$$

Step 2 Find the GCF.

$$\frac{6}{16} \text{ (GCF, 2)}$$

Step 3 Divide the numerator and denominator by the GCF.

$$\frac{6}{2} = 3, \ \frac{16}{2} = 8$$

$\frac{3}{8}$ of the freshwater is not frozen.

Practice Problem A bicycle rider is going 15 km/h for $\frac{4}{9}$ of his ride, 10 km/h for $\frac{2}{9}$ of his ride, and 8 km/h for the remainder of the ride. How much of his ride is he going over 8 km/h?

Reduce Fractions

$$\frac{66 \div 6}{90 \div 6} = \frac{11}{15}$$

Add and Subtract Fractions

$$\frac{4}{9} + \frac{2}{9} = \frac{6}{9}$$

$$\frac{6 \div 3}{9 \div 3} = \frac{2}{3}$$

Math Skill Handbook

MATH SKILL HANDBOOK E ◆ 173

Unlike Denominators

Problem 1

$1 \times 5 = 5, 8 \times 5 = 40$

$1 \times 4 = 4, 10 \times 4 = 40$

$\frac{5}{40} + \frac{4}{40} = \frac{9}{40}$

Problem 2

If $\frac{7}{10}$ are involuntary, the remainder are voluntary.

$\frac{10}{10} - \frac{7}{10} = \frac{3}{10}$

Unlike Denominators To add or subtract fractions with unlike denominators, first find the least common denominator (LCD). This is the smallest number that is a common multiple of both denominators. Rename each fraction with the LCD, and then add or subtract. Find the simplest form if necessary.

Example 1 A chemist makes a paste that is $\frac{1}{2}$ table salt (NaCl), $\frac{1}{3}$ sugar ($C_6H_{12}O_6$), and the rest water (H_2O). How much of the paste is a solid?

Step 1 Find the LCD of the fractions.

$\frac{1}{2} + \frac{1}{3}$ (LCD, 6)

Step 2 Rename each numerator and each denominator with the LCD.

$1 \times 3 = 3,\ 2 \times 3 = 6$

$1 \times 2 = 2,\ 3 \times 2 = 6$

Step 3 Add the numerators.

$\frac{3}{6} + \frac{2}{6} = \frac{(3 + 2)}{6} = \frac{5}{6}$

$\frac{5}{6}$ of the paste is a solid.

Example 2 The average precipitation in Grand Junction, CO, is $\frac{7}{10}$ inch in November, and $\frac{3}{5}$ inch in December. What is the total average precipitation?

Step 1 Find the LCD of the fractions.

$\frac{7}{10} + \frac{3}{5}$ (LCD, 10)

Step 2 Rename each numerator and each denominator with the LCD.

$7 \times 1 = 7,\ 10 \times 1 = 10$

$3 \times 2 = 6,\ 5 \times 2 = 10$

Step 3 Add the numerators.

$\frac{7}{10} + \frac{6}{10} = \frac{(7 + 6)}{10} = \frac{13}{10}$

$\frac{13}{10}$ inches total precipitation, or $1\frac{3}{10}$ inches.

Practice Problem On an electric bill, about $\frac{1}{8}$ of the energy is from solar energy and about $\frac{1}{10}$ is from wind power. How much of the total bill is from solar energy and wind power combined?

Example 3 In your body, $\frac{7}{10}$ of your muscle contractions are involuntary (cardiac and smooth muscle tissue). Smooth muscle makes $\frac{3}{15}$ of your muscle contractions. How many of your muscle contractions are made by cardiac muscle?

Step 1 Find the LCD of the fractions.

$\frac{7}{10} - \frac{3}{15}$ (LCD, 30)

Step 2 Rename each numerator and each denominator with the LCD.

$7 \times 3 = 21,\ 10 \times 3 = 30$

$3 \times 2 = 6,\ 15 \times 2 = 30$

Step 3 Subtract the numerators.

$\frac{21}{30} - \frac{6}{30} = \frac{(21 - 6)}{30} = \frac{15}{30}$

Step 4 Find the GCF.

$\frac{15}{30}$ (GCF, 15)

$\frac{1}{2}$

$\frac{1}{2}$ of all muscle contractions are cardiac muscle.

Example 4 Tony wants to make cookies that call for $\frac{3}{4}$ of a cup of flour, but he only has $\frac{1}{3}$ of a cup. How much more flour does he need?

Step 1 Find the LCD of the fractions.

$\frac{3}{4} - \frac{1}{3}$ (LCD, 12)

Step 2 Rename each numerator and each denominator with the LCD.

$3 \times 3 = 9,\ 4 \times 3 = 12$

$1 \times 4 = 4,\ 3 \times 4 = 12$

Step 3 Subtract the numerators.

$\frac{9}{12} - \frac{4}{12} = \frac{(9 - 4)}{12} = \frac{5}{12}$

$\frac{5}{12}$ of a cup of flour.

Practice Problem Using the information provided to you in Example 3 above, determine how many muscle contractions are voluntary (skeletal muscle).

Math Skill Handbook
174 ◆ E STUDENT RESOURCES

Multiply Fractions To multiply with fractions, multiply the numerators and multiply the denominators. Find the simplest form if necessary.

Example Multiply $\frac{3}{5}$ by $\frac{1}{3}$.

Step 1 Multiply the numerators and denominators.

$$\frac{3}{5} \times \frac{1}{3} = \frac{(3 \times 1)}{(5 \times 3)} = \frac{3}{15}$$

Step 2 Find the GCF.

$$\frac{3}{15} \text{ (GCF, 3)}$$

Step 3 Divide the numerator and denominator by the GCF.

$$\frac{3}{3} = 1, \quad \frac{15}{3} = 5$$

$$\frac{1}{5}$$

$\frac{3}{5}$ multiplied by $\frac{1}{3}$ is $\frac{1}{5}$.

Practice Problem Multiply $\frac{3}{14}$ by $\frac{5}{16}$.

Find a Reciprocal Two numbers whose product is 1 are called multiplicative inverses, or reciprocals.

Example Find the reciprocal of $\frac{3}{8}$.

Step 1 Inverse the fraction by putting the denominator on top and the numerator on the bottom.

$$\frac{8}{3}$$

The reciprocal of $\frac{3}{8}$ is $\frac{8}{3}$.

Practice Problem Find the reciprocal of $\frac{4}{9}$.

Divide Fractions To divide one fraction by another fraction, multiply the dividend by the reciprocal of the divisor. Find the simplest form if necessary.

Example 1 Divide $\frac{1}{9}$ by $\frac{1}{3}$.

Step 1 Find the reciprocal of the divisor.

The reciprocal of $\frac{1}{3}$ is $\frac{3}{1}$.

Step 2 Multiply the dividend by the reciprocal of the divisor.

$$\frac{\frac{1}{9}}{\frac{1}{3}} = \frac{1}{9} \times \frac{3}{1} = \frac{(1 \times 3)}{(9 \times 1)} = \frac{3}{9}$$

Step 3 Find the GCF.

$$\frac{3}{9} \text{ (GCF, 3)}$$

Step 4 Divide the numerator and denominator by the GCF.

$$\frac{3}{3} = 1, \quad \frac{9}{3} = 3$$

$$\frac{1}{3}$$

$\frac{1}{9}$ divided by $\frac{1}{3}$ is $\frac{1}{3}$.

Example 2 Divide $\frac{3}{5}$ by $\frac{1}{4}$.

Step 1 Find the reciprocal of the divisor.

The reciprocal of $\frac{1}{4}$ is $\frac{4}{1}$.

Step 2 Multiply the dividend by the reciprocal of the divisor.

$$\frac{\frac{3}{5}}{\frac{1}{4}} = \frac{3}{5} \times \frac{4}{1} = \frac{(3 \times 4)}{(5 \times 1)} = \frac{12}{5}$$

$\frac{3}{5}$ divided by $\frac{1}{4}$ is $\frac{12}{5}$ or $2\frac{2}{5}$.

Practice Problem Divide $\frac{3}{11}$ by $\frac{7}{10}$.

Multiply Fractions

$$\frac{3}{14} \times \frac{5}{16} = \frac{(3 \times 5)}{(14 \times 16)} = \frac{15}{224}$$

Find a Reciprocal

$$\frac{9}{4}$$

Divide Fractions

The reciprocal of $\frac{7}{10}$ is $\frac{10}{7}$.

$$\frac{3}{11} \times \frac{10}{7} = \frac{(3 \times 10)}{(11 \times 7)} = \frac{30}{77}$$

Math Skill Handbook

MATH SKILL HANDBOOK E ◆ 175

Use Ratios

$\frac{100 \text{ cm}}{144 \text{ cm}} = \frac{100 \div 4}{144 \div 4} = \frac{25}{36}$

25:36

Add or Subtract Decimals

$$\begin{array}{r} \scriptstyle 1 \\ 1.245 \\ +\ 3.842 \\ \hline 5.087 \end{array}$$

Use Ratios

When you compare two numbers by division, you are using a ratio. Ratios can be written 3 to 5, 3:5, or $\frac{3}{5}$. Ratios, like fractions, also can be written in simplest form.

Ratios can represent probabilities, also called odds. This is a ratio that compares the number of ways a certain outcome occurs to the number of outcomes. For example, if you flip a coin 100 times, what are the odds that it will come up heads? There are two possible outcomes, heads or tails, so the odds of coming up heads are 50:100. Another way to say this is that 50 out of 100 times the coin will come up heads. In its simplest form, the ratio is 1:2.

Example 1 A chemical solution contains 40 g of salt and 64 g of baking soda. What is the ratio of salt to baking soda as a fraction in simplest form?

Step 1 Write the ratio as a fraction.

$$\frac{\text{salt}}{\text{baking soda}} = \frac{40}{64}$$

Step 2 Express the fraction in simplest form.

The GCF of 40 and 64 is 8.

$$\frac{40}{64} = \frac{40 \div 8}{64 \div 8} = \frac{5}{8}$$

The ratio of salt to baking soda in the sample is 5:8.

Example 2 Sean rolls a 6-sided die 6 times. What are the odds that the side with a 3 will show?

Step 1 Write the ratio as a fraction.

$$\frac{\text{number of sides with a 3}}{\text{number of sides}} = \frac{1}{6}$$

Step 2 Multiply by the number of attempts.

$$\frac{1}{6} \times 6 \text{ attempts} = \frac{6}{6} \text{ attempts} = 1 \text{ attempt}$$

1 attempt out of 6 will show a 3.

Practice Problem Two metal rods measure 100 cm and 144 cm in length. What is the ratio of their lengths in simplest form?

Use Decimals

A fraction with a denominator that is a power of ten can be written as a decimal. For example, 0.27 means $\frac{27}{100}$. The decimal point separates the ones place from the tenths place.

Any fraction can be written as a decimal using division. For example, the fraction $\frac{5}{8}$ can be written as a decimal by dividing 5 by 8. Written as a decimal, it is 0.625.

Add or Subtract Decimals When adding and subtracting decimals, line up the decimal points before carrying out the operation.

Example 1 Find the sum of 47.68 and 7.80.

Step 1 Line up the decimal places when you write the numbers.

$$\begin{array}{r} 47.68 \\ +\ 7.80 \\ \hline \end{array}$$

Step 2 Add the decimals.

$$\begin{array}{r} 47.68 \\ +\ 7.80 \\ \hline 55.48 \end{array}$$

The sum of 47.68 and 7.80 is 55.48.

Example 2 Find the difference of 42.17 and 15.85.

Step 1 Line up the decimal places when you write the number.

$$\begin{array}{r} 42.17 \\ -15.85 \\ \hline \end{array}$$

Step 2 Subtract the decimals.

$$\begin{array}{r} 42.17 \\ -15.85 \\ \hline 26.32 \end{array}$$

The difference of 42.17 and 15.85 is 26.32.

Practice Problem Find the sum of 1.245 and 3.842.

Math Skill Handbook

176 ◆ E STUDENT RESOURCES

Multiply Decimals To multiply decimals, multiply the numbers like any other number, ignoring the decimal point. Count the decimal places in each factor. The product will have the same number of decimal places as the sum of the decimal places in the factors.

Example Multiply 2.4 by 5.9.

Step 1 Multiply the factors like two whole numbers.
$24 \times 59 = 1416$

Step 2 Find the sum of the number of decimal places in the factors. Each factor has one decimal place, for a sum of two decimal places.

Step 3 The product will have two decimal places.
14.16

The product of 2.4 and 5.9 is 14.16.

Practice Problem Multiply 4.6 by 2.2.

Divide Decimals When dividing decimals, change the divisor to a whole number. To do this, multiply both the divisor and the dividend by the same power of ten. Then place the decimal point in the quotient directly above the decimal point in the dividend. Then divide as you do with whole numbers.

Example Divide 8.84 by 3.4.

Step 1 Multiply both factors by 10.
$3.4 \times 10 = 34,\ 8.84 \times 10 = 88.4$

Step 2 Divide 88.4 by 34.

$$\begin{array}{r} 2.6 \\ 34\overline{)88.4} \\ -68 \\ \hline 204 \\ -204 \\ \hline 0 \end{array}$$

8.84 divided by 3.4 is 2.6.

Practice Problem Divide 75.6 by 3.6.

Use Proportions

An equation that shows that two ratios are equivalent is a proportion. The ratios $\frac{2}{4}$ and $\frac{5}{10}$ are equivalent, so they can be written as $\frac{2}{4} = \frac{5}{10}$. This equation is a proportion.

When two ratios form a proportion, the cross products are equal. To find the cross products in the proportion $\frac{2}{4} = \frac{5}{10}$, multiply the 2 and the 10, and the 4 and the 5. Therefore $2 \times 10 = 4 \times 5$, or $20 = 20$.

Because you know that both proportions are equal, you can use cross products to find a missing term in a proportion. This is known as solving the proportion.

Example The heights of a tree and a pole are proportional to the lengths of their shadows. The tree casts a shadow of 24 m when a 6-m pole casts a shadow of 4 m. What is the height of the tree?

Step 1 Write a proportion.
$$\frac{\text{height of tree}}{\text{height of pole}} = \frac{\text{length of tree's shadow}}{\text{length of pole's shadow}}$$

Step 2 Substitute the known values into the proportion. Let h represent the unknown value, the height of the tree.
$$\frac{h}{6} = \frac{24}{4}$$

Step 3 Find the cross products.
$h \times 4 = 6 \times 24$

Step 4 Simplify the equation.
$4h = 144$

Step 5 Divide each side by 4.
$$\frac{4h}{4} = \frac{144}{4}$$
$h = 36$

The height of the tree is 36 m.

Practice Problem The ratios of the weights of two objects on the Moon and on Earth are in proportion. A rock weighing 3 N on the Moon weighs 18 N on Earth. How much would a rock that weighs 5 N on the Moon weigh on Earth?

Multiply Decimals

Multiply 4.6 and 2.2 by 10.

$46 \times 22 = 1012$

Each factor had one decimal place.

10.12

Divide Decimals

Multiply both factors by 10.

Divide 756 by 36.

$$\begin{array}{r} 21 \\ 36\overline{)756} \\ 72 \\ \hline 36 \\ 36 \\ \hline 0 \end{array}$$

Use Proportions

$$\frac{3}{18} = \frac{5}{w}$$

$w \times 3 = 5 \times 18$

$$\frac{3w}{3} = \frac{90}{3}$$

$w = 30$

Math Skill Handbook

MATH SKILL HANDBOOK E ◆ 177

Use Percentages

The word *percent* means "out of one hundred." It is a ratio that compares a number to 100. Suppose you read that 77 percent of the Earth's surface is covered by water. That is the same as reading that the fraction of the Earth's surface covered by water is $\frac{77}{100}$. To express a fraction as a percent, first find the equivalent decimal for the fraction. Then, multiply the decimal by 100 and add the percent symbol.

Example Express $\frac{13}{20}$ as a percent.

Step 1 Find the equivalent decimal for the fraction.

$$\begin{array}{r} 0.65 \\ 20\overline{)13.00} \\ \underline{12\ 0} \\ 1\ 00 \\ \underline{1\ 00} \\ 0 \end{array}$$

Step 2 Rewrite the fraction $\frac{13}{20}$ as 0.65.

Step 3 Multiply 0.65 by 100 and add the % sign.

$0.65 \times 100 = 65 = 65\%$

So, $\frac{13}{20} = 65\%$.

This also can be solved as a proportion.

Example Express $\frac{13}{20}$ as a percent.

Step 1 Write a proportion.

$\frac{13}{20} = \frac{x}{100}$

Step 2 Find the cross products.

$1300 = 20x$

Step 3 Divide each side by 20.

$\frac{1300}{20} = \frac{20x}{20}$

$65\% = x$

Practice Problem In one year, 73 of 365 days were rainy in one city. What percent of the days in that city were rainy?

Solve One-Step Equations

A statement that two things are equal is an equation. For example, $A = B$ is an equation that states that A is equal to B.

An equation is solved when a variable is replaced with a value that makes both sides of the equation equal. To make both sides equal the inverse operation is used. Addition and subtraction are inverses, and multiplication and division are inverses.

Example 1 Solve the equation $x - 10 = 35$.

Step 1 Find the solution by adding 10 to each side of the equation.

$x - 10 = 35$

$x - 10 + 10 = 35 + 10$

$x = 45$

Step 2 Check the solution.

$x - 10 = 35$

$45 - 10 = 35$

$35 = 35$

Both sides of the equation are equal, so $x = 45$.

Example 2 In the formula $a = bc$, find the value of c if $a = 20$ and $b = 2$.

Step 1 Rearrange the formula so the unknown value is by itself on one side of the equation by dividing both sides by b.

$a = bc$

$\frac{a}{b} = \frac{bc}{b}$

$\frac{a}{b} = c$

Step 2 Replace the variables a and b with the values that are given.

$\frac{a}{b} = c$

$\frac{20}{2} = c$

$10 = c$

Step 3 Check the solution.

$a = bc$

$20 = 2 \times 10$

$20 = 20$

Both sides of the equation are equal, so $c = 10$ is the solution when $a = 20$ and $b = 2$.

Practice Problem In the formula $h = gd$, find the value of d if $g = 12.3$ and $h = 17.4$.

Math Skill Handbook

Use Percentages

$\frac{73}{365} = \frac{x}{100}$

$\frac{7300}{365} = \frac{365x}{365}$

$20\% = x$

Solve One-Step Equations

$h = gd$

$\frac{17.4}{12.3} = \frac{12.3d}{12.3}$

$1.41 = d$

Math Skill Handbook

178 ◆ E STUDENT RESOURCES

Use Statistics

The branch of mathematics that deals with collecting, analyzing, and presenting data is statistics. In statistics, there are three common ways to summarize data with a single number—the mean, the median, and the mode.

The **mean** of a set of data is the arithmetic average. It is found by adding the numbers in the data set and dividing by the number of items in the set.

The **median** is the middle number in a set of data when the data are arranged in numerical order. If there were an even number of data points, the median would be the mean of the two middle numbers.

The **mode** of a set of data is the number or item that appears most often.

Another number that often is used to describe a set of data is the range. The **range** is the difference between the largest number and the smallest number in a set of data.

A **frequency table** shows how many times each piece of data occurs, usually in a survey. **Table 2** below shows the results of a student survey on favorite color.

<table>
<tr><th colspan="3">Table 2 Student Color Choice</th></tr>
<tr><th>Color</th><th>Tally</th><th>Frequency</th></tr>
<tr><td>red</td><td>||||</td><td>4</td></tr>
<tr><td>blue</td><td>𝍸</td><td>5</td></tr>
<tr><td>black</td><td>||</td><td>2</td></tr>
<tr><td>green</td><td>|||</td><td>3</td></tr>
<tr><td>purple</td><td>𝍸 ||</td><td>7</td></tr>
<tr><td>yellow</td><td>𝍸 |</td><td>6</td></tr>
</table>

Based on the frequency table data, which color is the favorite?

Example The speeds (in m/s) for a race car during five different time trials are 39, 37, 44, 36, and 44.

To find the mean:

Step 1 Find the sum of the numbers.
$39 + 37 + 44 + 36 + 44 = 200$

Step 2 Divide the sum by the number of items, which is 5.
$200 \div 5 = 40$

The mean is 40 m/s.

To find the median:

Step 1 Arrange the measures from least to greatest.
36, 37, 39, 44, 44

Step 2 Determine the middle measure.
36, 37, 39, 44, 44

The median is 39 m/s.

To find the mode:

Step 1 Group the numbers that are the same together.
44, 44, 36, 37, 39

Step 2 Determine the number that occurs most in the set.
44, 44, 36, 37, 39

The mode is 44 m/s.

To find the range:

Step 1 Arrange the measures from largest to smallest.
44, 44, 39, 37, 36

Step 2 Determine the largest and smallest measures in the set.
44, 44, 39, 37, 36

Step 3 Find the difference between the largest and smallest measures.
$44 - 36 = 8$

The range is 8 m/s.

Practice Problem Find the mean, median, mode, and range for the data set 8, 4, 12, 8, 11, 14, 16.

Math Skill Handbook

Use Statistics

mean
$8 + 4 + 12 + 8 + 11 + 14 + 16 = 73$
$73 \div 7 = 10.4$

median
4, 8, 8, 11, 12, 14, 16

mode
4, 8, 8, 11, 12, 14, 16

range
4, 8, 8, 11, 12, 14, 16
$16 - 4 = 12$

Math Skill Handbook

MATH SKILL HANDBOOK E ◆ 179

Use Geometry

The branch of mathematics that deals with the measurement, properties, and relationships of points, lines, angles, surfaces, and solids is called geometry.

Perimeter The **perimeter** (P) is the distance around a geometric figure. To find the perimeter of a rectangle, add the length and width and multiply that sum by two, or $2(l + w)$. To find perimeters of irregular figures, add the length of the sides.

Example 1 Find the perimeter of a rectangle that is 3 m long and 5 m wide.

Step 1 You know that the perimeter is 2 times the sum of the width and length.

$P = 2(3\text{ m} + 5\text{ m})$

Step 2 Find the sum of the width and length.

$P = 2(8\text{ m})$

Step 3 Multiply by 2.

$P = 16\text{ m}$

The perimeter is 16 m.

Example 2 Find the perimeter of a shape with sides measuring 2 cm, 5 cm, 6 cm, 3 cm.

Step 1 You know that the perimeter is the sum of all the sides.

$P = 2 + 5 + 6 + 3$

Step 2 Find the sum of the sides.

$P = 2 + 5 + 6 + 3$

$P = 16$

The perimeter is 16 cm.

Practice Problem Find the perimeter of a rectangle with a length of 18 m and a width of 7 m.

Practice Problem Find the perimeter of a triangle measuring 1.6 cm by 2.4 cm by 2.4 cm.

Area of a Rectangle The **area** (A) is the number of square units needed to cover a surface. To find the area of a rectangle, multiply the length times the width, or $l \times w$. When finding area, the units also are multiplied. Area is given in square units.

Example Find the area of a rectangle with a length of 1 cm and a width of 10 cm.

Step 1 You know that the area is the length multiplied by the width.

$A = (1\text{ cm} \times 10\text{ cm})$

Step 2 Multiply the length by the width. Also multiply the units.

$A = 10\text{ cm}^2$

The area is 10 cm^2.

Practice Problem Find the area of a square whose sides measure 4 m.

Area of a Triangle To find the area of a triangle, use the formula:

$$A = \frac{1}{2}(\text{base} \times \text{height})$$

The base of a triangle can be any of its sides. The height is the perpendicular distance from a base to the opposite endpoint, or vertex.

Example Find the area of a triangle with a base of 18 m and a height of 7 m.

Step 1 You know that the area is $\frac{1}{2}$ the base times the height.

$A = \frac{1}{2}(18\text{ m} \times 7\text{ m})$

Step 2 Multiply $\frac{1}{2}$ by the product of 18×7. Multiply the units.

$A = \frac{1}{2}(126\text{ m}^2)$

$A = 63\text{ m}^2$

The area is 63 m^2.

Practice Problem Find the area of a triangle with a base of 27 cm and a height of 17 cm.

Math Skill Handbook

Perimeter

Problem 1

$P = 2(18\text{ m} + 7\text{ m})$

$P = 2(25\text{ m})$

$P = 50\text{ m}$

Problem 2

$P = 1.6\text{ cm} + 2.4\text{ cm} + 2.4\text{ cm}$

$P = 6.4\text{ cm}$

Area of a Rectangle

$A = (4\text{ m} \times 4\text{ m})$

$A = 16\text{ m}^2$

Area of a Triangle

$A = \frac{1}{2}(27\text{ cm} \times 17\text{ cm})$

$A = \frac{1}{2}(459\text{ cm}^2)$

$A = 229.5\text{ cm}^2$

Math Skill Handbook

180 ◆ E STUDENT RESOURCES

Circumference of a Circle The **diameter** (d) of a circle is the distance across the circle through its center, and the **radius** (r) is the distance from the center to any point on the circle. The radius is half of the diameter. The distance around the circle is called the **circumference** (C). The formula for finding the circumference is:

$$C = 2\pi r \text{ or } C = \pi d$$

The circumference divided by the diameter is always equal to 3.1415926... This nonterminating and nonrepeating number is represented by the Greek letter π (pi). An approximation often used for π is 3.14.

Example 1 Find the circumference of a circle with a radius of 3 m.

Step 1 You know the formula for the circumference is 2 times the radius times π.
$C = 2\pi(3)$

Step 2 Multiply 2 times the radius.
$C = 6\pi$

Step 3 Multiply by π.
$C = 19$ m

The circumference is 19 m.

Example 2 Find the circumference of a circle with a diameter of 24.0 cm.

Step 1 You know the formula for the circumference is the diameter times π.
$C = \pi(24.0)$

Step 2 Multiply the diameter by π.
$C = 75.4$ cm

The circumference is 75.4 cm.

Practice Problem Find the circumference of a circle with a radius of 19 cm.

Area of a Circle The formula for the area of a circle is:
$A = \pi r^2$

Example 1 Find the area of a circle with a radius of 4.0 cm.

Step 1 $A = \pi(4.0)^2$

Step 2 Find the square of the radius.
$A = 16\pi$

Step 3 Multiply the square of the radius by π.
$A = 50 \text{ cm}^2$

The area of the circle is 50 cm^2.

Example 2 Find the area of a circle with a radius of 225 m.

Step 1 $A = \pi(225)^2$

Step 2 Find the square of the radius.
$A = 50625\pi$

Step 3 Multiply the square of the radius by π.
$A = 158962.5$

The area of the circle is 158,962 m^2.

Example 3 Find the area of a circle whose diameter is 20.0 mm.

Step 1 You know the formula for the area of a circle is the square of the radius times π, and that the radius is half of the diameter.
$A = \pi\left(\frac{20.0}{2}\right)^2$

Step 2 Find the radius.
$A = \pi(10.0)^2$

Step 3 Find the square of the radius.
$A = 100\pi$

Step 4 Multiply the square of the radius by π.
$A = 314 \text{ mm}^2$

The area is 314 mm^2.

Practice Problem Find the area of a circle with a radius of 16 m.

Circumference of a Circle

$C = 2\pi r$

$C = 2\pi(19)$

$C = 38\pi$

$C = 119.3$

Area of a Circle

$A = \pi r^2$

$A = \pi(16 \text{ m})^2$

$A = \pi\, 256 \text{ m}^2$

$A = 803.8 \text{ m}^2$

Math Skill Handbook

MATH SKILL HANDBOOK E ◆ 181

Volume

Problem 1

$V = 8\text{ m} \times 4\text{ m} \times 4\text{ m}$

$V = 128\text{ m}^3$

Problem 2

$V = \pi r^2 \times \text{height}$

$V = \left[\pi\left(\frac{1}{2} \times 7\right)^2\right] \times 16$

$V = [\pi(3.5)^2] \times 16$

$V = [\pi(12.25)] \times 16$

$V = 38.46 \times 16$

$V = 615.36$

Math Skill Handbook

Math Skill Handbook

Volume The measure of space occupied by a solid is the **volume** (V). To find the volume of a rectangular solid multiply the length times width times height, or $V = l \times w \times h$. It is measured in cubic units, such as cubic centimeters (cm^3).

Example Find the volume of a rectangular solid with a length of 2.0 m, a width of 4.0 m, and a height of 3.0 m.

Step 1 You know the formula for volume is the length times the width times the height.

$V = 2.0\text{ m} \times 4.0\text{ m} \times 3.0\text{ m}$

Step 2 Multiply the length times the width times the height.

$V = 24\text{ m}^3$

The volume is 24 m^3.

Practice Problem Find the volume of a rectangular solid that is 8 m long, 4 m wide, and 4 m high.

To find the volume of other solids, multiply the area of the base times the height.

Example 1 Find the volume of a solid that has a triangular base with a length of 8.0 m and a height of 7.0 m. The height of the entire solid is 15.0 m.

Step 1 You know that the base is a triangle, and the area of a triangle is $\frac{1}{2}$ the base times the height, and the volume is the area of the base times the height.

$V = \left[\frac{1}{2}(b \times h)\right] \times 15$

Step 2 Find the area of the base.

$V = \left[\frac{1}{2}(8 \times 7)\right] \times 15$

$V = \left(\frac{1}{2} \times 56\right) \times 15$

Step 3 Multiply the area of the base by the height of the solid.

$V = 28 \times 15$

$V = 420\text{ m}^3$

The volume is 420 m^3.

Example 2 Find the volume of a cylinder that has a base with a radius of 12.0 cm, and a height of 21.0 cm.

Step 1 You know that the base is a circle, and the area of a circle is the square of the radius times π, and the volume is the area of the base times the height.

$V = (\pi r^2) \times 21$

$V = (\pi 12^2) \times 21$

Step 2 Find the area of the base.

$V = 144\pi \times 21$

$V = 452 \times 21$

Step 3 Multiply the area of the base by the height of the solid.

$V = 9490\text{ cm}^3$

The volume is 9490 cm^3.

Example 3 Find the volume of a cylinder that has a diameter of 15 mm and a height of 4.8 mm.

Step 1 You know that the base is a circle with an area equal to the square of the radius times π. The radius is one-half the diameter. The volume is the area of the base times the height.

$V = (\pi r^2) \times 4.8$

$V = \left[\pi\left(\frac{1}{2} \times 15\right)^2\right] \times 4.8$

$V = (\pi 7.5^2) \times 4.8$

Step 2 Find the area of the base.

$V = 56.25\pi \times 4.8$

$V = 176.63 \times 4.8$

Step 3 Multiply the area of the base by the height of the solid.

$V = 847.8$

The volume is 847.8 mm^3.

Practice Problem Find the volume of a cylinder with a diameter of 7 cm in the base and a height of 16 cm.

Math Skill Handbook

Science Applications

Measure in SI

The metric system of measurement was developed in 1795. A modern form of the metric system, called the International System (SI), was adopted in 1960 and provides the standard measurements that all scientists around the world can understand.

The SI system is convenient because unit sizes vary by powers of 10. Prefixes are used to name units. Look at **Table 3** for some common SI prefixes and their meanings.

Table 3 Some SI Prefixes

Prefix	Symbol	Meaning	
kilo-	k	1,000	thousand
hecto-	h	100	hundred
deka-	da	10	ten
deci-	d	0.1	tenth
centi-	c	0.01	hundredth
milli-	m	0.001	thousandth

Example How many grams equal one kilogram?

Step 1 Find the prefix *kilo* in **Table 3.**

Step 2 Using **Table 3,** determine the meaning of *kilo.* According to the table, it means 1,000. When the prefix *kilo* is added to a unit, it means that there are 1,000 of the units in a "*kilo*unit."

Step 3 Apply the prefix to the units in the question. The units in the question are grams. There are 1,000 grams in a kilogram.

Practice Problem Is a milligram larger or smaller than a gram? How many of the smaller units equal one larger unit? What fraction of the larger unit does one smaller unit represent?

Dimensional Analysis

Convert SI Units In science, quantities such as length, mass, and time sometimes are measured using different units. A process called dimensional analysis can be used to change one unit of measure to another. This process involves multiplying your starting quantity and units by one or more conversion factors. A conversion factor is a ratio equal to one and can be made from any two equal quantities with different units. If 1,000 mL equal 1 L then two ratios can be made.

$$\frac{1{,}000 \text{ mL}}{1 \text{ L}} = \frac{1 \text{ L}}{1{,}000 \text{ mL}} = 1$$

One can covert between units in the SI system by using the equivalents in **Table 3** to make conversion factors.

Example 1 How many cm are in 4 m?

Step 1 Write conversion factors for the units given. From **Table 3,** you know that 100 cm = 1 m. The conversion factors are

$$\frac{100 \text{ cm}}{1 \text{ m}} \text{ and } \frac{1 \text{ m}}{100 \text{ cm}}$$

Step 2 Decide which conversion factor to use. Select the factor that has the units you are converting from (m) in the denominator and the units you are converting to (cm) in the numerator.

$$\frac{100 \text{ cm}}{1 \text{ m}}$$

Step 3 Multiply the starting quantity and units by the conversion factor. Cancel the starting units with the units in the denominator. There are 400 cm in 4 m.

$$4 \cancel{\text{m}} \times \frac{100 \text{ cm}}{1 \cancel{\text{m}}} = 400 \text{ cm}$$

Practice Problem How many milligrams are in one kilogram? (Hint: You will need to use two conversion factors from **Table 3.**)

Measure in SI

smaller; 1000; one thousandth

Dimensional Analysis

$$x \text{ mg} = 1 \cancel{\text{kg}} \times \frac{1000 \cancel{\text{g}}}{1 \cancel{\text{kg}}} \times \frac{1000 \text{ mg}}{1 \cancel{\text{g}}} =$$

1,000,000 mg

1,000,000 mg = 1 kg

Math Skill Handbook

Convert Between Unit Systems

$$\frac{(1\text{ in})^3}{(2.54\text{ cm})^3} = \frac{1\text{ in} \times 1\text{ in} \times 1\text{ in}}{2.54\text{ cm} \times 2.54\text{ cm} \times 2.54\text{ cm}} = \frac{1\text{ in}^3}{16.39\text{ cm}^3}$$

Table 4 Unit System Equivalents

Type of Measurement	Equivalent
Length	1 in = 2.54 cm 1 yd = 0.91 m 1 mi = 1.61 km
Mass and Weight*	1 oz = 28.35 g 1 lb = 0.45 kg 1 ton (short) = 0.91 tonnes (metric tons) 1 lb = 4.45 N
Volume	1 in^3 = 16.39 cm^3 1 qt = 0.95 L 1 gal = 3.78 L
Area	1 in^2 = 6.45 cm^2 1 yd^2 = 0.83 m^2 1 mi^2 = 2.59 km^2 1 acre = 0.40 hectares
Temperature	$°C = \frac{(°F - 32)}{1.8}$ K = °C + 273

*Weight is measured in standard Earth gravity.

Convert Between Unit Systems **Table 4** gives a list of equivalents that can be used to convert between English and SI units.

Example If a meterstick has a length of 100 cm, how long is the meterstick in inches?

Step 1 Write the conversion factors for the units given. From **Table 4,** 1 in = 2.54 cm.

$$\frac{1\text{ in}}{2.54\text{ cm}} \quad \textit{and} \quad \frac{2.54\text{ cm}}{1\text{ in}}$$

Step 2 Determine which conversion factor to use. You are converting from cm to in. Use the conversion factor with cm on the bottom.

$$\frac{1\text{ in}}{2.54\text{ cm}}$$

Step 3 Multiply the starting quantity and units by the conversion factor. Cancel the starting units with the units in the denominator. Round your answer based on the number of significant figures in the conversion factor.

$$100\ \cancel{\text{cm}} \times \frac{1\text{ in}}{2.54\ \cancel{\text{cm}}} = 39.37\text{ in}$$

The meterstick is 39.4 in long.

Practice Problem A book has a mass of 5 lbs. What is the mass of the book in kg?

Practice Problem Use the equivalent for in and cm (1 in = 2.54 cm) to show how 1 in^3 = 16.39 cm^3.

Math Skill Handbook

184 ◆ E STUDENT RESOURCES

Precision and Significant Digits

When you make a measurement, the value you record depends on the precision of the measuring instrument. This precision is represented by the number of significant digits recorded in the measurement. When counting the number of significant digits, all digits are counted except zeros at the end of a number with no decimal point such as 2,050, and zeros at the beginning of a decimal such as 0.03020. When adding or subtracting numbers with different precision, round the answer to the smallest number of decimal places of any number in the sum or difference. When multiplying or dividing, the answer is rounded to the smallest number of significant digits of any number being multiplied or divided.

Example The lengths 5.28 and 5.2 are measured in meters. Find the sum of these lengths and record your answer using the correct number of significant digits.

Step 1 Find the sum.

5.28 m	2 digits after the decimal
+ 5.2 m	1 digit after the decimal
10.48 m	

Step 2 Round to one digit after the decimal because the least number of digits after the decimal of the numbers being added is 1.

The sum is 10.5 m.

Practice Problem How many significant digits are in the measurement 7,071,301 m? How many significant digits are in the measurement 0.003010 g?

Practice Problem Multiply 5.28 and 5.2 using the rule for multiplying and dividing. Record the answer using the correct number of significant digits.

Scientific Notation

Many times numbers used in science are very small or very large. Because these numbers are difficult to work with scientists use scientific notation. To write numbers in scientific notation, move the decimal point until only one non-zero digit remains on the left. Then count the number of places you moved the decimal point and use that number as a power of ten. For example, the average distance from the Sun to Mars is 227,800,000,000 m. In scientific notation, this distance is 2.278×10^{11} m. Because you moved the decimal point to the left, the number is a positive power of ten.

The mass of an electron is about 0.000 000 000 000 000 000 000 000 000 000 911 kg. Expressed in scientific notation, this mass is 9.11×10^{-31} kg. Because the decimal point was moved to the right, the number is a negative power of ten.

Example Earth is 149,600,000 km from the Sun. Express this in scientific notation.

Step 1 Move the decimal point until one non-zero digit remains on the left.
1.496 000 00

Step 2 Count the number of decimal places you have moved. In this case, eight.

Step 3 Show that number as a power of ten, 10^8.

The Earth is 1.496×10^8 km from the Sun.

Practice Problem How many significant digits are in 149,600,000 km? How many significant digits are in 1.496×10^8 km?

Practice Problem Parts used in a high performance car must be measured to 7×10^{-6} m. Express this number as a decimal.

Practice Problem A CD is spinning at 539 revolutions per minute. Express this number in scientific notation.

Precision and Significant Digits

Problem 1
7; 4

Problem 2
$5.28 \times 5.2 = 27.456$

5.28 has 3 significant digits.

5.2 has 2 significant digits.

When multiplying and dividing, the answer is rounded to the smallest number of significant digits of the numbers being multiplied or divided—in this case, 2.

27.456 is rounded to 27.

Scientific Notation

Problem 1
4; 4

Problem 2
0.000007

Problem 3
5.39×10^2

Math Skill Handbook

Line Graph

x	y
3	52
6	72
9	83
12	86

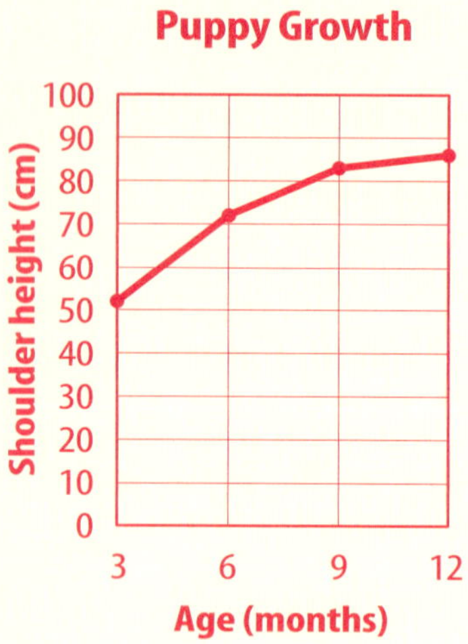

Make and Use Graphs

Data in tables can be displayed in a graph—a visual representation of data. Common graph types include line graphs, bar graphs, and circle graphs.

Line Graph A line graph shows a relationship between two variables that change continuously. The independent variable is changed and is plotted on the *x*-axis. The dependent variable is observed, and is plotted on the *y*-axis.

Example Draw a line graph of the data below from a cyclist in a long-distance race.

Table 5 Bicycle Race Data

Time (h)	Distance (km)
0	0
1	8
2	16
3	24
4	32
5	40

Step 1 Determine the *x*-axis and *y*-axis variables. Time varies independently of distance and is plotted on the *x*-axis. Distance is dependent on time and is plotted on the *y*-axis.

Step 2 Determine the scale of each axis. The *x*-axis data ranges from 0 to 5. The *y*-axis data ranges from 0 to 40.

Step 3 Using graph paper, draw and label the axes. Include units in the labels.

Step 4 Draw a point at the intersection of the time value on the *x*-axis and corresponding distance value on the *y*-axis. Connect the points and label the graph with a title, as shown in **Figure 20.**

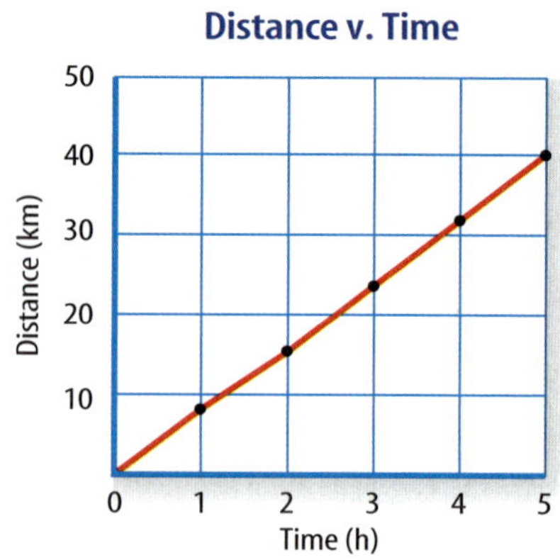

Figure 20 This line graph shows the relationship between distance and time during a bicycle ride.

Practice Problem A puppy's shoulder height is measured during the first year of her life. The following measurements were collected: (3 mo, 52 cm), (6 mo, 72 cm), (9 mo, 83 cm), (12 mo, 86 cm). Graph this data.

Find a Slope The slope of a straight line is the ratio of the vertical change, rise, to the horizontal change, run.

$$\text{Slope} = \frac{\text{vertical change (rise)}}{\text{horizontal change (run)}} = \frac{\text{change in } y}{\text{change in } x}$$

Example Find the slope of the graph in **Figure 20.**

Step 1 You know that the slope is the change in *y* divided by the change in *x*.

$$\text{Slope} = \frac{\text{change in } y}{\text{change in } x}$$

Step 2 Determine the data points you will be using. For a straight line, choose the two sets of points that are the farthest apart.

$$\text{Slope} = \frac{(40\text{–}0)\text{ km}}{(5\text{–}0)\text{ hr}}$$

Step 3 Find the change in *y* and *x*.

$$\text{Slope} = \frac{40\text{ km}}{5\text{h}}$$

Step 4 Divide the change in *y* by the change in *x*.

$$\text{Slope} = \frac{8\text{ km}}{\text{h}}$$

The slope of the graph is 8 km/h.

186 ◆ E STUDENT RESOURCES

Bar Graph To compare data that does not change continuously you might choose a bar graph. A bar graph uses bars to show the relationships between variables. The *x*-axis variable is divided into parts. The parts can be numbers such as years, or a category such as a type of animal. The *y*-axis is a number and increases continuously along the axis.

Example A recycling center collects 4.0 kg of aluminum on Monday, 1.0 kg on Wednesday, and 2.0 kg on Friday. Create a bar graph of this data.

Step 1 Select the *x*-axis and *y*-axis variables. The measured numbers (the masses of aluminum) should be placed on the *y*-axis. The variable divided into parts (collection days) is placed on the *x*-axis.

Step 2 Create a graph grid like you would for a line graph. Include labels and units.

Step 3 For each measured number, draw a vertical bar above the *x*-axis value up to the *y*-axis value. For the first data point, draw a vertical bar above Monday up to 4.0 kg.

Aluminum Collected During Week

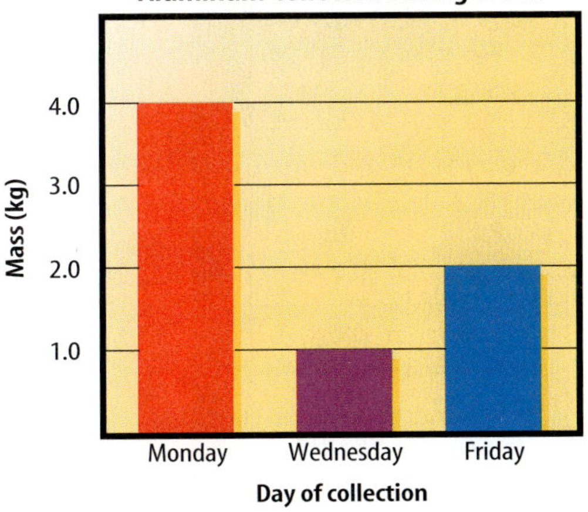

Practice Problem Draw a bar graph of the gases in air: 78% nitrogen, 21% oxygen, 1% other gases.

Circle Graph To display data as parts of a whole, you might use a circle graph. A circle graph is a circle divided into sections that represent the relative size of each piece of data. The entire circle represents 100%, half represents 50%, and so on.

Example Air is made up of 78% nitrogen, 21% oxygen, and 1% other gases. Display the composition of air in a circle graph.

Step 1 Multiply each percent by 360° and divide by 100 to find the angle of each section in the circle.

$$78\% \times \frac{360°}{100} = 280.8°$$

$$21\% \times \frac{360°}{100} = 75.6°$$

$$1\% \times \frac{360°}{100} = 3.6°$$

Step 2 Use a compass to draw a circle and to mark the center of the circle. Draw a straight line from the center to the edge of the circle.

Step 3 Use a protractor and the angles you calculated to divide the circle into parts. Place the center of the protractor over the center of the circle and line the base of the protractor over the straight line.

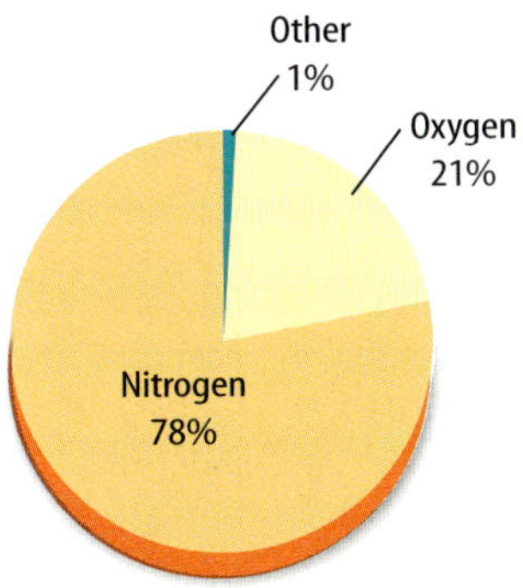

Practice Problem Draw a circle graph to represent the amount of aluminum collected during the week shown in the bar graph to the left.

Math Skill Handbook

Bar Graph

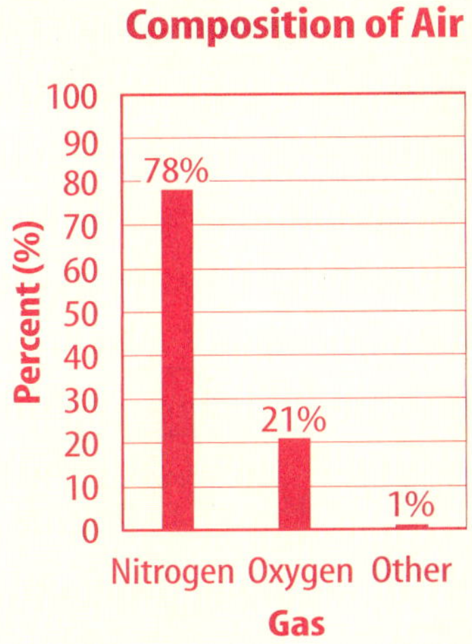

Circle Graph

The total amount of aluminum collected is:

$$4.0\text{ kg} + 1.0\text{ kg} + 2.0\text{ kg} = 7.0\text{ kg}$$

$$\frac{4.0\text{ kg}}{7.0\text{ kg}} = \frac{x}{360°};\ x = 206°$$

$$\frac{1.0\text{ kg}}{7.0\text{ kg}} = \frac{x}{360°};\ x = 51°$$

$$\frac{2.0\text{ kg}}{7.0\text{ kg}} = \frac{x}{360°};\ x = 103°$$

Math Skill Handbook

MATH SKILL HANDBOOK E ◆ 187

PERIODIC TABLE OF THE ELEMENTS

Columns of elements are called groups. Elements in the same group have similar chemical properties.

Gas
Liquid
Solid
Synthetic

Element — Hydrogen
Atomic number — 1
Symbol — H
Atomic mass — 1.008
State of matter

The first three symbols tell you the state of matter of the element at room temperature. The fourth symbol identifies elements that are not present in significant amounts on Earth. Useful amounts are made synthetically.

	1	2	3	4	5	6	7	8	9
1	Hydrogen 1 H 1.008								
2	Lithium 3 Li 6.941	Beryllium 4 Be 9.012							
3	Sodium 11 Na 22.990	Magnesium 12 Mg 24.305							
4	Potassium 19 K 39.098	Calcium 20 Ca 40.078	Scandium 21 Sc 44.956	Titanium 22 Ti 47.867	Vanadium 23 V 50.942	Chromium 24 Cr 51.996	Manganese 25 Mn 54.938	Iron 26 Fe 55.845	Cobalt 27 Co 58.933
5	Rubidium 37 Rb 85.468	Strontium 38 Sr 87.62	Yttrium 39 Y 88.906	Zirconium 40 Zr 91.224	Niobium 41 Nb 92.906	Molybdenum 42 Mo 95.94	Technetium 43 Tc (98)	Ruthenium 44 Ru 101.07	Rhodium 45 Rh 102.906
6	Cesium 55 Cs 132.905	Barium 56 Ba 137.327	Lanthanum 57 La 138.906	Hafnium 72 Hf 178.49	Tantalum 73 Ta 180.948	Tungsten 74 W 183.84	Rhenium 75 Re 186.207	Osmium 76 Os 190.23	Iridium 77 Ir 192.217
7	Francium 87 Fr (223)	Radium 88 Ra (226)	Actinium 89 Ac (227)	Rutherfordium 104 Rf (261)	Dubnium 105 Db (262)	Seaborgium 106 Sg (266)	Bohrium 107 Bh (264)	Hassium 108 Hs (277)	Meitnerium 109 Mt (268)

The number in parentheses is the mass number of the longest-lived isotope for that element.

Rows of elements are called periods. Atomic number increases across a period.

The arrow shows where these elements would fit into the periodic table. They are moved to the bottom of the table to save space.

Lanthanide series	Cerium 58 Ce 140.116	Praseodymium 59 Pr 140.908	Neodymium 60 Nd 144.24	Promethium 61 Pm (145)	Samarium 62 Sm 150.36
Actinide series	Thorium 90 Th 232.038	Protactinium 91 Pa 231.036	Uranium 92 U 238.029	Neptunium 93 Np (237)	Plutonium 94 Pu (244)

Metal

Metalloid

Nonmetal

The color of an element's block tells you if the element is a metal, nonmetal, or metalloid.

Science Online

Visit booke.msscience.com for updates to the periodic table.

10	11	12	13	14	15	16	17	18
								Helium 2 He 4.003
			Boron 5 B 10.811	Carbon 6 C 12.011	Nitrogen 7 N 14.007	Oxygen 8 O 15.999	Fluorine 9 F 18.998	Neon 10 Ne 20.180
			Aluminum 13 Al 26.982	Silicon 14 Si 28.086	Phosphorus 15 P 30.974	Sulfur 16 S 32.065	Chlorine 17 Cl 35.453	Argon 18 Ar 39.948
Nickel 28 Ni 58.693	Copper 29 Cu 63.546	Zinc 30 Zn 65.409	Gallium 31 Ga 69.723	Germanium 32 Ge 72.64	Arsenic 33 As 74.922	Selenium 34 Se 78.96	Bromine 35 Br 79.904	Krypton 36 Kr 83.798
Palladium 46 Pd 106.42	Silver 47 Ag 107.868	Cadmium 48 Cd 112.411	Indium 49 In 114.818	Tin 50 Sn 118.710	Antimony 51 Sb 121.760	Tellurium 52 Te 127.60	Iodine 53 I 126.904	Xenon 54 Xe 131.293
Platinum 78 Pt 195.078	Gold 79 Au 196.967	Mercury 80 Hg 200.59	Thallium 81 Tl 204.383	Lead 82 Pb 207.2	Bismuth 83 Bi 208.980	Polonium 84 Po (209)	Astatine 85 At (210)	Radon 86 Rn (222)
Darmstadtium 110 Ds (281)	Roentgenium 111 Rg (272)	Ununbium * 112 Uub (285)		Ununquadium * 114 Uuq (289)				

* The names and symbols for elements 112 and 114 are temporary. Final names will be selected when the elements' discoveries are verified.

Europium 63 Eu 151.964	Gadolinium 64 Gd 157.25	Terbium 65 Tb 158.925	Dysprosium 66 Dy 162.500	Holmium 67 Ho 164.930	Erbium 68 Er 167.259	Thulium 69 Tm 168.934	Ytterbium 70 Yb 173.04	Lutetium 71 Lu 174.967
Americium 95 Am (243)	Curium 96 Cm (247)	Berkelium 97 Bk (247)	Californium 98 Cf (251)	Einsteinium 99 Es (252)	Fermium 100 Fm (257)	Mendelevium 101 Md (258)	Nobelium 102 No (259)	Lawrencium 103 Lr (262)

Use and Care of a Microscope

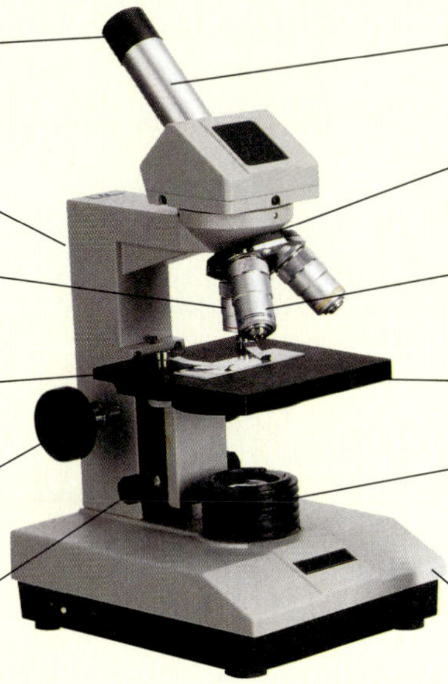

Caring for a Microscope

1. Always carry the microscope holding the arm with one hand and supporting the base with the other hand.
2. Don't touch the lenses with your fingers.
3. The coarse adjustment knob is used only when looking through the lowest-power objective lens. The fine adjustment knob is used when the high-power objective is in place.
4. Cover the microscope when you store it.

Using a Microscope

1. Place the microscope on a flat surface that is clear of objects. The arm should be toward you.
2. Look through the eyepiece. Adjust the diaphragm so light comes through the opening in the stage.
3. Place a slide on the stage so the specimen is in the field of view. Hold it firmly in place by using the stage clips.
4. Always focus with the coarse adjustment and the low-power objective lens first. After the object is in focus on low power, turn the nosepiece until the high-power objective is in place. Use ONLY the fine adjustment to focus with the high-power objective lens.

Making a Wet-Mount Slide

1. Carefully place the item you want to look at in the center of a clean, glass slide. Make sure the sample is thin enough for light to pass through.
2. Use a dropper to place one or two drops of water on the sample.
3. Hold a clean coverslip by the edges and place it at one edge of the water. Slowly lower the coverslip onto the water until it lies flat.
4. If you have too much water or a lot of air bubbles, touch the edge of a paper towel to the edge of the coverslip to draw off extra water and draw out unwanted air.

Diversity of Life: Classification of Living Organisms

A six-kingdom system of classification of organisms is used today. Two kingdoms—Kingdom Archaebacteria and Kingdom Eubacteria—contain organisms that do not have a nucleus and that lack membrane-bound structures in the cytoplasm of their cells. The members of the other four kingdoms have a cell or cells that contain a nucleus and structures in the cytoplasm, some of which are surrounded by membranes. These kingdoms are Kingdom Protista, Kingdom Fungi, Kingdom Plantae, and Kingdom Animalia.

Kingdom Archaebacteria

one-celled; some absorb food from their surroundings; some are photosynthetic; some are chemosynthetic; many are found in extremely harsh environments including salt ponds, hot springs, swamps, and deep-sea hydrothermal vents

Kingdom Eubacteria

one-celled; most absorb food from their surroundings; some are photosynthetic; some are chemosynthetic; many are parasites; many are round, spiral, or rod-shaped; some form colonies

Kingdom Protista

Phylum Euglenophyta one-celled; photosynthetic or take in food; most have one flagellum; euglenoids

Kingdom Eubacteria
Bacillus anthracis

Phylum Chlorophyta
Desmids

Phylum Bacillariophyta one-celled; photosynthetic; have unique double shells made of silica; diatoms

Phylum Dinoflagellata one-celled; photosynthetic; contain red pigments; have two flagella; dinoflagellates

Phylum Chlorophyta one-celled, many-celled, or colonies; photosynthetic; contain chlorophyll; live on land, in freshwater, or salt water; green algae

Phylum Rhodophyta most are many-celled; photosynthetic; contain red pigments; most live in deep, saltwater environments; red algae

Phylum Phaeophyta most are many-celled; photosynthetic; contain brown pigments; most live in saltwater environments; brown algae

Phylum Rhizopoda one-celled; take in food; are free-living or parasitic; move by means of pseudopods; amoebas

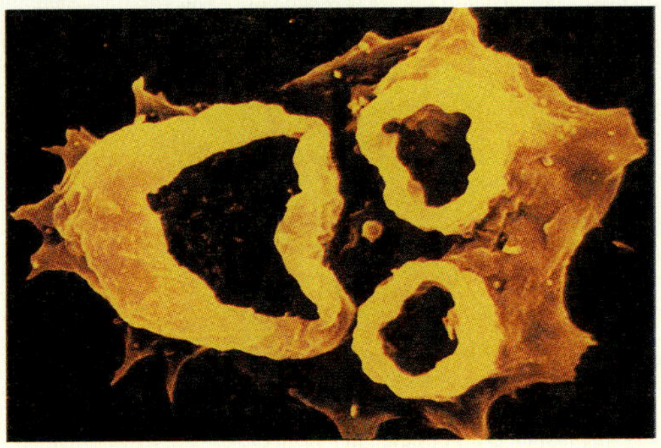

Amoeba

Phylum Zoomastigina one-celled; take in food; free-living or parasitic; have one or more flagella; zoomastigotes

Phylum Ciliophora one-celled; take in food; have large numbers of cilia; ciliates

Phylum Sporozoa one-celled; take in food; have no means of movement; are parasites in animals; sporozoans

Phylum Myxomycota
Slime mold

Phylum Oomycota
Phytophthora infestans

Phyla Myxomycota and Acrasiomycota one- or many-celled; absorb food; change form during life cycle; cellular and plasmodial slime molds

Phylum Oomycota many-celled; are either parasites or decomposers; live in freshwater or salt water; water molds, rusts and downy mildews

Kingdom Fungi

Phylum Zygomycota many-celled; absorb food; spores are produced in sporangia; zygote fungi; bread mold

Phylum Ascomycota one- and many-celled; absorb food; spores produced in asci; sac fungi; yeast

Phylum Basidiomycota many-celled; absorb food; spores produced in basidia; club fungi; mushrooms

Phylum Deuteromycota members with unknown reproductive structures; imperfect fungi; *Penicillium*

Phylum Mycophycota organisms formed by symbiotic relationship between an ascomycote or a basidiomycote and green alga or cyanobacterium; lichens

Lichens

Kingdom Plantae

Divisions Bryophyta (mosses), **Anthocerophyta** (hornworts), **Hepaticophyta** (liverworts), **Psilophyta** (whisk ferns) many-celled nonvascular plants; reproduce by spores produced in capsules; green; grow in moist, land environments

Division Lycophyta many-celled vascular plants; spores are produced in conelike structures; live on land; are photosynthetic; club mosses

Division Arthrophyta vascular plants; ribbed and jointed stems; scalelike leaves; spores produced in conelike structures; horsetails

Division Pterophyta vascular plants; leaves called fronds; spores produced in clusters of sporangia called sori; live on land or in water; ferns

Division Ginkgophyta deciduous trees; only one living species; have fan-shaped leaves with branching veins and fleshy cones with seeds; ginkgoes

Division Cycadophyta palmlike plants; have large, featherlike leaves; produces seeds in cones; cycads

Division Coniferophyta deciduous or evergreen; trees or shrubs; have needlelike or scalelike leaves; seeds produced in cones; conifers

Division Anthophyta
Tomato plant

Division Gnetophyta shrubs or woody vines; seeds are produced in cones; division contains only three genera; gnetum

Division Anthophyta dominant group of plants; flowering plants; have fruits with seeds

Kingdom Animalia

Phylum Porifera aquatic organisms that lack true tissues and organs; are asymmetrical and sessile; sponges

Phylum Cnidaria radially symmetrical organisms; have a digestive cavity with one opening; most have tentacles armed with stinging cells; live in aquatic environments singly or in colonies; includes jellyfish, corals, hydra, and sea anemones

Phylum Platyhelminthes bilaterally symmetrical worms; have flattened bodies; digestive system has one opening; parasitic and free-living species; flatworms

Division Bryophyta
Liverwort

Phylum Platyhelminthes
Flatworm

Phylum Chordata

Phylum Nematoda round, bilaterally symmetrical body; have digestive system with two openings; free-living forms and parasitic forms; roundworms

Phylum Mollusca soft-bodied animals, many with a hard shell and soft foot or footlike appendage; a mantle covers the soft body; aquatic and terrestrial species; includes clams, snails, squid, and octopuses

Phylum Annelida bilaterally symmetrical worms; have round, segmented bodies; terrestrial and aquatic species; includes earthworms, leeches, and marine polychaetes

Phylum Arthropoda largest animal group; have hard exoskeletons, segmented bodies, and pairs of jointed appendages; land and aquatic species; includes insects, crustaceans, and spiders

Phylum Echinodermata marine organisms; have spiny or leathery skin and a water-vascular system with tube feet; are radially symmetrical; includes sea stars, sand dollars, and sea urchins

Phylum Chordata organisms with internal skeletons and specialized body systems; most have paired appendages; all at some time have a notochord, nerve cord, gill slits, and a post-anal tail; include fish, amphibians, reptiles, birds, and mammals

Cómo usar el glosario en español:
1. Busca el término en inglés que desees encontrar.
2. El término en español, junto con la definición, se encuentran en la columna de la derecha.

Pronunciation Key

Use the following key to help you sound out words in the glossary.

a	**back** (BAK)	**ew**	**food** (FEWD)
ay	**day** (DAY)	**yoo**	**pure** (PYOOR)
ah	**father** (FAH thur)	**yew**	**few** (FYEW)
ow	**flower** (FLOW ur)	**uh**	**comma** (CAH muh)
ar	**car** (CAR)	**u** (+ con)	**rub** (RUB)
e	**less** (LES)	**sh**	**shelf** (SHELF)
ee	**leaf** (LEEF)	**ch**	**nature** (NAY chur)
ih	**trip** (TRIHP)	**g**	**gift** (GIHFT)
i (i + con + e)	**idea** (i DEE uh)	**j**	**gem** (JEM)
oh	**go** (GOH)	**ing**	**sing** (SING)
aw	**soft** (SAWFT)	**zh**	**vision** (VIH zhun)
or	**orbit** (OR buht)	**k**	**cake** (KAYK)
oy	**coin** (COYN)	**s**	**seed, cent** (SEED, SENT)
oo	**foot** (FOOT)	**z**	**zone, raise** (ZOHN, RAYZ)

English — A — Español

abiotic: nonliving, physical features of the environment, including air, water, sunlight, soil, temperature, and climate. (p. 36)

abiótico: características inertes y físicas del medio ambiente, incluyendo el aire, el agua, la luz solar, el suelo, la temperatura y el clima. (p. 36)

acid precipitation: precipitation with a pH below 5.6—which occurs when air pollutants from the burning of fossil fuels react with water in the atmosphere to form strong acids—that can pollute water, kill fish and plants, and damage soils. (p. 103)

lluvia ácida: precipitación con un pH menor de 5.6—lo cual ocurre cuando los contaminantes del aire provenientes de la quema de combustibles fósiles reaccionan con el agua en la atmósfera para formar ácidos fuertes—que puede contaminar el agua, matar peces y plantas, y dañar los suelos. (p. 103)

acid rain: forms when sulfur dioxide and nitrogen oxide from industries and car exhausts combine with water vapor in the air; can wash nutrients from soil and damage trees and aquatic life. (p. 135)

lluvia ácida: se forma cuando el dióxido de azufre y el óxido de nitrógeno derivados de la industria y de los escapes de los automóviles se combinan con vapor de agua en el aire; puede arrastrar nutrientes del suelo y causar daño a los árboles y a la vida acuática. (p. 135)

atmosphere: air surrounding Earth; is made up of gases, including 78 percent nitrogen, 21 percent oxygen, and 0.03 percent carbon dioxide. (p. 37)

atmósfera: aire que rodea a la Tierra; está compuesta de gases, incluyendo 78% de nitrógeno, 21% de oxígeno y 0.03% de dióxido de carbono. (p. 37)

B

biodiversity: variety of life in an ecosystem, most commonly measured by the number of species that live in a given area. (p. 126)

biodiversidad: variedad de vida en un ecosistema, comúnmente cuantificada mediante el número de especies que viven en un área determinada. (p. 126)

biomes (BI ohmz): large geographic areas with similar climates and ecosystems; includes tundra, taiga, desert, temperate deciduous forest, temperate rain forest, tropical rain forest, and grassland. (p. 68)

biomas: grandes áreas geográficas con climas y ecosistemas similares; incluyen la tundra, la taiga, el desierto, el bosque caducifolio templado, el bosque lluvioso templado, la selva húmeda tropical y los pastizales. (p. 68)

biosphere: part of Earth that supports life, including the top portion of Earth's crust, the atmosphere, and all the water on Earth's surface. (p. 8)

biosfera: capa de la Tierra que alberga la vida, incluyendo la porción superior de la corteza terrestre, la atmósfera y toda el agua de la superficie terrestre. (p. 8)

biotic (bi AH tihk): features of the environment that are alive or were once alive. (p. 36)

biótico: características del ambiente que tienen o alguna vez tuvieron vida. (p. 36)

C

captive population: population of organisms that is cared for by humans. (p. 142)

población cautiva: población de organismos bajo el cuidado de los seres humanos. (p. 142)

carbon cycle: model describing how carbon molecules move between the living and nonliving world. (p. 49)

ciclo del carbono: modelo que describe cómo se mueven las moléculas de carbono entre el mundo vivo y el mundo inerte. (p. 49)

carrying capacity: largest number of individuals of a particular species that an ecosystem can support over time. (p. 15)

capacidad de carga: el mayor número de individuos de una especie en particular que un ecosistema puede albergar en un periodo de tiempo. (p. 15)

chemosynthesis (kee moh SIN thuh sus): process in which producers make energy-rich nutrient molecules from chemicals. (p. 51)

quimiosíntesis: proceso a través del cual los productores fabrican moléculas ricas en energía a partir de agentes químicos. (p. 51)

climate: average weather conditions of an area over time, including wind, temperature, and rainfall or other types of precipitation such as snow or sleet. (p. 41)

clima: condiciones meteorológicas promedio de un área durante un periodo de tiempo; incluye viento, temperatura y precipitación pluvial u otros tipos de precipitación como la nieve o el granizo. (p. 41)

climax community: stable, end stage of ecological succession in which balance is in the absence of disturbance. (p. 67)

clímax comunitario: etapa final estable de la sucesión ecológica en la cual se da un equilibrio en ausencia de alteraciones. (p. 67)

commensalism: a type of symbiotic relationship in which one organism benefits and the other organism is not affected. (p. 22)

comensalismo: tipo de relación simbiótica en la que un organismo se beneficia sin afectar al otro. (p. 22)

community: all the populations of different species that live in an ecosystem. (p. 10)

comunidad: todas las poblaciones de diferentes especies que viven en un mismo ecosistema. (p. 10)

condensation: process that takes place when a gas changes to a liquid. (p. 45)

condensación: proceso que tiene lugar cuando un gas cambia a estado líquido. (p. 45)

conservation biology: study of methods for protecting Earth's biodiversity; uses strategies such as reintroduction programs and habitat restoration and works to preserve threatened and endangered species. (p. 138)

biología de la conservación: estudio de los métodos para proteger la biodiversidad de la Tierra; utiliza estrategias tales como programas de reintroducción y restauración de hábitat y busca preservar especies amenazadas o en peligro de extinción. (p. 138)

consumer: organism that cannot create energy-rich molecules but obtains its food by eating other organisms. (p. 21)

consumidor: organismo que no puede fabricar moléculas ricas en energía por lo que debe obtener su alimento ingiriendo otros organismos. (p. 21)

coral reef: diverse ecosystem formed from the calcium carbonate shells secreted by corals. (p. 81)

arrecife de coral: ecosistema diverso conformado de caparazones de carbonato de calcio secretados por los corales. (p. 81)

D

desert: driest biome on Earth with less than 25 cm of rain each year; has dunes or thin soil with little organic matter, where plants and animals are adapted to survive extreme conditions. (p. 74)

desierto: el bioma más seco sobre la Tierra con menos de 25 centímetros cúbicos de lluvia al año; tiene dunas o un suelo delgado con muy poca materia orgánica y aquí las plantas y animales están adaptados para sobrevivir en condiciones extremosas. (p. 74)

E

ecology: study of the interactions that take place among organisms and their environment. (p. 9)

ecosystem: all the living organisms that live in an area and the nonliving features of their environment. (p. 9)

endangered species: species that is in danger of becoming extinct. (p. 131)

energy pyramid: model that shows the amount of energy available at each feeding level in an ecosystem. (p. 53)

erosion: movement of soil from one place to another. (p. 109)

estuary: extremely fertile area where a river meets an ocean; contains a mixture of freshwater and saltwater and serves as a nursery for many species of fish. (p. 82)

evaporation: process that takes place when a liquid changes to a gas. (p. 44)

extinct species: species that was once present on Earth but has died out. (p. 130)

ecología: estudio de las interacciones que se dan entre los organismos y su medio ambiente. (p. 9)

ecosistema: conjunto de organismos vivos que habitan en un área y las características de su medio ambiente. (p. 9)

especie en peligro de extinción: especies que se encuentran en peligro de quedar extintas. (p. 131)

pirámide de energía: modelo que muestra la cantidad de energía disponible en cada nivel alimenticio de un ecosistema. (p. 53)

erosión: movimiento del suelo de un lugar a otro. (p. 109)

estuario: área extremadamente fértil donde un río desemboca en el océano; contiene una mezcla de agua dulce y salada y sirve como vivero para muchas especies de peces. (p. 82)

evaporación: proceso que tiene lugar cuando un líquido cambia a estado gaseoso. (p. 44)

especies extintas: especies que alguna vez estuvieron presentes en la Tierra pero que han desaparecido. (p. 130)

F

food web: model that shows the complex feeding relationships among organisms in a community. (p. 52)

fossil fuels: nonrenewable energy resources—coal, oil, and natural gas—that formed in Earth's crust over hundreds of millions of years. (p. 96)

cadena alimenticia: modelo que muestra las complejas relaciones alimenticias entre los organismos de una comunidad. (p. 52)

combustibles fósiles: recursos energéticos no renovables—carbón, petróleo y gas natural—que se formaron en la corteza terrestre durante cientos de millones de años. (p. 96)

G

geothermal energy: heat energy within Earth's crust, available only where natural geysers or volcanoes are located. (p. 99)

energía geotérmica: energía calórica en el interior de la corteza terrestre disponible sólo donde existen géiseres o volcanes. (p. 99)

grasslands: temperate and tropical regions with 25 cm to 75 cm of precipitation each year that are dominated by climax communities of grasses; ideal for growing crops and raising cattle and sheep. (p. 75)

greenhouse effect: heat-trapping feature of the atmosphere that keeps Earth warm enough to support life. (p. 104)

pastizales: regiones tropicales y templadas con 25 a 75 centímetros cúbicos de lluvia al año; son dominadas por el clímax comunitario de los pastos e ideales para la cría de ganado y ovejas. (p. 75)

efecto de invernadero: característica de la atmósfera que le permite atrapar calor y mantener la Tierra lo suficientemente caliente para favorecer la vida. (p. 104)

H

habitat: place where an organism lives and that provides the types of food, shelter, moisture, and temperature needed for survival. (p. 11)

habitat restoration: process of bringing a damaged habitat back to a healthy condition. (p. 141)

hazardous wastes: waste materials, such as pesticides and leftover paints, that are harmful to human health or poisonous to living organisms. (p. 110)

hydroelectric power: electricity produced when the energy of falling water turns the blades of a generator turbine. (p. 97)

hábitat: lugar donde vive un organismo y que le proporciona los tipos de alimento, refugio, humedad y temperatura necesarios para su supervivencia. (p. 11)

restauración de hábitat: proceso de restaurar la condiciones favorables de un hábitat alterado. (p. 141)

desperdicios peligrosos: materiales de desecho como los pesticidas y residuos de pintura nocivos para la salud humana o dañinos para los organismos vivos. (p. 110)

energía hidroeléctrica: electricidad producida cuando la energía generada por la caída del agua hace girar las aspas de una turbina generadora. (p. 97)

I

intertidal zone: part of the shoreline that is under water at high tide and exposed to the air at low tide. (p. 82)

introduced species: species that moves into an ecosystem as a result of human actions. (p. 134)

zona litoral: parte de la línea costera que está bajo el agua durante la marea alta y expuesta al aire durante la marea baja. (p. 82)

especies introducidas: especies que ingresan en un ecosistema como resultado de las actividades humanas. (p. 134)

L

limiting factor: anything that can restrict the size of a population, including living and nonliving features of an ecosystem, such as predators or drought. (p. 14)

factor limitante: cualquier factor que pueda restringir el tamaño de una población, incluyendo las características biológicas y no biológicas de un ecosistema, tales como los depredadores o las sequías. (p. 14)

M

mutualism: a type of symbiotic relationship in which both organisms benefit. (p. 22)

mutualismo: tipo de relación simbiótica en la que ambos organismos se benefician. (p. 22)

N

native species: original organisms in an ecosystem. (p. 134)

natural resources: parts of Earth's environment that supply materials useful or necessary for the survival of living organisms. (p. 94)

niche: in an ecosystem, refers to the unique ways an organism survives, obtains food and shelter, and avoids danger. (p. 23)

nitrogen cycle: model describing how nitrogen moves from the atmosphere to the soil, to living organisms, and then back to the atmosphere. (p. 46)

nitrogen fixation: process in which some types of bacteria in the soil change nitrogen gas into a form of nitrogen that plants can use. (p. 46)

nonrenewable resources: natural resources, such as petroleum, minerals, and metals, that are used more quickly than they can be replaced by natural processes. (p. 95)

nuclear energy: energy produced from the splitting apart of billions of uranium nuclei by a nuclear fission reaction. (p. 98)

especies nativas: organismos originales de un ecosistema. (p. 134)

recursos naturales: partes del medio ambiente terrestre que proporcionan materiales útiles o necesarios para la supervivencia de los organismos vivos. (p. 94)

nicho: en un ecosistema, se refiere a las formas únicas en las que un organismo sobrevive, obtiene alimento, refugio y evita el peligro. (p. 23)

ciclo del nitrógeno: modelo que describe cómo se mueve el nitrógeno de la atmósfera al suelo, a los organismos vivos y de nuevo a la atmósfera. (p. 46)

fijación del nitrógeno: proceso en el cual algunos tipos de bacterias en el suelo transforman el nitrógeno gaseoso en una forma de nitrógeno que las plantas pueden usar. (p. 46)

recursos no renovables: recursos naturales, como el petróleo, los minerales y los metales, que son utilizados más rápidamente de lo que pueden ser reemplazados mediante procesos naturales. (p. 95)

energía nuclear: energía producida a partir del fraccionamiento de billones de núcleos de uranio mediante una reacción de fisión nuclear. (p. 98)

O

ozone depletion: thinning of Earth's ozone layer caused by chlorofluorocarbons (CFCs) leaking into the air and reacting chemically with ozone, breaking the ozone molecules apart. (pp. 105, 136)

agotamiento del ozono: reducción de la capa de ozono causada por clorofluorocarbonos (CFCs) que se liberan al aire y reaccionan químicamente con el ozono descomponiendo sus moléculas. (pp. 105, 136)

P

parasitism: a type of symbiotic relationship in which one organism benefits and the other organism is harmed. (p. 22)

petroleum: nonrenewable resource formed over hundreds of millions of years mostly from the remains of microscopic marine organisms buried in Earth's crust. (p. 95)

pioneer species: first organisms to grow in new or disturbed areas. (p. 64)

pollutant: substance that contaminates any part of the environment. (p. 102)

population: all the organisms that belong to the same species living in a community. (p. 10)

parasitismo: tipo de relación simbiótica en la que un organismo se beneficia y el otro es perjudicado. (p. 22)

petróleo: recurso no renovable formado durante cientos de millones de años, en su mayoría a partir de los restos de organismos marinos microscópicos sepultados en la corteza terrestre. (p. 95)

especies pioneras: primeros organismos que crecen en áreas nuevas o alteradas. (p. 64)

contaminante: sustancia que contamina cualquier parte del medio ambiente. (p. 102)

población: todos los organismos que pertenecen a la misma especie dentro de una comunidad. (p. 10)

producer: organism, such as a green plant or alga, that uses an outside source of energy like the Sun to create energy-rich food molecules. (p. 20)

productor: organismo, como una planta o un alga verde, que utiliza una fuente externa de energía, como la luz solar, para producir moléculas de nutrientes ricas en energía. (p. 20)

R

recycling: conservation method that is a form of reuse and requires changing or reprocessing an item or natural resource. (p. 113)

reintroduction program: conservation strategy that returns organisms to an area where the species once lived and may involve seed banks, captive populations, and relocation. (p. 142)

renewable resources: natural resources, such as water, sunlight, and crops, that are constantly being recycled or replaced by nature. (p. 95)

reciclaje: método de conservación como una forma de reutilización y que requiere del cambio o reprocesamiento del producto o recurso natural. (p. 113)

programa de reintroducción: estrategia de conservación que devuelve a los organismos a un área en la que la especie vivió alguna vez, pudiendo involucrar bancos de semillas, poblaciones cautivas y reubicación. (p. 142)

recursos renovables: recursos naturales, como el agua, la luz solar y los cultivos, que son reciclados o reemplazados constantemente por la naturaleza. (p. 95)

S

soil: mixture of mineral and rock particles, the remains of dead organisms, air, and water that forms the topmost layer of Earth's crust and supports plant growth. (p. 38)

succession: natural, gradual changes in the types of species that live in an area; can be primary or secondary. (p. 64)

symbiosis: any close relationship between species, including mutualism, commensalism, and parasitism. (p. 22)

suelo: mezcla de partículas minerales y rocas, restos de organismos muertos, aire y del agua que forma la capa superior de la corteza terrestre y favorece el crecimiento de las plantas. (p. 38)

sucesión: cambios graduales y naturales en los tipos de especies que viven en un área; puede ser primaria o secundaria. (p. 64)

simbiosis: cualquier relación estrecha entre especies, incluyendo mutualismo, comensalismo y parasitismo. (p. 22)

T

taiga (TI guh): world's largest biome, located south of the tundra between 50° N and 60° N latitude; has long, cold winters, precipitation between 35 cm and 100 cm each year, cone-bearing evergreen trees, and dense forests. (p. 70)

temperate deciduous forest: biome usually having four distinct seasons, annual precipitation between 75 cm and 150 cm, and climax communities of deciduous trees. (p. 71)

temperate rain forest: biome with 200 cm to 400 cm of precipitation each year, average temperatures between 9°C and 12°C, and forests dominated by trees with needlelike leaves. (p. 71)

taiga: el bioma más grande del mundo, localizado al sur de la tundra entre 50° y 60° de latitud norte; tiene inviernos prolongados y fríos, una precipitación que alcanza entre 35 y 100 centímetros cúbicos al año, coníferas perennifolias y bosques espesos. (p. 70)

bosque caducifolio templado: bioma que generalmente tiene cuatro estaciones distintas, con una precipitación anual entre 75 y 150 centímetros cúbicos y un clímax comunitario de árboles caducifolios. (p. 71)

bosque lluvioso templado: bioma con 200 a 400 centímetros cúbicos de precipitación al año; tiene una temperatura promedio entre 9 y 12°C y bosques dominados por árboles de hojas aciculares. (p. 71)

threatened species: species that is likely to become endangered in the near future. (p. 131)

tropical rain forest: most biologically diverse biome; has an average temperature of 25°C and receives between 200 cm and 600 cm of precipitation each year. (p. 72)

tundra: cold, dry, treeless biome with less than 25 cm of precipitation each year, a short growing season, permafrost, and winters that can be six to nine months long. Tundra is separated into two types: arctic tundra and alpine tundra. (p. 69)

especies amenazadas: especies susceptibles de verse amenazadas en un futuro cercano. (p. 131)

selva húmeda tropical: el bioma más diverso biológicamente; tiene una temperatura promedio de 25°C y recibe entre 200 y 600 centímetros cúbicos de precipitación al año. (p. 72)

tundra: bioma sin árboles, frío y seco, con menos de 25 centímetros cúbicos de precipitación al año; tiene una estación corta de crecimiento y permafrost e inviernos que pueden durar entre 6 y 9 meses. La tundra se divide en dos tipos: tundra ártica y tundra alpina. (p. 69)

W

water cycle: model describing how water moves from Earth's surface to the atmosphere and back to the surface again through evaporation, condensation, and precipitation. (p. 45)

wetland: a land region that is wet most or all of the year. (p. 79)

ciclo del agua: modelo que describe cómo se mueve el agua de la superficie de la Tierra hacia la atmósfera y nuevamente hacia la superficie terrestre a través de la evaporación, la condensación y la precipitación. (p. 45)

zona húmeda: región lluviosa la mayor parte del año. (p. 79)

Italic numbers = *illustration/photo* **Bold numbers = vocabulary term**
lab = *indicates a page on which the entry is used in a lab*
act = *indicates a page on which the entry is used in an activity*

A

B

C

Index

D

E

F

Index

Index

Index

W

Y

Credits

Magnification Key: Magnifications listed are the magnifications at which images were originally photographed.
LM–Light Microscope
SEM–Scanning Electron Microscope
TEM–Transmission Electron Microscope

Acknowledgments: Glencoe would like to acknowledge the artists and agencies who participated in illustrating this program: Absolute Science Illustration; Andrew Evansen; Argosy; Articulate Graphics; Craig Attebery represented by Frank & Jeff Lavaty; CHK America; John Edwards and Associates; Gagliano Graphics; Pedro Julio Gonzalez represented by Melissa Turk & The Artist Network; Robert Hynes represented by Mendola Ltd.; Morgan Cain & Associates; JTH Illustration; Laurie O'Keefe; Matthew Pippin represented by Beranbaum Artist's Representative; Precision Graphics; Publisher's Art; Rolin Graphics, Inc.; Wendy Smith represented by Melissa Turk & The Artist Network; Kevin Torline represented by Berendsen and Associates, Inc.; WILDlife ART; Phil Wilson represented by Cliff Knecht Artist Representative; Zoo Botanica.

Photo Credits

Cover Darrell Gulin/Getty Images; **i ii** Darrell Gulin/Getty Images; **iv** (bkgd)John Evans, (inset)Darrell Gulin/Getty Images; **v** (t)PhotoDisc, (b)John Evans; **vi** (l)John Evans, (r)Geoff Butler; **vii** (l)John Evans, (r)PhotoDisc; **viii** PhotoDisc; **ix** Aaron Haupt Photography; **x** Lynn M. Stone/DRK Photo; **xi** Hal Beral/Visuals Unlimited; **xii** (l)Michael P. Gadomski/Photo Researchers, (r)Zig Leszczynski/Earth Scenes; **1** (l)Rob & Ann Simpson/Visuals Unlimited, (r)Steve Wolper/DRK Photo; **2** (t)Darren Bennett/Animals Animals, (b)Collection of Glenbow Museum, Calgary, Canada; **3** (t)Mathew Cavanaugh/AP/Wide World Photos, (b)Helen Hardin 1971; **4** (t)Grant Heilman Photography, Inc., (b)Rick Poley/Visuals Unlimited, Inc.; **5** (t)Jeff Cooper/The Salina Journal/Associated Press, (b)Fletcher & Baylus/Photo Researchers; **6–7** Joe McDonald/Visuals Unlimited; **8** (tr)Richard Kolar/Animals Animals, (l)Adam Jones/Photo Researchers, (c)Tom Van Sant/Geosphere Project, Santa Monica/Science Photo Library/Photo Researchers, (br)G. Carleton Ray/Photo Researchers; **9** (t)John W. Bova/Photo Researchers, (b)David Young/Tom Stack & Assoc.; **11** (l)Zig Leszczynski/Animals Animals, (r)Gary W. Carter/Visuals Unlimited; **14** Mitsuaki Iwago/Minden Pictures; **15** Joel Sartore from Grant Heilman; **17** (t)Norm Thomas/Photo Researchers, (b)Maresa Pryor/Earth Scenes; **18** (tl)Wyman P. Meinzer, (r)Bud Neilson/Words & Pictures/PictureQuest, (bl)Wyman P. Meinzer; **20** (l)Michael Abbey/Photo Researchers, (r)OSF/Animals Animals, (b)Michael P. Gadomski/Photo Researchers; **21** (tl)William J. Weber, (tlc)William J. Weber, (tcr)Lynn M. Stone, (tr)William J. Weber, (bl)Larry Kimball/Visuals Unlimited, (blc)George D. Lepp/Photo Researchers, (bcr)Stephen J. Krasemann/Peter Arnold, Inc., (br)Mark Steinmetz; **22** (t)Milton Rand/Tom Stack & Assoc., (c)Marian Bacon/Animals Animals, (b)Sinclair Stammers/Science Photo Library/Photo Researchers; **23** (tl)Raymond A. Mendez/Animals Animals, (bl)Donald Specker/Animals Animals, (br)Joe McDonald/Animals Animals; **24** Ted Levin/Animals Animals; **25** Richard L. Carlton/Photo Researchers; **26** (t)Jean Claude Revy/PhotoTake, NYC, (b)OSF/Animals Animals; **27** Runk/Schoenberger from Grant Heilman; **28** Eric Larravadieu/Stone/Getty Images; **29** (l)C.K. Lorenz/Photo Researchers, (r)Hans Pfletschinger/Peter Arnold, Inc.; **30** CORBIS; **32** (l)Michael P. Gadomski/Photo Researchers, (r)William J. Weber; **34–35** Ron Thomas/Getty Images; **36** Kenneth Murray/Photo Researchers; **37** (t)Jerry L. Ferrara/Photo Researchers, (b)Art Wolfe/Photo Researchers; **38** (t)Telegraph Colour Library/FPG/Getty Images, (b)Hal Beral/Visuals Unlimited; **39** (l)Fritz Polking/Visuals Unlimited, (r)R. Arndt/Visuals Unlimited; **40** Tom Uhlman/Visuals Unlimited; **44** Jim Grattan; **47** (t)Rob & Ann Simpson/Visuals Unlimited, (c b)Runk/Schoenberger from Grant Heilman; **50** WHOI/Visuals Unlimited; **54** Gerald and Buff Corsi/Visuals Unlimited; **55** Jeff J. Daly/Visuals Unlimited; **56** Gordon Wiltsie/Peter Arnold, Inc.; **57** (l)Soames Summerhay/Photo Researchers, (r)Tom Uhlman/Visuals Unlimited; **62–63** William Campbell/CORBIS Sygma; **64** Jeff Greenberg/Visuals Unlimited; **65** Larry Ulrich/DRK Photo; **66** (bkgd)Craig Fujii/Seattle Times, (l)Kevin R. Morris/CORBIS, (tr br)Jeff Henry; **67** Rod Planck/Photo Researchers; **69** (t)Steve McCutcheon/Visuals Unlimited, (bl)Pat O'Hara/DRK Photo, (br)Erwin & Peggy Bauer/Tom Stack & Assoc.; **70** (tl)Peter Ziminski/Visuals Unlimited, (c)Leonard Rue III/Visuals Unlimited, (bl)C.C. Lockwood/DRK Photo, (br)Larry Ulrich/DRK Photo; **71** (t)Fritz Polking/Visuals Unlimited, (b)William Grenfell/Visuals Unlimited; **72** Lynn M. Stone/DRK Photo; **74** (l)McDonald Wildlife Photgraphy, Inc./DRK Photo, (r)Steve Solum/Bruce Coleman, Inc.; **75** Kevin Schafer; **77** W. Banaszewski/Visuals Unlimited; **78** (l)Dwight Kuhn, (r)Mark E. Gibson/Visuals Unlimited; **79** James R. Fisher/DRK Photo; **80** D. Foster/WHOI/Visuals Unlimited; **81** (l)C.C. Lockwood/Bruce Coleman, Inc., (r)Steve Wolper/DRK Photo; **82** (tl)Dwight Kuhn, (tr)Glenn Oliver/Visuals Unlimited, (b)Stephen J. Krasemann/DRK Photo; **83** (l)John Kaprielian/Photo Researchers, (r)Jerry Sarapochiello/Bruce Coleman, Inc.; **84** (t)Dwight Kuhn, (b)John Gerlach/DRK Photo; **85** Fritz Polking/Bruce Coleman, Inc.; **86** courtesy Albuquerque Public Schools; **87** (l)James P. Rowan/DRK Photo, (r)John Shaw/Tom Stack & Assoc.; **91** (l)Leonard Rue III/Visuals Unlimited, (r)Joe McDonald/DRK Photo; **92–93** Grant Heilman Photography; **94** (l)Keith Lanpher/Liaison Agency/Getty Images, (r)Richard Thatcher/David R. Frazier Photolibrary; **95** (t)Solar Cookers International, (bl)Brian F. Peterson/The Stock Market/CORBIS, (br)Ron Kimball Photography; **96** Larry Mayer/Liaison Agency/Getty Images; **99** (tr)Torleif Svenson/The Stock Market/CORBIS, (bl)Rob Williamson, (br)Les Gibbon/Cordaiy Photo Library Ltd./CORBIS; **100** Sean Justice; **101** (t)Lowell Georgia/Science Source/Photo Researchers, (cl)NASA, (c)CORBIS, (cr)Sean Sprague/Impact Visuals/PictureQuest, (bl)Lee Foster/Bruce Coleman, Inc., (br)Robert Perron; **102** Philippe Renault/Liaison Agency/Getty Images; **103** (l)NYC Parks Photo Archive/Fundamental Photographs, (r)Kristen Brochmann/Fundamental Photographs; **107** (l)Jeremy Walker/Science Photo Library/Photo Researchers, (c)John Colwell from Grant Heilman, (r)Telegraph Colour Library/FPG/Getty Images; **108** Wilford Haven/Liaison Agency/Getty Images; **109** (tl)Larry Mayer/Liaison Agency/Getty Images, (tr) ChromoSohm/The Stock Market/CORBIS, (cr)David R. Frazier Photolibrary, (br)Inga Spence/Visuals Unlimited; **110** (r)Andrew Holbrooke/The Stock Market/CORBIS, (Paint Cans)Amanita Pictures, (Turpantine, Paint thinner, epoxy)Icon Images, (Batteries)Aaron Haupt; **112** Paul A. Souders/CORBIS;

113 Icon Images; **115** Larry Lefever from Grant Heilman; **116** (t)Howard Buffett from Grant Heilman, (b)Solar Cookers International; **117** John D. Cunningham/Visuals Unlimited; **118** Frank Cezus/FPG/Getty Images; **120** Robert Cameron/Stone/Getty Images; **121** (l)Steve McCutcheon/Visuals Unlimited, (r)James N. Westwater; **122** David R. Frazier Photolibrary; **124–125** Jeffrey Greenberg/Photo Researchers; **126** (l)Andy Sacks/Stone/Getty Images, (r)Erwin and Peggy Bauer/Bruce Coleman, Inc.; **127** (tr)Leonard Lee Rue III/Photo Researchers, (l)Chuck Pefley/Stone/Getty Images, (br)Charles W. Mann/Photo Researchers; **128** (l)Ray Pfortner/Peter Arnold, Inc., (c)Gilbert S. Grant/Photo Researchers, (r)Rexford Lord/Photo Researchers; **129** William D. Adams; **131** (t)Joe McDonald/Tom Stack & Assoc., (b)Len Rue, Jr./Animals Animals; **132** (tl)Joel Sartore, (tr)William H. Amos, (c)Joseph Van Wormer/Bruce Coleman, Inc./PictureQuest, (bl)Kennan Word/CORBIS, (br)Erwin & Peggy Bauer/Bruce Coleman, Inc./PictureQuest; **133** (l)Jim Baron/The Image Finders, (r)Melissa Hathaway/Ohio Dept of Natural Resources Division of Wildlife; **134** (t)University of Minnesota Sea Grant Program, (b)Marcia Griffen/Earth Scenes; **135** SuperStock; **137** A. Greth/BIOS/Peter Arnold, Inc.; **138** Michael Pogany/Columbus Zoo; **139** (l)Bill Pogue/Stone/Getty Images, (r)Daniel J Cox/Stone/Getty Images; **141** Save The Bay® - People for Narragansett Bay; **142** Leonard L.T. Rhodes/Animals Animals; **143** Hal Beral/Visuals Unlimited; **144** Zig Leszczynski/Earth Scenes; **146** (t)Jacques Jangoux/Photo Researchers, (b)Michael Fogden/DRK Photo; **147** (l)Inga Spence/Visuals Unlimited, (tr)Juan Manuel Renjifo/Earth Scenes, (br)Bogart Photography; **148** Kenneth W. Fink/Bruce Coleman, Inc.; **150** Gustav Verderber/Visuals Unlimited; **151** Linda Saville-Rath; **152** PhotoDisc; **154** Tom Pantages; **158** Michell D. Bridwell/PhotoEdit, Inc.; **159** (t)Mark Burnett, (b)Dominic Oldershaw; **160** StudiOhio; **161** Timothy Fuller; **162** Aaron Haupt; **164** KS Studios; **165** Matt Meadows; **166** Rod Planck/Photo Researchers; **169** Amanita Pictures; **170** Bob Daemmrich; **172** Davis Barber/PhotoEdit, Inc.; **190** Matt Meadows; **191** (l)Dr. Richard Kessel, (c)NIBSC/Science Photo Library/Photo Researchers, (r)David John/Visuals Unlimited; **192** (t)Runk/Schoenberger from Grant Heilman, (bl)Andrew Syred/Science Photo Library/Photo Researchers, (br)Rich Brommer; **193** (tr)G.R. Roberts, (l)Ralph Reinhold/Earth Scenes, (br)Scott Johnson/Animals Animals; **194** Martin Harvey/DRK Photo.

PERIODIC TABLE OF THE ELEMENTS

Columns of elements are called groups. Elements in the same group have similar chemical properties.

Element — Hydrogen
Atomic number — 1
Symbol — H
Atomic mass — 1.008
State of matter

Gas
Liquid
Solid
Synthetic

The first three symbols tell you the state of matter of the element at room temperature. The fourth symbol identifies elements that are not present in significant amounts on Earth. Useful amounts are made synthetically.

	1	2	3	4	5	6	7	8	9
1	Hydrogen 1 **H** 1.008								
2	Lithium 3 **Li** 6.941	Beryllium 4 **Be** 9.012							
3	Sodium 11 **Na** 22.990	Magnesium 12 **Mg** 24.305							
4	Potassium 19 **K** 39.098	Calcium 20 **Ca** 40.078	Scandium 21 **Sc** 44.956	Titanium 22 **Ti** 47.867	Vanadium 23 **V** 50.942	Chromium 24 **Cr** 51.996	Manganese 25 **Mn** 54.938	Iron 26 **Fe** 55.845	Cobalt 27 **Co** 58.933
5	Rubidium 37 **Rb** 85.468	Strontium 38 **Sr** 87.62	Yttrium 39 **Y** 88.906	Zirconium 40 **Zr** 91.224	Niobium 41 **Nb** 92.906	Molybdenum 42 **Mo** 95.94	Technetium 43 **Tc** (98)	Ruthenium 44 **Ru** 101.07	Rhodium 45 **Rh** 102.906
6	Cesium 55 **Cs** 132.905	Barium 56 **Ba** 137.327	Lanthanum 57 **La** 138.906	Hafnium 72 **Hf** 178.49	Tantalum 73 **Ta** 180.948	Tungsten 74 **W** 183.84	Rhenium 75 **Re** 186.207	Osmium 76 **Os** 190.23	Iridium 77 **Ir** 192.217
7	Francium 87 **Fr** (223)	Radium 88 **Ra** (226)	Actinium 89 **Ac** (227)	Rutherfordium 104 **Rf** (261)	Dubnium 105 **Db** (262)	Seaborgium 106 **Sg** (266)	Bohrium 107 **Bh** (264)	Hassium 108 **Hs** (277)	Meitnerium 109 **Mt** (268)

The number in parentheses is the mass number of the longest-lived isotope for that element.

Rows of elements are called periods. Atomic number increases across a period.

The arrow shows where these elements would fit into the periodic table. They are moved to the bottom of the table to save space.

Lanthanide series	Cerium 58 **Ce** 140.116	Praseodymium 59 **Pr** 140.908	Neodymium 60 **Nd** 144.24	Promethium 61 **Pm** (145)	Samarium 62 **Sm** 150.36
Actinide series	Thorium 90 **Th** 232.038	Protactinium 91 **Pa** 231.036	Uranium 92 **U** 238.029	Neptunium 93 **Np** (237)	Plutonium 94 **Pu** (244)